Environmental

Biotreatment

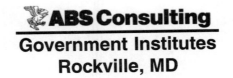

Technologies for Air, Water, Soil, and Wastes

Catherine N. Mulligan, PhD

ABS Consulting
Government Institutes
Rockville, MD

ABS Consulting

Government Institutes
4 Research Place, Rockville, Maryland 20850, USA
Phone: (301) 921-2300
Fax: (301) 921-0373
Email: giinfo@govinst.com
Internet: http://www.govinst.com

        Library of Congress Cataloging-in-Publication Data
Mulligan, Catherine N.
        Environmental biotreatment: technologies for air, water, soil, and waste / Catherine N. Mulligan.
        p. cm.
        Includes bibliographical references and index.
        ISBN: 0-86587-890-0
        1. Bioremediation. I. Title.

TD192.5 .M85 2001
628.5--dc21

                                                      2001040712

Printed in the United States of America

Dedication

The author would like to dedicate this book to Dr. Bernard F. Gibbs, formerly Senior Scientist at BRI, National Research Council of Canada, with whom she has collaborated for many years. He is currently Principal Scientist/Research Fellow at MDS Pharma Services, and Professor at McGill University in Montreal, Canada. She thanks him for his guidance, motivation, encouragement, inspiration, and support throughout all these years.

Summary of Contents

Contents

List of Illustrations

Boxes

Figures

Tables

Foreword

The availability of biological techniques for treatment of soil solids and water, air, and waste materials has provided geoenvironmental scientists and engineers with almost unlimited opportunities for bioremediation treatment of contaminated materials, effluents, and contaminated sites. The range of techniques covers such procedures as bioventing, biostripping, biorestoration, biostimulation, biodegradation, landfarming, biofilters, and agitated reactors. Most of the procedures involve processes that include microbial degradation, hydrolysis, substitution, aerobic and anaerobic transformation, degradations, biotic redox reactions, mineralization, and volatilization. To achieve effective remediation treatment, it is necessary to have a good working knowledge not only of the types of wastes and contaminants but also of: (a) the basic principles involved in the biological degradation of organic chemicals and how these are manifested in a soil and/or water system, (b) the nature of wastes and contaminants and the nature of contaminant environment, (c) the partitioning of contaminants in a contaminated site, and (d) the interactions and bonding relationships established between soil solids and contaminants and also within the porewater.

This book not only develops the basic fundamental principles and processes involved in application of biological treatments but also provides the reader with an understanding of the various treatment techniques and available technologies. In combination with the evaluation of the merits and deficiencies of these various techniques and their applicability, this book is a timely and valuable resource textbook for geoenvironmental and environmental scientists and engineers.

Raymond N. Yong, CQ, Ph. D., FRSC
Geoenvironmental Research Centre
University of Wales, Cardiff, UK

Preface

The purpose of this book is to provide a complete state-of-the-art review, description, and evaluation of the technologies available for the biological treatment of air, water, soil, and solid wastes. It will be of interest to environmental consultants, engineering firms, and students and academics in chemical, civil, and environmental engineering programs. This book examines the various technologies available; provides process descriptions and conditions used in various case studies, performance data, scale-up issues, and site characteristics affecting performance; and weighs the advantages and disadvantages of each process and how they compare to traditional processes, such as incineration in terms of time requirements, safety issues, and other factors.

What is unique about this book is that processes for the treatment of air, water, soil, and wastes are described in one book. This is important since pollutants can be found in all media and can also be transferred from one to another. In addition, information on technologies developed for a type of medium can often be applied to another. This book will provide the professional with potential solutions to the remediation of these media. Biological processes are of increasing interest since they do not produce toxic by-products and are less expensive. Many advancements have been made in recent years. However, there are several misconceptions that exist concerning biological treatment processes that can inhibit their acceptance by practitioners. For example, they are believed to be too slow, unreliable, difficult to control, and subject to upsets. Many times, these problems are due to poor design and lack of understanding of the process. Real cases are presented in which the biological technologies have been successful. Flow diagrams are used extensively to assist in understanding the process descriptions.

The organization of the book is as follows. In the first chapter, the near- and long-term prospects for biological treatment processes are discussed. Aspects include what further process developments (e.g., faster degradation rates, smaller reactors) are required so that biological processes will be competitive with conventional processes and what other considerations must be overcome (e.g., public perception, regulatory aspects).

The second chapter describes the basic principles of the biological degradation of chemicals by bacteria, yeast, and fungus. The types of chemicals that can be degraded or transformed, the pathways of conversion, the potential for genetic improvement, the general conditions used, and the importance of acclimatization are discussed. The general criteria that must be met for biological technologies to be considered as a choice is examined. Some of these include the toxicity of the chemical to be degraded, the availability of the component to the microorganism, favourable growth conditions, and low process cost.

The third chapter describes the biological technologies available for air pollution treatment. The components to be treated are either odorous or volatile organic chemicals. Industrial applications, such as in the printing, flexography, and pharmaceutical industries, are included. This chapter consists of three sections—biofiltration, bioscrubbers, and trickling filters, which are the main technologies. The composition, concentration, and types of chemicals that can be biologically treated are examined. Examples from Germany, Netherlands, and other countries are included and compared to conventional technologies, such as activated carbon filtration and incineration.

The fourth chapter, which is on water treatment, includes aerobic and anaerobic processes for industrial and municipal applications. Aerobic processes such as activated sludge, aerated lagoon and trickling filters, and anaerobic ones such as upflow sludge blanket reactor and others are described. Each type of reactor is illustrated with the type and concentration of components treated, the water flow rate treated, and the conditions used. Processes for conversion or adsorption of organics or inorganics such as metals, nitrogen, or phosphorus compounds are covered.

The next chapter details the processes available for biological soil and groundwater treatment. *In situ* processes include various methods for the addition of microorganisms, oxygen, nutrients, and water, the anaerobic processes for chlorinated organics, and the use of plants (phytoremediation) for metal accumulation. *Ex situ* processes where treatment parameters such as mixing, temperature, and aeration can be more easily controlled are also described. Examples of technologies for treatment of excavated soil and sediments are landfarming, biopiles, and slurry reactors. Parameters such as the time schedule available, concentration of contaminants, treatment costs, and availability of components to be treated are considered in the choice of technology.

Chapter six contains a description of the biological treatment of solid wastes in terms of what types and concentrations of wastes can be treated in industrial and municipal applications, what factors are important in choosing a process, and which processes are available for treatment or conversion to higher value products. Anaerobic digestion, composting, leaching of mining wastes, and conversion of food wastes to animal feed are some examples. Various processes are described through the use of illustrations and performance data.

This is followed by a glossary of terms and abbreviations and a directory of vendors that should be of specific interest to engineering firms, consultants, and environmental managers who must choose appropriate technologies to solve pollution problems in all media. After reading this book the reader will have a better understanding of the design of biological treatment processes, thus avoiding process failure.

About the Author

Catherine Mulligan has worked for the past 15 years in the field of biotechnology in research, industrial, and academic environments. After six years in the production of biosurfactants by fermentation at McGill University and the Biotechnology Research Institute, she joined the SNC Research Corporation in 1989, a subsidiary of SNC-Lavalin Group, where she has conducted various projects in the development of environmental processes, bioconversions, fermentation, process evaluation, and technico-economic studies. She was involved in the development of the anaerobic treatment of various types of industrial wastewater and air, in addition to the development of bioleaching of mining residues. She has recently joined Concordia University, Department of Building Civil and Environmental Engineering, as a professor where her research interests include biosurfactant-washing of contaminated soils, treatment of metal-contaminated soils and wastes, bioremediation, the biological treatment of wastewater, and the biological treatment of air. She has taught courses in Site Remediation, Environmental Engineering, Fate and Transport of Contaminants in the Environment and Geoenvironmental Engineering. She earned B. Eng. and M. Eng. degrees in Chemical Engineering and a Ph. D. in Civil Engineering at McGill University, Montreal, Canada. She is a member of Order of Engineers of Québec, Canadian Society of Chemical Engineering, American Institute of Chemical Engineering, Air and Waste Management Association, Association for the Environmental Health of Soils, Canadian Society of Civil Engineering, American Chemical Society, and the Canadian Geotechnical Society. She has authored over 30 articles and presented papers at numerous conferences.

Acknowledgments

The author thanks the publisher, Government Institutes, and in particular Charlene P. Ikonomou, Acquisitions Editor, and Patti Koch, Production Editor, for their assistance and patience throughout the preparation and publishing of this book. She would also like to thank Alexander Padro, Associate Publisher, for the initial work on the book. It has been a pleasure to work with them. Finally, she would like to thank Mr. Asif Choudhury for assisting in the preparation of some of the figures.

Overview and Future Direction of Biological Treatment

Introduction

The biological treatment of contamination in air, water, soil, groundwater, wastes, and sludges can address major industrial contamination problems. Numerous private companies and government agencies currently employ biological treatment processes. The use of microorganisms is relatively new, but there have been numerous full-scale applications, particularly within the last decade. The key to the success of a biological degradation process is the biodegradability of the wastes. All biological processes require biodegradable contaminants or another process to enhance the biodegradability of the waste. Optimal design of a biological degradation system requires that adequate nutrients and electron acceptors be delivered to the microorganisms. A multidisciplinary approach is required due to various aspects that must be addressed. Some of the disciplines include microbiology, chemistry, ecology, geology, hydrology, and engineering. Successful bioremediation results from the establishment of the appropriate environmental factors for the appropriate microorganisms. Multiphasic and heterogeneous environments increase the complexity of biological treatment processes.

Biotreatment processes are less expensive than other waste treatment processes, and energy requirements for biotreatment are fewer. Furthermore, treatments can be done onsite, wastes can be degraded completely, and other chemical and physical treatments can enhance the process. However, some chemicals cannot be removed by bioremediation, heavy metals cannot be degraded, toxic chemicals can be produced, and remediation, particularly in the case of contaminated soil, is highly site-specific. It may also be difficult to make the conditions ideal for the biotreatment to occur. Characterization of the contaminants and treatability studies are required in this case. A balance, however, must be maintained between sampling and analysis costs and the information obtained. Extensive work remains in the areas of microbial ecology, physiology, and genetic expression. The engineering of the various processes requires a stronger scientific basis.

Before biological treatment can be initiated, it is necessary to thoroughly understand the regulations. The site or process should be thoroughly characterized for the concentration and quantity of contaminants and any other interferences. Treatment or feasibility studies are often the best method to determine if the process will work and under what conditions. The design and engineering of the process can then be based on the studies, increasing the likelihood of success.

The environmental parameters that are the most important to determine during the tests are the energy sources, electron acceptors, nutrients, pH, and temperature. The presence of inhibitory substrates or metabolites can also influence the process. Other factors depend on the properties of the contaminants. Slow mass transfer to the microorganism can decrease the availability of the substrate to the microorganism. The addition of biological or synthetic surfactants may be necessary to increase the bioavailability of the substrate.

Regulatory Factors Influencing Biological Treatment

There are several non-technical issues that influence the choice of a biological treatment process. More contractors and consultants now recommend biotreatment. Regulatory issues determine the type of contaminant that must be treated, treatment levels, and what methods must be used. Approval for the use of genetically engineered microorganisms has been extremely difficult. For example, in the United States, there are several environmental laws that control bioremediation:

- Toxic Substances Control Act of 1976;
- Federal Water Pollution Control Act of 1971, Clean Water Act of 1977, and the Water Quality Act of 1987;
- Solid Waste Disposal Act of 1965, Resource Conservation and Recovery Act (RCRA) of 1976, and the Hazardous and Solid Waste Amendments of 1984;
- Comprehensive Environmental Response, Compensation and Liability Act of 1980, Superfund Amendments, and Reauthorization Act of 1986;
- Clean Air Act of 1970 and CAA amendments of 1977 and 1990; and National Environmental Policy Act of 1969.

Some of these regulations have been changed to allow reuse of cleaned soil after excavation and bioremediation. Other regulations, however, impede bioremediation by being very conservative in setting cleanup levels and restricting the injection of nutrients that assist biological processes. Setting levels that protect human health and the environment would be a more useful approach for detoxifying, degrading, or immobilizing the target chemicals. Government agencies such as the U.S. Environmental Protection Agency (U.S. EPA) enable the development of innovative technologies such as biological treatment processes by promoting pilot testing, demonstrations, training seminars, and symposia.

Research and Technical Factors

Although petroleum hydrocarbons, solvents, and alcohols are readily biodegradable, other contaminants such as pesticides, chlorinated compounds, coal tars, polycyclic aromatic hydrocarbons (PAHs), and polychlorinated biphenyls (PCBs) are not very degradable. Research into co-metabolism, white rot fungi, and anaerobic processes will enhance this area significantly. Therefore, more research is needed, which requires funding—funding that has decreased significantly in recent years. In addition, since each case is unique, it requires particular attention and development work during pilot testing.

Human Resources

There is still a lack of trained personnel due to the newness of the technology. Knowledge in many fields must be integrated including engineering, geology, hydrogeology, soil science, microbiology, and project management. Previously, on-the-job training was the only way to obtain the required expertise; however, the number of university programs in environmental engineering that include bioremediation, both at the undergraduate and graduate levels, is increasing.

Economic and Liability Factors

Clients and regulatory agencies tend to view biological treatment processes as experimental and are less likely to invest in the technology. Thus, performance standards are usually tougher. There are also liability issues if the standards are not met. Government policies also tend to favor proven technologies. Regulations are often complex, and changing regulations in view of new technologies can take significant periods of time. There has been some change in regulations to remove some of the restrictions, in particular, for excavating the soil, bioremediating it, and reusing the soil at the site. There are still many limitations on the types of nutrients that can be injected for *in situ* use. New developments are continuing with new applications. Regulations will continue to tighten for industrial processes. Even with the adoption of ISO 14001, which features clean production processes and closed water cycles, end of pipe treatment processes are vital.

Status and Specific Needs

All media are presented in this book because chemicals, in many cases, can easily be transferred from one media to another or are released into more than one media at a time. More than one treatment method may then be required. For example during bioventing, biological degradation occurs in the subsurface. However, volatilization of the components may require that biofiltration be used to capture and treat this contaminant above the surface. In addition, many of the principles required for biological treatment in one media are the same for another since the degradability of the contaminant is the main issue in many cases. The following sections illustrate specific needs for the improvement of biological treatment processes for each media.

Air Technologies

Recently, the development of biological techniques for the treatment of volatile organic compounds (VOCs) has progressed significantly. There are now many applications in the area of odor and VOC control. These applications can lead to savings in energy, catalysts, and capital and operating costs compared with many of the physical and chemical processes currently used. Simplicity and biodegradation to carbon dioxide are other advantages of these processes. Some limitations (particularly with biofilters) include large installation area requirements, pH control when sulfides or chlorinated compounds are present, inability to treat variable and high concentrations of contaminants, high-temperature gas streams, maintenance of adequate moisture levels, and process stability. Improper maintenance of biofilters has led to system failures. These problems are in the process of being solved through the development of fully enclosed and controlled systems. Since many industries are conservative and do not run the risks associated with full-scale development, it is important to cooperate fully with the industry and monitor the applications extensively to solve the development problems. In addition, education of potential users regarding the benefits and limitations of biological treatment technologies by carrying out reliable demonstrations will enable biological treatment technologies to compete with traditional technologies. More full-scale demonstrations are needed to enable competition with established technologies.

It is expected that large gas flows of greater than 1 million m^3/h will be able to be treated in the near future, particularly for the treatment of SO_2 (Pâques 1997). Further developments will be made in NO_x removal by coordinating physical, chemical, and biological processes. Recently it was demonstrated on a laboratory scale that the fungi *Rhinocladiella* sp. and *Acremonium* sp. were able to remove 93% of NO (250 mg/m^3) in the presence of 14.4 L/min of toluene in the air in less than one minute (Woertz 1999). The toluene served as the carbon and energy source.

Other areas of future development include the treatment of compounds of low water solubility, the use of fungi for other applications and specific bacterial species, and the treatment of thermophilic gas streams up to 250°C for sulfate reduction and sulfite oxidation (Stewart et al. 2000). For example, Bio-Reaction Industries has developed a fungal-based system with pre-adapted, compost media. Multiple beds are used in series to prevent overloads and enable polishing. Up to 95.1% total VOC removal for emissions of 90,000 ppm from a glycol dehydrator can be obtained.

Since compounds must be water-soluble to be treated by bacteria, methods to overcome this limitation include the use of two liquid phases in liquid-impelled loop reactors or by the addition of activated carbon as an absorbent. Some low water soluble compounds include alkanes, alkenes, benzenes, etc. Membrane reactors or using fungi in dry biofilms are other alternatives under development. New cultures can be isolated from contaminated gas streams such as styrene and toluene treatment by fungus. Pilot tests (Kraakman et al. 1997) have shown that the fungus *Exaphiala jeanselmei* on ceramic packing could treat a load of styrene of 25 to 30 g/m^3-h (100 to 900 mg/m^3 concentration) with a 90% removal efficiency. A buffer unit with activated carbon and a humidifying unit preceded the biological unit.

Water Technologies

Biological treatment is commonly used as a secondary wastewater treatment. The major groups of biological processes include aerobic, anaerobic, and a combination of both. The systems are divided into suspended or attached growth processes for the removal of biochemical oxygen demand (BOD), nitrification, denitrification, stabilization, and phosphorus removal. Aerobic processes, including activated sludge, trickling filters, aerated lagoons, and rotating biological contactors, have been used extensively. However, the supply of air is expensive. In addition, large amounts of sludge must be sent for disposal. Recently, significant developments have been made in the area of anaerobic water treatment. Treatment is now reliable and retention times are low. A net production of methane also makes these systems attractive.

Biotreatment systems for wastewater contain living organisms such as bacteria, algae, fungi, or plants. The requirements for these organisms vary according to the organism present. In general, temperature, light, and environmental conditions such as pH, oxygen, and metal concentrations have the most effect on treatment efficiencies. With proper analysis and environmental control, many types of wastewater can be treated biologically. Changes in the environment must allow for the organisms to adapt, or the effects may be highly detrimental. In the future, research must focus on the development of microorganisms and systems that can increase the rate of the treatment process to decrease retention times and, subsequently, decrease reactor volumes. There is usually resistance by engineers and waste treatment plant operators, among others, to employ biological augmentation. Experts in design, operation, and biological processes will need to combine their efforts to enhance system performance, particularly for the treatment of recalcitrant compounds. Because experience is fairly limited in the biological treatment of toxic compounds, it is difficult to predict their fate and effect in bioreactors. Systems such as wetlands are highly complex. Research is needed to determine specific mechanisms for toxicity reduction.

Soil and Groundwater Technologies

Contaminated groundwater and soil are still a major problem with up to 60,000 sites in need of urgent treatment in Europe (Anonymous 1999). Leaking underground storage tanks (UST) continue to be a major cause of soil contamination. In Germany, the short term treatment market is in the order of $15 to $20 billion; it is $30 to $60 billion in the Netherlands. The development of improved technologies is, therefore, of utmost importance. Bioremediation has been effective at many sites in the United States, Canada, and throughout the world. In 1995, the bioremediation market was $228 million in the United States. In 1999, it grew to $1.5 to 2 billion compared to $1 billion in Europe (Zechendorf 1999), and the market for environmental biotechnology worldwide was estimated at $11.5 billion. This market is expected to grow at a rate of 16%. Early in the 21st Century, the U.S. market is expected to reach $2.8 billion. Bioremediation is a small part of this market but is growing.

Since bioremediation is still a relatively new process, there are still many requirements for further improvement. Minimizing environmental risk and maintaining low costs are two of the most important factors. Research needs to focus on halogenated compounds, metals, explosives, and nitrates. Much work has been done on co-metabolism for the treatment of halogenated solvents and PCBs, but research still is required to ensure that unwanted by-products do not accumulate. Such accumulation would reduce the large ratios of co-substrates to contaminants in ratios from 5:1 to 20:1 and would reduce the huge oxygen demand. Some other needs include optimization of bioreactor design; isolation and characterization of organisms capable of degrading numerous compounds such as creosote; degradation of complex mixtures; anaerobic degradation; evolution of methods for *in situ* monitoring of degradation processes; and determination of the effect of environmental factors such as soil type, oxygen levels, temperature, redox potential, etc. Other needs involve the development of methods to deliver nutrients for bioventing and biosparging processes for unsaturated zones. The use of gaseous nitrogen and phosphorus may be the solution.

When bioremediation was first employed, other technologies were combined with it to assist in the treatment. For example, chemical and physical pretreatments were used prior to landfarming. In the future, combinations of anaerobic/aerobic bioremediation could be used for the treatment of chlorinated compounds, in particular. Chemical oxidation could also be used as a pretreatment.

In situ bioremediation processes have been successful for VOCs and other contaminants when excavation is restricted. *In situ* bioremediation requires minimal worker exposure making it safe for workers. The technology is becoming more and more publicly acceptable. Bioremediation technologies for soil remediation can be limited if the soil is clayey, highly heterogeneous, or of low permeability. Highly-absorbed contaminants such as PAHs or oils that are highly weathered can also be difficult to remediate.

A knowledge of the sorbing soil fractions can significantly enhance remediation efficiency. An example of the disruption of the bonds between pollutants and soil fractions is achieved when surfactants are added to a hydrophobic organic-contaminated soil by reducing the surface tension at the soil-water interface (Mulligan et al. 1999a). The solubility of the contaminant is also increased, which can enhance *in situ* biodegradation. Further examination of the use of surfactants and biosurfactants, in particular, is valuable to their potential in assisting biodegradation and enhancing metal removal (Mulligan et al. 1999b). Furthermore, the development of techniques such as selective sequential extraction to determine the heavy metals desorbed from the various soil fractions can enhance the design of remediation techniques (Mulligan et al. 2001).

Delivery of oxygen, moisture, and nutrients to the microorganisms and optimal conditions of pH and temperature are of utmost importance for efficient remediation processes, yet these are the most difficult aspects to resolve. Inadequate delivery can lead to non-achievement of remediation goals. Gaseous forms of nutrients (nitrogen and phosphorus) may remediate this problem. Low soil permeability, high contaminant levels, and limitations in oxygen and nutrients constrain the effectiveness of natural remediation processes. It can also be difficult to prove that remediation is actually taking place. Tools for designing, predicting, and monitoring bioremediation processes have been limited, particularly in heterogeneous environments. Although cost effective, the monitoring requirements by regulatory agencies for natural bioremediation can be extensive. Regulators are increasing their acceptance, however, of natural attenuation since minimal site disruption is necessary. Engineered bioremediation processes are more effective since oxygen and nutrient delivery can be optimized.

A variety of bioremediation processes have been developed. Landfarming techniques are inexpensive and simple. However, air emissions can become a concern when the contaminants are volatile. Extensive land areas may also be required. Landfarming is usually not applicable to clay soils unless bulking agents are added. There seems to be some evidence that the slow release oxygen compounds (calcium or magnesium hydrogen peroxide complexes) can increase soil permeability. Bioventing is useful for the *in situ* remediation of volatile components although this is not applicable if the soil is less than 0.6 m from the surface or the water table is less than 3 m. While low permeability clay soils are usually difficult to remediate, some success has been achieved in clayey soils that is not understood. These technologies have been effective for crude oil, jet and diesel fuels, gasoline, waste oil, and BTEX. Chlorinated compounds have been difficult to remediate.

There are many unknowns regarding air sparging, a new bioremediation technique, including the influence of clay lenses, water table level, channeling of air bubbles, and other phenomenon which decrease the effectiveness of this remediation process. Nutrient delivery systems can become inefficient if soil pores clog, which could be caused by iron precipitation after oxygen addition to high iron aquifers, excessive biological growth (particularly at the point of injection), or nutrient solutions that are incompatible with components in the ground water. Further research is needed concerning microbial transport and enabling the survival of desirable microorganisms.

Ex situ biological processes have been effective for excavated soils. Windrow composting has been used for explosive-contaminated soils; it is cost-effective for this application and can be easily monitored for moisture content, oxygen level, and temperature. Slurry bioreactors can provide high rates of biodegradation but can require extensive supervision or be limited by high contaminant concentrations. Closed reactors or *in situ* lagoons can be used. Various agents such as neutralizing agents, surfactants, and dispersants can be used to enhance the biodegradation process. Off-gases should also be controlled, which is difficult if open lagoons are used. Wood preserving or oil refinery wastes have been successfully treated by this method. Bioslurry and windrow composting have been shown to be half the cost of incineration processes.

Phytoremediation is another emerging process. This technique has been mainly applicable for heavy metal removal, but levels of accumulation in plants have been fairly low. There has been uptake of some organic pollutants such as trichloroethylene (TCE) (Schnoor et al. 1997), but there is much research that needs to be done. This area is promising due to low costs, and it is asthetically pleasing. Hyperaccumulating plants are required to improve the efficiency of phytoremediation.

Waste Treatment Technologies

There are numerous types of industrial, agricultural, and municipal wastes. The most common practice until now has been to dispose of them in landfills. As many of these landfills are closing, biological treatment offers a clear method of reducing wastes going to landfills. In addition, biological processes can lead to the production of valuable products. Among the biological treatments, anaerobic digestion is frequently cost-effective due to the insignificant environmental impact and the energy production through methane. In Europe, there are 36,000 anaerobic digesters in operation, treating up to 50% of the sludges (Tilche and Malaspina 1998). Recent investigations have also indicated that emissions from anaerobic

Table 1.1 Estimation of Biogas Production in Europe (adapted from Tilche and Malaspina 1998)

Source	Current Biogas Production Rate (10^6 m³/day)
Suspended solids	1.7
Organic fraction of municipal waste	4.5
Industrial wastewater	0.8
Animal wastes	0.5
Total	7.5

digestion plants are much less than from composting plants (Mata-Alvarez et al. 2000). Estimates from different sources for methane production are shown in Table 1.1. Anaerobic digestion has a significant future in the treatment of organic wastes. Degradation of chlorinated compounds in particular, must be investigated further.

Composting can also significantly reduce the amount of waste, particularly waste derived from food, going to the landfill. A variety of composting systems are now available ranging from fairly simple to highly complex. Composting is applicable for home owners, individual institutions, and even communities and has increased in popularity significantly because of public interest in decreasing the amount of wastes going to landfills. Moisture and oxygen levels, pile temperature, and odors must be monitored throughout the process. Little is known, however, about the microbiology of composting. Carbon to nitrogen ratios are important factors that must be optimal to ensure the success of the process. Research is needed for hazardous wastes such as explosives and petroleum sludges. Composting can be used if the soil conditions or space requirements are not appropriate for landfarming. Vermicomposting has also shown potential and needs to be further developed.

Aerobic digestion is a highly stable process that can be operated so that nitrification also takes place. Due to the aeration requirements, it is more suitable for small and medium-sized plants. Aerobic digestion and nitrification are considered safer than anaerobic processes since there are no explosive gases produced. Supernatants are also of higher quality. Aerobic digestion tanks do not need to be heated or covered; therefore, they are less expensive to manufacture than anaerobic tanks. Aeration systems are similar to conventional aeration tanks and can be designed for spiral or cross roll aeration with diffused air equipment. Thermophilic digestion is an area in which developments will continue because this method requires less space than other methods, and it results in pathogen destruction and high sludge treatment rates.

Landfill bioreactors, a newly developed method for operating landfills, have several advantages over conventional landfills. Further efforts are required to optimize leachate recirculation, gas production and the degradation of recalcitrant compounds. There are still many challenges including the reluctance of regulators, the ability to wet the waste uniformly, and the availability of design criteria.

Landfarming is a low-cost, simple method of treating petroleum and wood preserving wastes. Common cultivation equipment is used. The process design is simple, and remediation can be performed in six to 24 months. There is little operating data available, which can present problems when designing systems. Tilling provides porosity to the soils, mixes the soil and additives, and enhances oxygen transfer. Frequency of tilling depends mainly on oxygen uptake. Leachate collection systems should be incorporated if mobile, halogenated solvents are to be treated. Space requirements for landfarming can be large, and air emissions and leachates should be monitored. Landfarming in cold climates can slow degradation rates significantly.

Slurry phase systems in the form of reactors or lagoons incorporate continuous mixing, thus decreasing the degradation times and reducing land requirements. Emissions and leachates can be easily controlled. Significant advances can be made in reducing retention times. The process design includes pretreatment, determination of optimal solids concentration, mixer design, and retention time. Pretreatments to enhance slurry degradation processes include particle-size fractionation and/or surfactant addition. Mixing is highly important in slurry reactors. Constant supervision is necessary to obtain optimal performance. Full-scale applications need to be expanded from petroleum and wood preserving sludges. With the appropriate conditions, bacterial leaching of mining ores (Mulligan and Galvez-Cloutier 2000), tailings, and other metal-containing residues can be performed in slurry reactors and have shown promise.

Agricultural wastes, wood residuals, yard debris, low-grade paper, and forest wastes can be used as feedstocks for ethanol production. Improvements are needed for the conversion of cellulose to sugars by enzymes or chemicals. In some cases, municipal wastes have been converted to ethanol. Other processes have also been developed for conversion of wastes to fertilizer or animal feed. The success of these processes will depend on the market for the products, process economics, and the quality of the final product.

Future Trends

An area that should continue to develop is metal bioremediation. Bacteria can be used to produce biosurfactants and other chelants of metals to enhance metal removal from contaminated sites (Mulligan et al. 1999c). This can be done *in situ* or in *ex situ* reactors. Anaerobic processes could also be developed for conversion of radionuclides to insoluble species (Tucker 1997). The biological remediation of nitrates and sulfates is another area that must continue as a result of recent significant advances in wastewater treatment.

Intrinsic anaerobic treatment in aquifers is another area where significant work needs to be done. Compounds such as halogenated solvents are suitable for anaerobic degradation, but such systems are anaerobic and difficult to aerate. Compounds such as perchloroethylene (PCE) do not seem to be aerobically dehalogenated. The anaerobic degradation produces Dichloroethylene (DCE) and vinyl chloride (VC). Aerobic degradation is needed to completely mineralize these compounds. Processes for anaerobic/aerobic treatment need to be developed and optimized for *in situ* and *ex situ* applications. Chemical oxidation processes could also be used as a pretreatment for biological treatment of compounds that are difficult to treat.

The use of acclimated or engineered microorganisms is an area with potential but with uncertainty. Novel microorganisms could produce the enzymes capable of degrading recalcitrant components by the insertion of genes to destroy particular compounds. Many of the genes with potential importance in biodegradation processes are unknown. Derepression techniques could also be developed to enhance acclimation of the bacteria. The ability of engineered microbes to complement indigenous microorganisms is a major issue since it has been shown that the natural organisms can sometimes grow better than the developed organisms. In addition, regulatory agencies do not favor the injection of foreign organisms; unless the agencies work together with researchers, this area will not be fully developed. Other considerations are the potential formation of toxic metabolites. The microorganisms could be plant pathogens, or there could be the development of other unexpected microbial mutants due to chemical-microbial interactions (Yong 2001).

The list of contaminants that have been treated by biological means is continually growing as is information on biological treatment processes. A more scientific and technical approach is necessary instead of a trial and error one for evaluating the potential for biotreatment of all media. This will lead to more reliable processes, particularly for recalcitrant compounds including chlorinated solvents, pesticides, explosives, and metals. Isolation and characterization of microorganisms capable of degrading these recalcitrant compounds are also essential.

References

Anonymous. 1999. Bioremediation. *Chemical Weekly*. 30 March.

Kraakman, N. J. R., J. W. van Groenstijn, B. Koers, and D. C. Heslinga. 1997. Styrene removal using a new type of bio-reactor with fungi. *Biological Gas Cleaning*. Maastricht, The Netherlands. W. L. Prins and J. Van Ham, eds. Düsseldorf: VDI Verlag. 28–29 April. 225–232.

Mata-Alvarez, J., S. Macé, and P. Llabrés. 2000. Anaerobic digestion of organic solid wastes. An overview of research achievements and perspectives. *Bioresource Technology*, 74:3–16.

Mulligan, C. N., and R. Galvez-Cloutier. 2000. Bioleaching of copper mining residues by *Aspergillus niger*. *Water Science and Technology*, 41(12):255–262.

Mulligan, C. N., R. N. Yong, and B. F. Gibbs. 2001. The use of selective extraction procedures for soil remediation. *Proceedings of the International Symposium On Suction, Swelling, Permeability and Structure of Clays*. Rotterdam: Balkema. 377–384.

———. 1999a. Metal removal from contaminated soil and sediments by the biosurfactant surfactin. *Environmental Science and Technology*, 33:3812–3820.

———. 1999b. A review of surfactant-enhanced remediation of contaminated soil. *Geoenvironmental Engineering-Ground Contamination: Pollutant Management and Remediation*. R. N. Yong and H. R. Thomas, eds. Thomas Telford: London. 441–449.

———. 1999c. On the use of biosurfactants for the removal of heavy metals from oil-contaminated soil. *Environmental Progress*, 18(1):50–54.

Pâques, J. H. J. 1997. Biological waste gas treatment: A competitive alternative. *Biological Gas Cleaning*. Maastricht, The Netherlands. 28–29 April. W. L. Prins and J. Van Ham, eds. Düsseldorf: VDI Verlag.1–4.

Schnoor, J. L. 1997. *Phytoremediation*. Technology Evaluation Report TE-98-01. Pittsburg, PA: GWRTAC.

Stewart, W. C., T. A. Barton, and R. R. Thom. 2000. High VOC loadings in biofilters: Petroleum and industrial applications. Presented at the *Air & Waste Management Association's 93rd Annual Meeting & Exhibition*. Salt Lake City, UT. 18–22 June.

Tilche, A., and F. Malaspina.1998. Biogas production in Europe. Paper presented at the *10th European Conference for Energy and Biomass*. Wurzburg, Germany. 8–10 June.

Tucker, M. D. 1997. Assessment of chemical and biological processes for treatment of mixed waste. Poster presented at *WERC/HSRC '97 Joint Conference on the Environment*. Albuquerque, NM. 22–24 April.

Woertz, J. 1999. Removal of nitric oxide in a fungal vapor-phase bioreactor. Paper F7-297. *Air & Waste Management Association's 92nd Annual Meeting & Exhibition*. St. Louis, MO. 20–24 June.

Yong, R. N. 2001. *Geoenvironmental Engineering: Contaminated Soils, Pollutant Fate and Mitigation*. Boca Raton: CRC Press.

Zechendorf, B. 1999. Sustainable development: How can biotechnology contribute? *Trends in Biotechnology*, 17:219–225.

Principles of Biological Treatment

Introduction

The main emphases of this chapter are the principles of biological treatment. After a short discussion of the sources of contaminants and their effects on human health, the various types of organisms involved are described. In particular, the growth and metabolism of bacteria are emphasized. The types of chemicals that can be biologically degraded or transformed, the pathways of conversion, the potential for genetic improvement, and the importance of acclimation are discussed. The toxicity and availability of several chemicals are also examined.

Understanding these aspects is very important because they will serve as the foundation of knowledge required for subsequent chapters on the biological treatment technologies for air, water, soil, and waste. Previously many people believed that biotreatment processes are unpredictable and unreliable. Many of these failures, however, have occurred due to a lack of understanding of the microbial requirements during the process. This chapter provides sufficient information to design biotreatment processes while considering the needs of the microorganisms. This enables readers to predict intermediate products and make informed choices about the most appropriate biotreatment method to use.

Sources of Contamination

Over the past decades the contamination of the environment through the water, land, and air has occurred due to the generation of waste by agricultural enterprises, households, industrial plants, and through the extraction and treatment of natural resources. Common contaminants include oil and grease, solvents, acids, cyanides, and metals (Figure 2.1). Some of the toxic effluents and their sources are shown in Table 2.1 while specific chemical wastes and their toxic ffects are shown in Table 2.2.

In 1984 the Office of Technology Assessment reported more than 200 substances in groundwater with highly variable concentrations, varying according to the place where the sample was taken and the time when it was taken (Protecting 1984). Of the substances, 175 were of organic nature, 37 were inorganic, and 19 radionuclides. Only a small percentage of the known substances have been thoroughly evaluated for their health hazards due to the extensive requirements to evaluate various concentration levels over time of exposure. Acute and chronic effects on humans have, in many cases, been extrapolated from animal studies. Some of the reported effects on humans, seen in Table 2.2, include eye and skin irritation, kidney damage, cancer, genetic mutation, and central nervous system, lung, and respiratory tract problems (Protecting 1984).

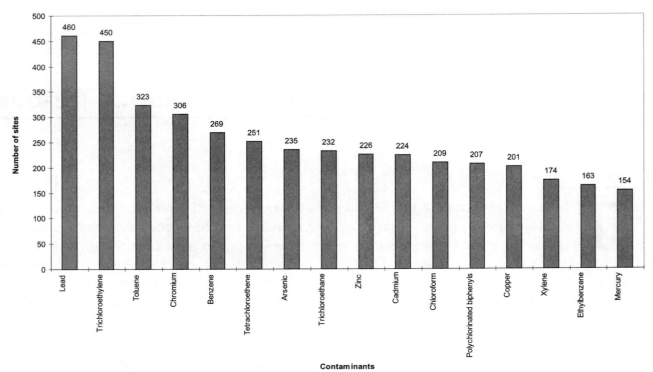

Figure 2.1 Common Contaminants at U.S. Superfund Sites

Table 2.1 Sources and Composition of Toxic Effluents from Selected Industries

Industry	Waste streams
Laboratories	acids, bases, heavy metals, inorganics, ignitable wastes, solvents
Flexography	acids, bases, heavy metals, inorganics, solvents, ink sludges, spent plating
Pesticides	metals, inorganics, pesticides, solvents
Construction	acids, bases, ignitable wastes, solvents
Metal finishing	acids, bases, reactives, heavy metals, inorganics, ignitable wastes, solvents
Formulators	acids, bases, reactives, heavy metals, inorganics, ignitable wastes, solvents, pesticides
Chemical manufacture	acids, bases, reactives, heavy metals, inorganics, ignitable wastes, solvents
Dry cleaning	dry clean filtration residue, solvents

Adapted from Yong et al. 1992

Table 2.2 Acute Effects of Some Hazardous Wastes on Human Health

Type of waste	Toxic effects					
	Nervous system	Gastric system	Neural system	Respiratory system	Skin	Death
Halogenated organic pesticide	X		X	X		X
Methyl bromide			X			
Halogenated organic phenoxy herbicide					X	
Organophosphorous pesticide	X		X	X		X
Organonitrogen herbicide		X				X
Carbamate insecticide	X		X	X		X
Dimethyldithiocarbonate fungicide				X		
Aluminum phosphide		X				
Polychlorinated biphenyls		X			X	
Cyanide wastes	X		X	X		X
Halogenated organics	X		X		X	
Non-halogenated volatile organics			X	X		
Zn, Cu, Se, Cr, Ni		X		X	X	
As		X			X	X
Organic lead compounds	X	X	X			
Hg	X	X	X			X
Cd		X		X		X

Adapted from: Governor's Office of Appropriate Technology, Toxic Waste Assessment Group, Calif. 1981

Treatment of Contaminants

Microorganisms—bacteria, protozoa, fungi, algae and viruses—are the key to the biological treatment of air, water, soil and wastes. These are smaller than plant and animal cells and are divided into two groups based on their cell structures: prokaryotes (simple, single cells less than 5 microns) and eukaryotes (single or multi-cells which are more complex and greater than 20 microns). Prokaryotes have a nuclear region not encompassed in a membrane and only a single strand of deoxyribonucleic acid (DNA) whereas eukaryotes have a nucleus surrounded by a membrane containing DNA molecules. Eukaryotes are subdivided into unicellular organisms that have multipurpose cells and multicellular organisms (plants and animals) with special-purpose cells.

Types of Organisms

Protozoa

Protozoa include pseudopods, flagellates, amoebas, ciliates, and parasitic protozoa. Their sizes can vary from 1 to 2,000 μm. They are aerobic, single-celled chemoheterotrophs and eukaryotes with no cell walls. In aerobic biological treatment processes, they are common, scavenging particles such as bacteria, yeasts, fungal spores, and other protozoa. Their concentrations can be as high as 5×10^4 protozoa per ml. Since they feed on single-celled bacteria, they are useful in creating clear water and reducing bacterial numbers. They can cause turbidity, though, if they are not agglomerated. Some protozoa such as *Entamoeba histolytica* and *Giardia lambia* can cause diseases such as amoebic dysentery and giardiasis, respectively. Protozoa are found in water and soil. Water is an absolute requirement for their survival. In soil, protozoa are useful in reducing bacterial numbers near injection wells. During bioremediation these can become clogged due to excessive growth (Sinclair et al. 1993). In general, though, protozoa cannot reduce contaminant concentrations.

Fungi

Fungi are aerobic, multicellular eukaryotes and chemoheterotrophs which use organic compounds as substrates for energy and carbon. They reproduce by formation of asexual spores. Compared to bacteria, they do not require as much nitrogen and are also more sensitive to changes in moisture levels. They are also larger than bacteria, and they grow more slowly and in a more acidic pH range (less than pH 5). Fungi mainly live in or on dead plants or in the soil; they are sometimes found in fresh water. Molds, yeasts, and mushrooms are all fungi. Molds (filamentous fungus) form mycelium for adsorbing nutrients from a substrate. They reproduce through formation of spores or spore sacs. *Mucor, Rhizopus,* and *Penicillium* are examples of different species of molds. Yeasts are unicellular organisms that are larger than bacteria (1 to 5 μm in diameter and 5 to 30 μm in length). They are shaped like eggs, spheres, or ellipsoids and reproduce by fission or formation of buds. *Neurospora* and *Candida* are two examples of yeasts. Yeasts are used industrially for the production of beer, wine, cheese, antibiotics, and many chemicals. They may or may not require oxygen. Mushrooms are composed of two main portions, the basidia, the fruiting body above the ground, and the mycelium, which is below the ground. The common mushroom is a well-known representative of this family of fungi.

Although fungi are useful for composting and in rotating biological contactors, they are detrimental to activated sludge processes due to their poor settling characteristics. Appropriate aeration of the soil is required for effective growth of the fungus. *Phaenaerochaete chrysosporium* (also known as white rot fungus) produces an enzyme, peroxidase, under nitrogen limitation, which has the potential for attacking chlorinated compounds such as dioxins (Aust. et al. 1988).

Algae

Algae are single-celled and multicellular microorganisms which are green, greenish-tan to golden brown, yellow to golden brown (marine), or red (marine). They grow in fresh and salt water, in the soil, and on trees. When they grow with fungi, they are called lichens. Seaweeds and kelps are examples of algae. Since they are photosynthetic, they can produce oxygen, new cells from carbon dioxide or bicarbonate (HCO_3^-), and dissolved nutrients including nitrogen and phosphorus. They use light of wavelengths between 300 and 700 nm. Red tides are an example of excessive growth of dinoflagellates in the sea. The green color in lakes and rivers is eutrophication and is due to the accumulation of nutrients such as fertilizers in the water. Although algae can cause problems in drinking water treatment due to the production of taste and odor, they can be useful in wastewater treatment since they can reduce inorganic nutrient concentrations and can produce oxygen for the bacteria required for the treatment process. Algae can also bioaccumulate hydrophobic compounds (Matsumura and Esaac 1979) or metals (Schiewer and Volesky 1998).

Viruses

Viruses are smaller than bacteria and require a living cell to reproduce. Their relationship to other organisms is not clear. Viruses interrupt the activity of the cell they invade so they can replicate. Within a protein coat, they consist of a strand of DNA and one of ribonucleic acid (RNA). A virus can only attack a specific host. For example, those that attack bacteria are called bacteriophages. In the field of remediation, there is a potential for introducing genes in bacteria by viruses, but this has not been fully explored. Viruses are also responsible for various diseases such as HIV, herpes, ebola, meningitis, gastroenteritis, hepatitis, and poliomyelitis. Thus, removal of these organisms is very important. Coagulation seems to remove viruses more effectively than sand filters. It is much more difficult to disinfect using chlorine and UV than when bacteria alone are present.

Animals (worms)

The most significant animals are millimeter size worms. Some nematodes can be useful in wastewater treatment systems since they can prevent excessive growth in percolating filters by detaching some of the biofilm on the support media. They are cylindrical in shape and are able to move within bacterial flocs. Nematodes cannot contribute to activated sludge treatment since their reproduction times are too long as compared to the wastewater residence time. They enter treatment plants from stormwater drains. Flatworms such as tapeworms, eel worms, roundworms, and threadworms, which are nematodes, can cause diseases such as roundworm, hookworm, and filarisis.

Plants

Plants have a very evolved cell structure. They are classified as eukaryotes and include aquatic plants, crop plants, macrophytes, seed plants, grasses, ferns, and mosses. Recently, plants have been used to remove, degrade, or accumulate chemical contaminants in the soil, sediment, groundwater, surface water, or air. Plants can be used to treat petroleum hydrocarbons, chlorinated solvents, pesticides, metals, radionuclides, explosives, and excessive nutrients (Chappell 1997). The mechanisms for contaminant treatment by the plants are shown in Table 2.3. Genetic engineering or selective breeding of the plants can lead to faster uptake and higher yields of contaminants. Some examples of plant species that accumulate metals include *Thlaspi calaminare* (zinc accumulation), *Alyssum* species (nickel accumulation), *Brassica juncea* (lead accumulation) (PRC Environmental 1997), and *Populus* (TCE remediation) (Chappell 1997).

Table 2.3 Mechanisms of Contaminant Removal by Plants

Method	Mechanism	Media
Rhizofiltration	Absorption of metals by plant roots	Surface and pumped water
Phytotransformation	Uptake and degradation of organics	Surface water, groundwater
Plant-assisted bioremediation	Enhanced microbial degradation in rhizosphere	Soils and groundwater in rhizosphere
Phytoextraction	Uptake and accumulation of metals into plant tissue	Soils
Phytostabilization	Excretion of components from the roots which precipitate metals	Soils, groundwater, mining residues
Phytovolatization	Evapotranspiration of selenium, mercury, and volatile organics	Soils, groundwater
Removal of organics from air	Uptake of volatile organics by leaves	Air
Vegetative caps	Evapotranspiration of rainwater to reduce contaminant leaching	Soils, solid wastes

Source: Modified from Chappell 1997

Bacteria

Bacteria are prokaryotes and reproduce by binary fission by dividing into two cells in about 20 minutes. The doubling time, the time it takes for one cell to become two, however, depends on the temperature and species. For example, the optimal doubling time for *Bacillus subtilis* (37°C) is 24 minutes while for *Nitrobacter agilis* (27°C) it is 20 hours. Classification is by shape, including rod-shaped (bacillus), spherical-shaped (coccus), and spiral-shaped (spirillum). Rods usually have a diameter of one-half to one micron and a length of three to five microns. The diameter of spherical cells can vary from 0.2 to two microns. Spiral-shaped cells can be 0.3 to five microns in diameter and six to 15 microns in length. The cells can be in clusters, chains, or as single cells and may or may not be motile. The substrate of the bacteria must be soluble. Bacteria in most cases are classified according to the genus and species such as *Pseudomonas aeruginosa* and *Bacillus subtilis*. Some of the most common species are *Pseudomonas, Arthrobacter, Bacillus, Acinetobacter, Micrococcus, Vibrio, Achromobacter, Brevibacterium, Flavobacterium,* and *Corynebacterium*. Various strains exist within species which can behave differently. Some strains can survive in certain conditions while others cannot. The ones that are better adapted will survive. Some strains, called mutants, originate due to a mistake in the genetic copying mechanisms. Some species are dependent on other species for survival. Degradation of chemicals to an intermediate stage by one species of bacteria may be required for the growth of another species that utilizes the intermediate.

Makeup of Bacterial Cells

A schematic of the bacterial cell is shown in Figure 2.2. The main components of the cell are the cytoplasm, the nucleus, the cell membrane, and the cell wall. Proteins, carbohydrates, and other complex materials are contained in the cytoplasm. The synthesis of numerous components such as proteins occurs here. The cytoplasm is a tightly-packed granular region containing ribosomes of

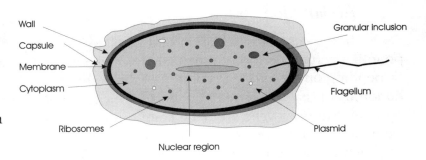

Figure 2.2 Schematic Diagram of Bacterial Cell

ribonucleic acid (RNA), proteins, enzymes for catalysis of chemical reactions, plasmids, glycogen and poly-ß-hydroxybutyrate (PHB) granular inclusions, polyphosphate (volutin granules for energy metabolism), and sulfur. *Acinetobacter* has been used to accumulate volutin for phosphorus removal processes (Metcalf and Eddy 1991). The nucleus controls the reproduction of the cell while the cell membranes, made of phospholipids and proteins, regulate what enters and leaves the cell. In cells that require oxygen, the electron transport system enzymes, producing adenosine triphosphate (ATP) and consuming oxygen, are attached to the cell membrane of 7 to 8 nm in thickness. The purpose of the cell wall is to maintain the shape of the cell and protect the interior of the cell from toxic materials and antibiotics. The slime layer or capsule can help maintain the moisture of the cell even in a dry environment. The flagella are found only in motile bacteria since they assist in movement. Enzymes can be endocellular (inside the cell) for synthesis of proteins or exocellular (on the cell wall) for transport or to break down large molecules so that they can enter the cell.

Growth of Microorganisms

The growth of microorganisms can be represented by several phases including lag, exponential or log growth, declining growth, stationary growth, and death (Figure 2.3). In the lag phase, the microorganisms acclimate to the substrate and environment (i.e., pH, temperature, soil type, etc.) before growth. During this time, bacteria must produce the appropriate enzymes for the new growth medium. There is no increase in the number of cells at the beginning, but as the metabolism of the cells increases, the number of cells starts to increase. This phase can be decreased by increasing both the amount and the age of the inoculum. This period can last between one and 20 hours or longer.

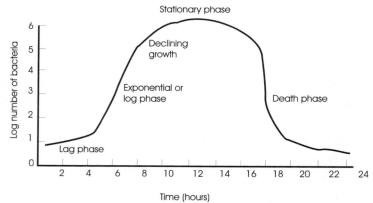

Figure 2.3 Bacterial Growth as a Function of Time

Acclimation Periods

In the case of inoculation of an isolated culture for degradation of a particular compound, a certain period of time called acclimation is required before metabolism can begin. Depending on the biodegradability of the compound, the length of time can be as long as hundreds of days. Substrates such as herbicides, insecticides, quaternary ammonium compounds, and polycyclic aromatic compounds are particularly difficult to acclimatize. For example, several aromatics can require ten to 30 hours for acclimation in soil, whereas others, such as the pesticide 2,4,5-T, can require four to ten weeks (Alexander 1994). The concentration and environmental factors such as pH, temperature, oxygen, and nutrient availability may also affect the length of time for acclimation to take place. This extended exposure to the chemical may be necessary to induce (which normally takes only a few hours) or to develop an enzyme system due to genetic modification of the microbial population. Mutation can occur in a population or from the transfer of genetic material between species. If the enzymes of the mutant can degrade a substrate better, this mutant will have an advantage over non-mutated organisms and will grow better. The time for acclimation may be due to the time required for a mutant or mutants to dominate (Alexander 1994).

Log Growth Phase

The maximum rate of growth and metabolism occurs during the log growth phase. It is called this since the log of the number of cells versus time is linear (Monod 1949). The following equation represents the growth:

$$\frac{dN}{dT} = kN$$

where N = number of cells per volume, t = time, and k is the growth rate in t^{-1}. Integration of this equation for time t = 0, when N = N_o to time t gives the equation:

$$N = N_o e^{kt}$$

At the doubling time, t_d, when N = $2N_0$, the specific growth rate can be determined during exponential growth as:

$$k = \frac{\ln 2}{t_d}$$

Although the doubling time can vary from minutes to days, usually it is from ten to 60 minutes. As the substrate, oxygen, or nutrient becomes limiting or is depleted (which takes place as the treatment process is nearing completion), pH shifts, or toxic components start to accumulate, the declining growth phase occurs. Those microorganisms capable of storing organics earlier in the growth phase, such as glycogen or poly-ß-hydroxybutyrate (PHB), will survive very well in this phase. The rate of growth decreases and microorganisms begin to die. The stationary phase occurs as the numbers of cells produced and dying are equal. This usually occurs over a 12- to 36-hour period.

When the number of microorganisms dying exceeds the number growing, this signifies the death phase. The cells may either be inactivated or start to degrade and break apart (lyse). An equation like the growth phase can be used as follows:

$$\frac{dN}{dt} = -bN$$

where b is the rate constant for the decrease in the cell number.

Classification of Microorganisms

Microorganisms are classified by energy or carbon requirements. Most of the bacteria used for treatment require organic substrates for energy and are called chemoorganotrophs. If they use the organics as a carbon source, they are also called heterotrophs. Those that use inorganic compounds as an energy source are named chemolithotrophs. Nitrifying bacteria (*Nitrosomonas* and *Nitrobacter*), however, use carbon dioxide for carbon instead of organic compounds and are named autotrophs, a class of chemolithotrophs. They are called nitrifying since they produce nitrite from ammonium ions, which is then followed by conversion to nitrate. Photolithotrophs use light as an energy source and inorganic compounds for respiration while photoorganotrophs use light with organic compounds for respiration. These organisms have chlorophyll to adsorb the light at specific wavelengths. In plants, blue (430 nm) and red light (680 nm) are adsorbed by chlorophyll. Purple and green bacteria have bacteriochlorophyll, adsorbing light at wavelengths of 720 to 780 nm, 850 nm, or 1,020 nm. Photoautotrophic organisms convert light to chemical energy via adenosine triphosphate (ATP) and then reduce carbon dioxide to cellular materials also known as carbon dioxide fixation. Electron donors can include sulfur, hydrogen sulfide, thiosulfate, hydrogen, or organic compounds.

Nutrition of Microorganisms

Microorganisms have specific nutritional needs. Carbon, hydrogen, oxygen, nitrogen, and phosphorus are the major elements required (Table 2.4). The most-used empirical formula for the cell is $C_5H_7NO_2$ (Porgues et al. 1953). The incorporation of phosphate leads to the formula $C_{42}H_{100}N_{11}O_{13}P$ (McCarty 1965). Sulfur, potassium, calcium, sodium, magnesium, and chlorine are some of the minor elements. Other substances such as iron, cobalt, copper, zinc, molybdenum, boron, and aluminum are required in trace amounts. Vitamins, essential amino acids, and other precursors may also be required for growth since these are not produced by the cells. Due to deficiencies in the soil, water, or waste that is to be treated by the microorganism, some of these components may have to be added so that the microorganism can grow and perform in an optimal manner.

Classification of Bacteria

Bacteria can also be classified depending on their requirements for oxygen. They can either require oxygen (aerobes) or not require oxygen (anaerobes) for respiration (Frobisher et al. 1974). Energy is obtained from the respiration process by the release of hydrogen from a donor enzymatically, which results in oxidation. When the hydrogen atom meets an acceptor, reduction occurs. Oxygen is the hydrogen acceptor during aerobic respiration. It is the best one since the most free energy is released this way. Thus, aerobic microbes can grow faster than anaerobic strains. Water is produced as the end product. Carbonate, iron, nitrate or sulfate ions, or an organic compound can serve as the hydrogen acceptor by formation of oxygen radicals in anaerobic respiration. Methane, ammonia, hydrogen sulfide, or a reduced organic compound are the final products depending on the hydrogen acceptor.

Table 2.4 Composition of Bacterial Cells on a Dry Basis

Element	Dry weight (percent)	Physiological function
Carbon	50	Component of cell
Oxygen	20	Component of cell and water
Nitrogen	14	For proteins, nucleic acids, coenzymes
Hydrogen	8	Component of water and cell
Phosphorus	3	Component of nucleic acids, phospholipids, coenzymes
Sulfur	1	Component of coenzymes and proteins
Potassium	1	Cation for cell processes
Sodium	1	Cation for cell processes
Calcium	0.5	Cation for cell processes and enzyme co-factor
Magnesium	0.5	Cation for cell processes and co-factor in ATP reactions
Chlorine	0.5	Anion for cell processes
Iron	0.2	Component of cytochromes, proteins, and enzymes

Adapted from Stanier et al. (1986)

Bacterial Metabolism of Nitrogen

In nitrate respiration, nitrate is reduced to NO_2^-, then N_2O, NO and finally, nitrogen gas. This reaction, known as denitrification, is a naturally occurring process that removes nitrogen from the soil and releases it into the atmosphere. This process is also called dissimulative nitrate reduction when it is performed by bacteria. When ammonia and then amino acids and proteins are produced for growth, this process is called assimilative nitrate reduction. Bacteria, fungi, and plants are able to perform this process. This latter process can occur when oxygen is present whereas the former process cannot. Microorganisms that can use oxygen for respiration if it is available, or other compounds if it is not, are called facultative anaerobes. Denitrifying bacteria are always facultative. These are commonly found in activated sludge processes. Obligate anaerobes do not require oxygen, which can also be toxic to some species of bacteria.

Bacterial Metabolism of Sulfur

Sulfate-reducing bacteria can convert sulfate to hydrogen sulfide, H_2S, or sulfur, S.

Sulfur can be used for cell synthesis in assimilatory reactions or as an electron acceptor for energy generation in dissimilatory reactions. In the latter case, hydrogen sulfide is the final gaseous product. Fewer than 12 species of bacteria are capable of performing this process. Less energy is obtained from sulfate than oxygen or nitrate. Acetate, lactate, or pyruvate can be used as organic substrates by some sulfate-reducing bacteria while hydrogen gas can be used by some lithotrophic bacteria.

Oxidation of hydrogen can be performed by hydrogen lithotrophic bacteria. If hydrogen is used to form cell material, the following reaction takes place:

$$6H_2 + 2O_2 + CO_2 \rightarrow CH_2O + 5H_2O$$

As in most cases, oxygen is required and carbon dioxide is used as the carbon source. CH_2O is the general formula of cell material. Some bacteria are facultative and can use organic material for energy if hydrogen is not available. Some can also use organic compounds as the carbon source but use hydrogen for energy, such as the bacteria *Escherichia coli,* which are used to indicate contamination in water.

Sulfur bacteria are facultative autotrophs that use sulfur, hydrogen sulfide, thiosulfate, and organic sulfides or organic materials for energy. Typical reactions are

$$H_2S + 2O_2 \rightarrow SO_4^{2-} + 2\,H^+$$

$$S^o + H_2O + 3/2O_2 \rightarrow SO_4^{2-} + 2\,H^+$$

$$S_2O_3^{2-} + H_2O + 2O_2 \rightarrow 2SO_4^{2-} + 2\,H^+$$

The conversion of hydrogen sulfide to sulfate is a multistep process in which sulfur is produced. As hydrogen sulfide is depleted in some species, the sulfur inside the cell can then be used as the energy source. Those microorganisms that use the very insoluble sulfur must grow on the sulfur solids, slowly oxidizing the gradually soluble sulfur. In all of these reactions, sulfuric acid is produced, lowering the pH to 1 in high concentrations. *Beggiatoa* is frequently found in sediments containing high levels of hydrogen sulfide and requires organic materials for carbon (Eweis et al. 1998).

Actinomycetes

Actinomycetes are mainly aerobic prokaryotes that are multicellular, filamentous, rod-shaped bacteria. Like fungus, they can produce spores or spore chains called conidia. They are commonly found in soil and water and often cause odor and taste problems in water. They can degrade lignocellulosics, phenols, aromatics, steroids, and chlorinated aromatics (U.S. EPA 1983). Actinomycetes are very resilient since they can resist desiccation and withstand an extensive range of pH and temperature values.

Environmental Effects on Microorganisms

Physical Factors

Environmental factors are important in the optimal performance of the microorganism. The most important physical factors are temperature, oxygen availability, presence of moisture, and osmotic pressure.

Temperature

There are three main temperature ranges in which the various classes of microorganisms can grow optimally. Psychrophiles grow at 0 to 10°C, mesophiles grow at 10 to 45°C, and thermophiles at 45 to 75°C. Mesophiles are the most common microorganisms, particularly for biological treatment processes. Few microorganisms can survive extended exposure to temperatures above 75°C. Capsule formation, however, can enable microorganisms to survive at reduced temperatures and grow as the temperature increases to more favorable conditions (Sims et al. 1990). In general, the U.S. EPA uses a rule that a 10°C increase in temperature can double the growth rate.

Water and Oxygen

Moisture is essential for the survival of microorganisms since it is the principal component of cell protoplasm and is required for nutrient transport into the cell. Although this is not usually a problem in wastewater treatment applications, it can become limiting in the event of frozen water in the winter or dryness of the biofilter, soil, or waste. In general, moisture levels of 50 to 75% are optimal (U.S. EPA 1985). Oxygen limitations can occur in the soil as a result of too much moisture causing the soil pores to become saturated with water. Transport of oxygen to bacteria in soil remediation sites can be a problem since oxygen solubility in water is low (9 mg/L at 20°C) and the soil is very heterogeneous. Since oxygen is extremely important for aerobic and facultative anaerobic bacteria, sufficient quantities must be available. In general, a concentration of approximately 2.0 mg/L is considered a good design value (Reynolds and Richards 1996).

Osmotic Pressure

Another factor that can influence microbial activity is osmotic pressure since the salt concentration affects the intake of nutrients by osmosis. A range of 500 to 35,000 mg/L is an acceptable value for salt concentrations.

Contaminant Availability

Availability of the contaminant to the microorganism is another important factor, particularly when the contaminant is sorbed onto a solid surface such as soil. Heavy petroleum components and others can sorb strongly onto soil. Surfactants have been added in soil flushing situations to enhance the desorption of the contaminants (Aronstein et al. 1991). For *ex situ* situations after excavation of the soil, mixing of the soil with water can enhance the desorption by breaking up larger particles into smaller ones, exposing more surface area to the water, increasing the scrubbing of the soil, and enabling the release of more components from the soil into the water. Components can also penetrate small micropores less than one micron in diameter where bacteria cannot easily grow or their enzymes cannot penetrate. Molecules can also form complexes with humic substances in soils. These complexes can be resistant to microorganisms and thus remain undegraded for extended periods of time (Alexander 1994).

Chemical Factors

Some of the major chemical factors which effect microbial activity include pH, toxicity, heavy metals, molecular structure, and co-metabolism.

Most microbes prefer a pH range of 6.5 to 9.0. Some microorganisms, however, can function at high or low pH values. For example, fungus grows at pH values below 5 while sulfur-oxidizing bacteria produce sulfuric acid which lowers the pH below 1 (e.g., H_2S $+2O_2 \rightarrow H_2SO_4$). Adjustment to obtain the preferred pH may be necessary before biological treatment, particularly in the case of industrially generated wastewaters or wastes. Once a neutral pH is established, generation of carbon dioxide by the microorganisms can help to buffer the system and maintain the neutral pH. In soil or water, pH can be raised through the addition of lime, calcium hydroxide, calcium carbonate, magnesium carbonate, etc. To lower soil pH, sulfur compounds such as sulfuric acid can be used (Dupont et al. 1988).

Toxicity

Toxicity is the inhibition of microbial growth or substrate utilization. The process of degradation of substances which are normally biodegradable can be slowed or stopped by the presence of a toxic component. Several mechanisms are possible. Inhibition of a single enzyme in a metabolic pathway can adversely affect microbial growth or substrate utilization. Sulfonylurea is known to specifically inhibit acetolactate synthase, which is required for the synthesis of amino acids, the basis of proteins (LaRossa and Schloss 1984). Mercuric, cupric, or uranyl metal ions can nonspecifically inhibit the functioning of the cell membrane by binding to the cell surface and causing leaks (Khovrychev et al. 1974). This can cause many problems in maintaining concentration gradients across the cell. Cyanide and azide are toxic in a different manner since they can bind to heme iron, present in cytochromes, which are required for many electron transport reactions (Brock et al. 1984). This binding eliminates the cytochromes from taking part in these reactions and subsequently inhibits cell energy-generating processes. Organic acids, ethers, aldehydes, phenol, chlorophenol, cyanide, ammonia compounds, antibiotics, and dyes can also be toxic, depending on their concentrations. Grease can inhibit respiration by coating the microorganism.

Heavy Metals

Numerous components can be toxic to microorganisms at low or high concentrations. Contact time, temperature, and concentration of these agents will influence their effect on the microbes. In general, lower concentrations at short contact times are less toxic. For example, 2,4-D will degrade much faster at 1 to 100 ppm than at higher concentrations where it can remain undegraded for years. Toxic compounds include acids such as benzoic acid, bases such as ammonium hydroxide, oxidizing and reducing agents, and halogens (chlorine, iodine, bromine, fluorine). Heavy metals such as mercury, arsenic, and lead are particularly toxic. Other toxic heavy metals in wastewaters include chromium, zinc, cadmium, copper, barium, and nickel.

Molecular Structure

Molecules that are too large to enter the microbial cell (e.g., high molecular weight or long chain lengths), of low water solubility, or that contain structural factors that inhibit metabolism can be difficult to degrade. Amine, methoxy, sulfonate and nitro groups, benzene with meta substitution, branched carbon chains, and ether linkages are some examples of structural factors that affect biodegradability. Biodegradability is dependent on nature, the amount of substitution, and the degree of saturation or branching. Water solubility is important since components enter the cell by water transportation. Increasing the saturation in a compound tends to decrease its solubility. The substrate to be degraded can become unavailable to the microorganism by sorption onto a solid substrate such as soil or a waste, or it may have limited water solubility.

A high level of branching can also decrease biodegradability (Evans et al. 1988). N-octane (a straight chain compound) is easier to biodegrade than its branched isomer, 3-propyl pentane. Replacement of carbon atoms with sulfur, phosphorus, or chlorine atoms can decrease biodegradability. Substitution with nitrogen atoms can also adversely affect degradability, with the exception of amino acids and proteins, which are vital for metabolism.

Co-metabolism

Co-metabolism is the degradation of a substrate in the presence of another substrate (a co-substrate). The co-substrate, which is not required for energy or nutrient purposes, is converted when the enzymes are induced for utilization of the principal substrate. The concentration of the co-substrate must not be too high, though, to cause competition for enzymes with the substance required for growth and thus inhibit its degradation. Pulsing of the co-substrate into the soil to assist biodegradation of a contaminant is one technique that has been used (McCarty 1988). Co-metabolism of difficult to degrade components such as trichloroethene (TCE), dioxin, and polychlorinated biphenyls (PCBs) has been performed only on a small scale (Eweis et al. 1998). Another situation that can occur is that the more easily degraded component is depleted before the other compound can be degraded. This is called diauxic growth.

Biological Factors

Microbial activity can be affected by competition from undesirable organisms. Growth of *Sphaerotilis* in activated sludge can cause poor settling due to its filamentous characteristics. Most of the bacteria in the activated sludge are single-celled bacteria that have formed an agglomerated floc from the slime layers of the cells. The presence of the multicellular *Sphaerotilis* causes the floc to be less compact and more difficult to settle.

Where Microorganisms Are Found

Although the numbers and types of microbes can vary widely depending on the pH, moisture content, soil type, and type of nutrients available, they are found everywhere in the soil, surface water, and groundwater. They can be attached to plant roots due to the availability of organic and inorganic nutrients. Since most of the plant roots are, by definition, found in the soil layer called the rhizosphere, this is where the highest microbial activity is found (Tate 1995). Contaminant degradation rates are higher here and acclimation periods are decreased. Nutrient contents are also higher. Protozoa, algae, and fungi are also found but in lower amounts than bacteria which are found at the rate of 10^7 to 10^{10} units per gram of soil. The numbers of bacteria, decrease to 10^6 to 10^7 units per gram of soil deeper into the soil since oxygen and nutrient levels decrease. Fungi, algae, and protozoa can still be found.

Substrate Utilization

The bacterial growth rate is dependent on the environmental conditions, such as the substrate concentration. The higher the substrate concentration, the higher the growth rate. The substrate can be a single compound or a mixture. In the latter case, the terms biochemical oxygen demand (BOD), chemical oxygen demand (COD), total organic carbon (TOC), or total petroleum hydrocarbons (TPH) can be used on a mass or liquid basis since these parameters are an indication of the total substrate concentration. The Monod equation (1949) is:

$$\mu = \frac{\mu_{max} S}{K_S + S}$$

where

μ = specific growth rate, time^{-1}

μ_{max} = the maximum specific growth rate, time^{-1}

S = substrate concentration, mass/volume

K_S = substrate concentration when the growth rate is half of the maximum growth rate, mass /volume.

A plot of the growth rate versus the substrate concentration is shown in Figure 2.4. Maximum growth, μ_{max}, occurs at the highest substrate concentration. As the substrate concentration decreases, the growth rate also decreases. It also can be seen that S is equal to K_S at one-half of the μ_{max} value. This value is dependent on pH, temperature, nutrient composition, and bacterial species. Lower K_S values mean that the microorganism has a higher affinity for the substrates. For example, K_S

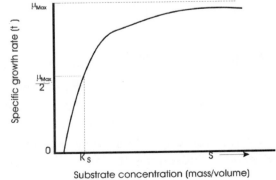

Figure 2.4 Effect of Substrate Concentration on the Rate of Growth According to the Monod Model

values for *Escherichia* for glucose and lactose of 4.0 and 20.0 mg/L, respectively, have been determined, which means that this microorganism has a higher affinity for glucose. In biological treatment processes, the substrate is the contaminant. This equation can be used to design wastewater treatment systems by determination of the retention time. Due to heterogeneity and complexity, soil treatment systems are much more difficult to model.

Biodegradation of Chemicals

Petroleum Hydrocarbons

Petroleum contains a wide variety of compounds ranging from very biodegradable to recalcitrant. As mentioned earlier, this is due to their various degrees of branching, chain lengths, molecular sizes, and substitution with nitrogen, oxygen, or sulfur atoms. A wide variety of microbial species would then be required for efficient degradation of these components. Usually aerobic bacteria are used initially (Atlas 1991), followed by anaerobic ones which can utilize nitrogen or sulfur compounds as electron acceptors (Bartha 1986).

Alkanes

The various classes of petroleum hydrocarbons are alkanes, cycloalkanes, aromatics, polycyclic aromatic hydrocarbons, asphaltenes, and resins. Although alkanes, represented by the formula C_nH_{2n+2}, where n is the number of carbons and 2n+2 is the number of hydrogens, can have many isomers as the number of carbons increases, relatively few exist in petroleum. Low molecular weight alkanes are the most easily degraded by microorganisms despite the harsh conditions required for chemical degradation (strong acids, high temperatures, ultraviolet light, etc.) (Morrison and Boyd 1973). As the chain length is increased to between C_{20} and C_{40}, these compounds become more hydrophobic and less soluble, decreasing the rate of biodegradation. Conversion of alkanes occurs mainly via formation of an alcohol using a monooxygenase enzyme, followed by oxidation to an aldehyde and then to a fatty acid (Pitter and Chudoba 1990). Further oxidation of the fatty acid occurs, leading to products which are less volatile than the original contaminants. The oxidation of the fatty acids is called ß-oxidation and is required for growth by many aerobic bacteria. Since oxygen is not required for this reaction, anaerobic bacteria such as sulfate-reducing bacteria are capable of degrading the fatty acids via this step (Widdel 1988).

Many of the alkanes found in petroleum are branched. Although bacteria that are capable of degrading n-alkanes cannot degrade branched ones (Higgins and Gilbert 1978), *Brevibacterium ethrogenes, Corynebacteria* sp., *Mycobacterium fortuitum, Mycobacterium smegmatis*, and *Nocardia* sp. have been shown to grow on branched alkanes. The first degradation step is the same as for the unbranched alkanes; however, the ß-oxidation is more difficult and less efficient (Pirnik 1977). In addition, in the presence of n-alkanes, the metabolism of the branched alkanes will be repressed, which will cause difficulties during the degradation of mixtures such as petroleum.

Alkenes

Alkenes, which contain a double bond between carbons, have not been extensively studied for biodegradation. It has been postulated, however, that those containing the double bond on the first carbon are more easily degradable than those alkenes with the double bond at another position (Pitter and Chudoba 1990). The products of oxidation of 1-alkenes can be either diols or the methyl group.

Cycloalkanes

Cycloalkanes are not as degradable as alkanes due to their cyclic structure. Their biodegradability also decreases as the number of rings increases, partly because of decreasing solubility (Pitter and Chudoba 1990). Bacteria that grow on cycloalkanes are not frequently found and are difficult to isolate. Specific enzymes are needed to cleave the ring. A species of *Nocardia* and *Pseudomonas* are able to use cyclohexane as a carbon source. Oxidation of the cycloalkanes with the oxidase enzyme leads to a cyclic alcohol and then formation of a ketone (Bartha 1986). The ketone is then converted to a lactone by insertion of an oxygen in the ring. In the case of cyclohexane as the initial substrate, the lactone (e-caprolactone) is hydrolyzed to adipic acid, which is a substrate for ß-oxidation (Perry 1984).

Aromatics

Aromatics, compounds containing benzene structures, are more difficult to degrade than cycloalkanes. Benzene, toluene, ethylbenzene, and xylene, commonly called BTEX, are volatile, water soluble, hazardous components of gasoline. Benzene is a known carcinogen. Degradation is catalyzed by oxygenases from fungi and other eukaroyotes to form transdihydrodiols. *Dihydrodiols*, formed by dioxygenases from bacteria, have been produced from naphthalene, anthracene, biphenyl, benzene, toluene, xylene, benzo(a)pyrene, and phenanthrene (Gibson 1988). The dihydrodiol is then converted to catechol by dehydrogenation. Catechol is then either converted by the ortho pathway to produce muconic acid (splitting of the ring between carbons with the hydroxyl groups) or by the meta pathway to produce *2-hydroxymuconic semialdehyde* (splitting of the ring at the carbon with a hydroxyl group) (Cerniglia 1984). Further reactions to produce easily degradable acids are shown in Figure 2.5.

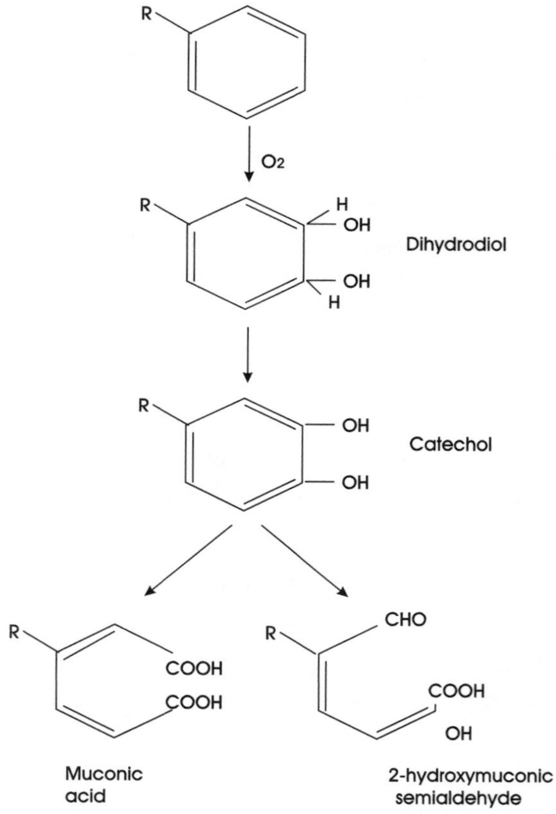

Figure 2.5 Oxidation of Aromatic Compounds

Table 2.5 Properties of Some Examples of PAHs

Compound	Molecular formula (molecular weight)	Molecular structure	Solubility (mg/L at 25°C)	Vapour pressure, mmHg (°C)
2-ring: Naphthalene	$C_{10}H_8$ (128)		31.7	0.082 (25)
3-ring: Anthracene	$C_{14}H_{10}$ (178)		0.045	1.9×10^{-4} (20)
Phenanthrene	$C_{14}H_{10}$ (178)		1.0	6.8×10^{-4} (20)
4-ring: Pyrene	$C_{16}H_{10}$ (202)		0.132	2.5×10^{-6} (25)
Benzo(a)anthracene	$C_{18}H_{12}$ (228)		9.4×10^{-3}	1.0×10^{-8} (20)
Chrysene	$C_{18}H_{12}$ (228)		1.8×10^{-3}	6.3×10^{-9} (25)
5-ring: Benzo(a)pyrene	$C_{20}H_{12}$ (252)		1.2×10^{-3}	5.6×10^{-9} (25)
6-ring: Benzo(g,h,i)perylene	$C_{22}H_{12}$ (276)		0.7×10^{-3}	1.0×10^{-10} (25)

Adapted from Eweis et al. 1998

Polycyclic or Polynuclear Aromatics Hydrocarbons (PAHs)

Polycyclic or polynuclear aromatic hydrocarbons are components of creosote and are produced during petroleum refining, coke production, and wood preservation (Park et al. 1990). Many are suspected to be carcinogens. A general form for PAHs is $C_{4n+2}H_{2n+4}$, where n is the number of rings. A four-ring PAH would be $C_{18}H_{12}$. Like cycloalkanes, as the number of rings increases, these compounds are more difficult to degrade due to decreasing volatility and solubility. Some examples are shown in Table 2.5. PAHs are also degraded in a manner similar to single ring aromatics since PAHs are degraded ring by ring. However, these simple aromatics are required to induce the enzymes necessary for PAH degradation (Atlas 1991). If enzymes are not induced, limited biodegradation of the PAHs may take place. Fungi, due to their oxidase production, can significantly contribute to biodegradation.

Halogenated Aliphatic Compounds

Examples of halogenated aliphatic compounds are pesticides such as ethylene dibromide (DBR) or $CHCl_3$, $CHCl_2Br$, and industrial solvents including methylene chloride and trichloroethylene. Aerobic degradation can be more difficult because of the lower energy and higher oxidation state of the compound due to the presence of the halogen. Thus, anaerobic biodegradation is more beneficial as the number of halogens in the compound increases. Aerobic degradation becomes more difficult. Methylene chloride, chlorophenol, and chlorobenzoate are the most aerobically biodegradable. The ease of replacement of the halogens in decreasing order is as follows due to decreasing electronegativity (the ability to accept electrons): iodine, bromine, chlorine, and fluorine. Removal of the halogen and replacement by a hydroxide group is often the first step of the degradation process, particularly when the carbon chain length is short.

$$R - X + H_2O \rightarrow R - OH + HX$$

Dichloromethane is such an example where formaldehyde, 2-chloroethanol, and 1,2-ethanediol are intermediates and carbon dioxide is the final product (Pitter and Chudoba 1990). If the chain is long, the methyl group at the end of the molecule will be oxidized to yield a halogenated alcohol, since the halogen has less influence on the reaction. Biodegradation of chlorinated ethenes involves formation of an epoxide and hydrolysis to carbon dioxide and hydrochloric acid. Reductive dehalogenation can occur anaerobically and involves replacement of the halogen with hydrogen or formation of a double bond when two adjacent halogens are removed (dihalo-elimination). Perchloroethylene (PCE) and trichloroethylene (TCE) can be reduced to form vinylidene and vinyl chloride, which are more toxic and volatile than the original compounds (Freeman and Gossett 1989). Another mechanism, induction of monooxygenase or dioxygenase enzymes, can lead to the co-metabolism of TCE by methanotrophs (Alvarez-Cohen and McCarty 1991).

Pesticides

The pesticides DDT, 2,4-D and 2,4,5-T, plasticizers, pentachlorophenol, and polychlorinated biphenyls, among others, are examples of halogenated aromatic compounds. Their stability and toxicity cause great concern for the environment and public health. The halogenated aliphatic compounds and the position and number of halogens are important in determining the ease and mechanism of biodegradation. In addition, the mechanisms of conversions are also similar and include hydrolysis (replacement of halogen with hydroxyl group), reductive dehalogenation (replacement of halogen with hydrogen), and oxidation (introduction of oxygen into the ring causing removal of halogen). Unlike the other two, the latter mechanism can take place only in the presence of oxygen. Reductive dehalogenation will more likely occur as the number of halogens increases.

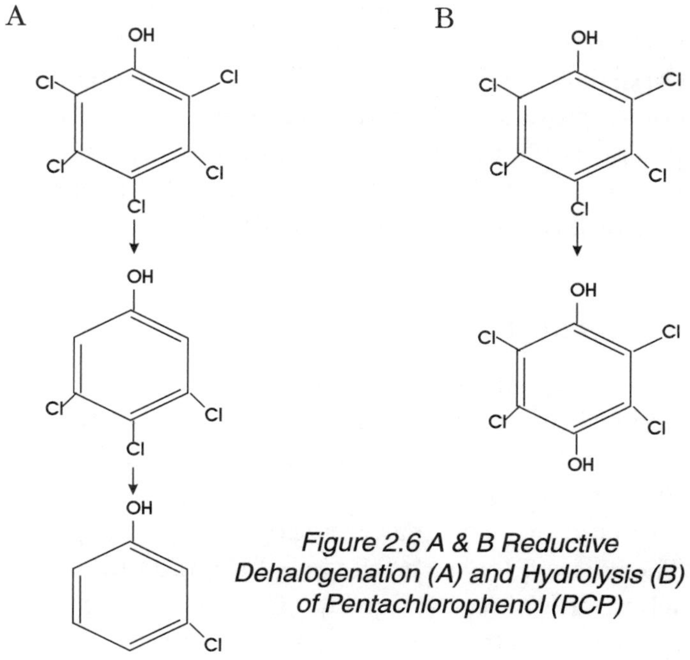

Figure 2.6 A & B Reductive Dehalogenation (A) and Hydrolysis (B) of Pentachlorophenol (PCP)

Ring cleavage could occur before oxidation, reduction, or substitution of the halogen.

Chlorophenoxy herbicides and chlorobenzene can be oxidized to halocatechols, which is then followed by ring cleavage (Pitter and Chudoba 1990). Bacterial strains of *Pseudomonas* sp., *Acinetobacter calcoaceticus,* and *Alkaligenes eutrophus* have been able to degrade aromatic halogenated compounds in this manner (Reineke and Knackmuss 1988). The cleavage for chlorobenzene can occur either at the ortho position to form chloromuconic acid or at the meta position to form chlorohydroxymuconic semialdehyde. Subsequent dehalogenation can be spontaneous (Reineke and Knackmuss 1988).

Chlorinated benzoates (Suflita et al. 1983), 2,4,5-T pesticides, PCBs (Thayer 1991), and 1,2,4-trichlorobenzenes (Reineke and Knackmuss 1988) are known to undergo reductive dehalogenation under anaerobic conditions. *Rhodococcus chlorophenolicus* (Apajalahti and Salkinoja-Salonen 1987) and *Flavobacterium* sp. (Steiert and Crawford 1986) can aerobically biodegrade pentachlorophenol while anaerobic degradation of 3 -chlorobenzoate and PCBs has been identified by methanogenic consortia (Shelton and Tiedje 1984; Nies and Vogel 1990). Pentachlorophenol (PCP) is an example of this type of reaction (shown in Figure 2.6A), which forms 3,4,5-trichlorophenol, 3,5-dichlorophenol, and 3-chlorophenol (Reineke and Knackmuss 1988). These products can be oxidized aerobically only.

Monochlorobenzoate, another product of this reaction, however, degrades aerobically to carbon dioxide. The position of the halogen with respect to other functional groups on the molecule can influence the ease of reductive dehalogenation. For monochlorophenol, an anaerobic sludge could remove the chlorine more easily from the *ortho*-position than *meta*- and *para*-positions (Boyd and Shelton 1984). In contrast, PCBs were dechlorinated at the *meta*- and *para*-positions first (Nies and Vogel 1990).

Pseudomonas sp. and *Arthrobacter* sp. (Reineke and Knackmuss 1988) are able to hydrolytically dehalogenate 4-chlorobenzoate while Rhodococcus chlorophenolicus (Apajalahti and Salkinoja-Salonen 1987) and *Flavobacterium* sp (Steiert and Crawford 1986) could degrade PCP in this manner. Displacement of the halogen is from *ortho-* and/or *para-*positions. Figure 2.6B is a diagram of hydrolysis of pentachlorophenol (PCP) (Reineke and Knackmuss 1988). In this reaction, the product, tetrachloro-*p*-hydroquinone, formed by substitution of the chlorine with a hydroxyl group, can only be further degraded anaerobically.

Nitroaromatics

Nitroaromatics in the form of 2,4,6-trinitrotoluene (TNT), 2,6- and 2,4- isomers of dinitrotoluene, nitrobenzene, 1,3-dinitrobenzene, nitrophenol isomers, and nitroaniline can be found in contaminated soils, groundwaters, wastewater, and leachates from the manufacture of many products such as explosives (Wujcik et al. 1992). The nitrogen group on an aromatic ring can make biodegradation very difficult and can also be toxic to the bacteria. There is evidence, however, of degradation by two different pathways: nitrogroup reduction, where the nitro (NO group) is converted to an amine ($-NH_2$) (Preuss et al. 1993), or oxidation of the nitrophenol such that the nitro is replaced by a hydroxyl group (Spain and Gibson 1991). In the first reaction, production of the amino aromatics may not be beneficial since the products can be toxic.

In summary, aerobic degradation of alkanes or aromatic compounds can lead to the formation of a fatty acid or catechol that is less volatile, more soluble, and more degradable. Biodegradation of PAHs may form more volatile, lower molecular weight products, particularly through fungal metabolism. Conversion of chlorinated compounds such as TCE by anaerobic reductive dehalogenation can lead to the formation of ethene, which can persist in the environment. Alternating aerobic and anaerobic conditions may be required to eliminate these types of compounds.

References

Alexander, M. 1994. *Biodegradation and Bioremediation*. San Diego: Academic Press.

Alvarez-Cohen, L., and P. L. McCarty. 1991. Effects of toxicity, aeration and reductant supply on trichloroethene transformation by a mixed methanotropic culture. *Applied and Environmental Microbiology*, 57(1):228–235.

Apajalahti, J. H. A., and M. S. Salkinoja-Salonen. 1987. Complete dechlorination of tetrahydroquinone by cell extracts of pentachloro-induced *Rhodococcus chlorophenolicus. Journal of Bacteriology*, 169:5125–5130.

Aronstein, B. N., Y. M. Calvillo, and M. Alexander. 1991. Effects of surfactants at low concentrations on the desorption and biodegradation of sorbed aromatic compounds in soil. *Environmental Science and Technology*, 25(10):1728–1731.

Atlas, R. M. 1991. Microbial hydrocarbon degradation: Bioremediation of oil spills. *Journal of Chemical Technology and Biotechnology*, 52:149–152.

Aust, S., T. Fernando, B. Brock, H. Tuisel, and J. Bumpus. 1988. Biological treatment of hazardous wastes by *Phanaerochaete chrysosporium*. In *Proceedings of Conference on Biotechnology Applications in Hazardous Waste Treatment*. G. Lewandowski, B. Blatiz, and P. Armenante, eds. Longboat Key, Florida. 30 October–4 November. New York: Engineering Foundation.

Bartha, R. 1986. Biotechnology of petroleum pollutant biodegradation. *Microbial Ecology*, 12:155–172.

Boyd, S. A., and D. R. Shelton. 1984. Anaerobic biodegradation of chlorophenols in fresh and acclimated sludge. *Applied and Environmental Microbiology*, 47:272–277.

Brock, T. D., D. W. Smith, and M. T. Madigan.1984. *Biology of Microorganisms, 4th ed.* Englewood Cliffs: Prentice-Hall.

Cerniglia, C. E. 1984. Microbial metabolism of polycyclic aromatic hydrocarbons. *Advances in Applied Microbiology*, 30:31–39.

Chappell, J. 1997. *Phytoremediation of TCE using Populus*. Report prepared for U.S. EPA Technology Innovation Office. August.

Dupont, R. R., R. C. Sims, J. L. Sims, and D. Sorensen. 1988. *In situ* biological treatment of hazardous waste-contaminated soils. *Biotreatment Systems, Vol. II*, D. L. Wise, ed. Boca Raton, FL: CRC Press.

Evans, W. C., and G. Fuchs. 1988. Anaerobic degradation of aromatic compounds. *Annual Review of Microbiology*, 42:289–317.

Eweis, J. B., S. J. Ergas, D. P. Y. Chang, and E. D. Schroeder. 1998. *Bioremediation Principles*. Boston: WCB McGraw-Hill.

Frobisher, M., R. D. Hinsdill, K. T. Crabtree, and C. R. Goodheart. 1974. *Fundamentals of Microbiology*. Philadelphia: Saunders.

Freeman, D. L. and J. M. Gossett. 1989. Biological reductive dechlorination of tetrachloroethylene and trichloroethene to ethylene under methanogenic conditions. *Applied and Environmental Microbiology*, 55(9):2144–2151.

Gibson, D. T. 1988. Microbial metabolism of aromatic hydrocarbons and the carbon cycle. *Microbial Metabolism and the Carbon Cycle*, S. R. Hagedorn, R. S. Handson, and D. A. Kunz, eds. New York: Harwood Academic Publishers.

Governor's Office of Appropriate Technology. 1981. Toxic Waste Assessment Group. California.

Higgins, I. J., and P. D. Gilbert. 1978. The biodegradation of hydrocarbons. *The Oil Industry and Microbioal Ecosystems*. London: Heyden and Son, Ltd. 80–115.

Khovrychev, M. P., I. I. Ivanova, and S. D. Taptykova. 1974. Chemical composition and physiological properties of *Candida utilis* in the presence of inhibition of growth by copper ions. *Microbiology*, 43:405–409.

LaRossa, R. A., and J. V. Schloss. 1984. The sulfonyl berbicie sulfometuron methyl is an extremely potent and selective inhibitor of acetolactate synthase in *Salmonella typhimurium*. *Journal of Biological Chemistry*, 259:8753–8757.

Matsumura, F., and E. G. Esaac. 1979. Degradation of pesticides by algae and microorganisms. *Pesticides and Xenobiotic Metabolism in Aquatic Organisms*, M. A. Q. Kahn, ed. American Chemical Society. Washington, D.C.: ACS Symposium Series 99.

McCarty, P. L. 1965. Thermodynamics of biological synthesis and growth. *Proceedings of Second International Conference on Water Pollution Research*. New York: Pergamon Press. 169.

———. 1988. Bioengineering issues related to *in situ* remediation of contaminated soils and groundwater. *Environmental Biotechnology*, G. S. Owen, ed. New York: Plenum Publishing Company. 143–162.

Metcalf and Eddy, Inc. 1991. *Water Engineering Treatment, Disposal and Reuse*. 3rd ed. New York: McGraw-Hill.

Monod, J. 1949. The bacterial cultures. *Annual Review of Microbiology*, 3:371.

Morrison, R. T., and R. N. Boyd. 1973. *Organic Chemistry*. Boston: Allyn and Bacon, Inc.

Nies, L., and T. M. Vogel. 1990. Effects of organic substrates on dechlorination of Arochlor 1242 in anaerobic sediments. *Applied and Environmental Microbiology*, 56:2612–2617.

Park, K. S., R. C. Sims, and R. R. Dupont. 1990. Transformation of PAHs in Soil Systems. *Journal of Environmental Engineering*, 116(3):632–640.

Perry, J. J. 1984. Microbial metabolism of cyclic alkanes. *Petroleum Microbiology*, R. M. Atlas, ed. New York: MacMillan Publishing Company. 61–97.

Pirnik, M. P. 1977. Microbial oxidation of methyl branched alkanes. *CRC Critical Reviews in Microbiology*, 5:413–422.

Pitter, P., and J. Chudoba 1990. *Biodegradability of Organic Substances in the Aquatic Environment*. Boca Raton, FL: CRC Press.

Porgues, N., L. Jaiswicz, and S. R. Hoover. 1953. Biological oxidation of dairy waste VII. *Proceedings of 24th Industrial Waste Conference*, Purdue University, West Lafayette, IN.

Preuss, A., J. Fimpel, and G. Diekart. 1993. Anaerobic transformation of 2,4,6-trinitrotoluene (TNT). *Archives of Microbiology*, 159:345–353.

Protecting the nation's groundwater from contamination, 1984. OTA-0-233. 1:19–60, Washington, D.C.: Office of Technology Assessment.

Reineke, W., and H. J. Knackmuss. 1988. *Annual Review of Microbiology*. 42:263–287.

Reynolds, T. D., and P. A. Richards. 1996. *Unit Operations and Processes in Environmental Engineering*. Boston: PWS Publishing Company.

Schiewer, S., and B. Volesky. 1998. Biosorption processes for heavy metal removal. In *Environmental Metal-Microbes Interactions*, D. Lovely, ed. New York: Chapman Hill.

Shelton, D. R., and J. M. Tiedje. 1984. Isolation and partial characterization of bacteria in an anaerobic consortium that mineralizes 3-chlorobenzoic acid. *Applied and Environmental Microbiology*, 48:840–848.

Sims, J. L., R. C. Sims, and J. E. Mathews. 1990. Approach to bioremediation of contaminated soil. *Hazardous Waste and Hazardous Materials*, 7(4):117–149.

Sinclair, J. L., D. H. Kampbell, L. Cook, and J. T. Wilson. 1993 Protozoa in subsurface sediments from sites contaminated with aviation gasoline or jet fuel. *Applied and Environmental Microbiology*, 59(2):46–472.

Spain, J. C., and D. T. Gibson. 1991. Pathway for biodegradation of p-nitrophenol in a *Moraxella* sp. *Applied Environmental Microbiology*, 57:812–819.

Stanier, R. Y., J. L. Ingraham, M. L. Wheelis, and P. R. Painter. 1986. *The Microbial World*. Englewood Cliffs, NJ: Prentice-Hall.

Steiert, J. G., and R. L. Crawford. 1986. Catabolism of pentachlorophenol by a *Flavobacterium* sp. *Biochemical and Biophysical Research Communications*, 141:825–830.

Suflita, J. M., J. A. Robinson, and J. M. Tiedje. 1983. Kinetics of microbial dehalogenation of haloaromatic substrates in methanogenic environments. *Applied and Environmental Microbiology*, 45:1466–1473.

Tate, R. L. 1995. *Soil Microbiology*. New York: Wiley.

Thayer, A. M. 1991. Bioremediation: Innovative technology for cleaning up hazardous waste. *Chemical and Engineering News*, 69(34):23–44.

U.S. Environmental Protection Agency (U.S. EPA). 1983 *Guidance Manual for POTW Pretreatment Program Development*. Washington, D.C.: U.S. EPA.

U.S. Environmental Protection Agency (U.S. EPA). 1985. *EPA Guide for Identifying Cleanup Alternatives at Hazardous Waste Sites and Spills: Biological Treatment*. EPA 600/3-83/063. Washington D.C.: U.S. EPA.

Widdel, F. 1988. Microbiology and ecology of sulfate- and sulfur-reducing bacteria. *Biology of Anaerobic Microorganisms*. J. B. Zehner, ed. New York: John Wiley and Sons. 469–585.

Wujcik, W. J., W. L. Lowe, and P. J. Marks. 1992. Granular activated carbon pilot treatment studies for explosives removal from contaminated groundwater. *Environmental Progress*, 11(3):178–196.

Yong, R. N., A. M. O. Mohamed, and B. P. Warkentin. 1992. *Principles of Contaminant Transport in Soils*. Amsterdam: Elsevier Science Publishers B.V.

Biological Air Treatment

Introduction

This chapter describes the biological technologies available for air pollution treatment. The components to be treated are primarily odorous or volatile organic chemicals (VOCs). Air streams of low concentrations (less than 1 g/m³) can be generated by many types of sources such as wastewater treatment plants, gasoline and solvents from contaminated vadose zones, composting, landfill leachates, food industry, agriculture, and chemical processes including printing, flexography, and pharmaceutical industries (Edwards and Nirmalakhandan 1996). This chapter covers the three main technologies: biofiltration, trickling filters, and bioscrubbers. In addition to airflow rates, the composition, concentration, and types of chemicals that can be biologically treated are examined. For more than 20 years, biological treatment of air has been used in Germany and the Netherlands, because regulations there allow for innovation and cost savings in design. Examples from these countries and other countries are described and compared to conventional technologies including activated carbon filtration and thermal or catalytic oxidation. In comparison to conventional processes, these biological air treatment processes have become increasingly popular in the last decade since their advantages include low cost, simple operation, high efficiencies (greater than 90%), and production of nontoxic by-products. In addition, energy and raw materials requirements are lower. The major disadvantage is a lack of confidence in their performance, but this is improving as more full-scale demonstrations are performed.

Biofiltration

Description

A biofilter is a bed (open or closed) of material consisting of peat, compost, soil, plastics, ceramics, activated carbon, or other materials. The oldest biofilters consisting of soil beds were used to remove hydrogen sulfide (H_2S) from the gases in wastewater treatment processes in the 1920s (Leson and Winer 1991). In the 1980s, the treatment of many types of low-concentration soluble volatile suspended solids (VOCs) began.

Figure 3.1 shows a schematic of a biofilter. The air containing the contaminants is passed through the bed after humidification. Microorganisms are attached to the material as a biofilm which also contains extracellular polysaccharides and water. A liquid is required to surround the microbes to enable them to transport substrates and nutrients into the cells. The compounds to be treated are absorbed by the bed and biologically converted by the microorganisms to carbon dioxide, water, nitrate (NO_3^-), and sulfate (SO_4^{2-}) ions. The fibrous portion of the bed such as peat or compost contains the microbes and has a very large specific surface area and, thus, high activity. The purpose of the coarse fraction is to support the filter and prevent high pressure drops. Materials like wood chips, wood bark, or polystyrene can be used.

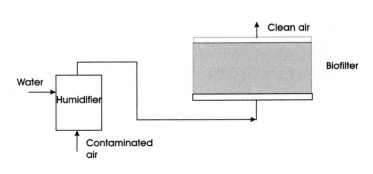

Figure 3.1 Schematic of a Biofilter

Box 3-1 Overview of Biofiltration

Applications	Suitable for treatment of odorous compounds, aromatic compounds, aldehydes, alcohols, mercaptans, amides, hydrogen sulfide, ammonia and other water soluble, biodegradable compounds of low concentration in air of flows 33,000 to 170,000 m^3/h and low particulate levels (less than 10 mg/m^3)
Cost	Operating: $0.12 to $0.36/1,000 m^3, Capital: $1.2 to $13.0 per m^3/h
Advantages	Simple to start-up and operate
	Low capital and operating costs
	High gas/liquid surface area
Disadvantages	Removal rates greater than 99% difficult to achieve
	Slow to adapt to fluctuating concentrations
	Large area requirement
Concentrations treated	5 to 1,000 mg/m^3 with typical loadings of 100 to 200 m^3/m^3-h
Other considerations:	Highly accepted in Europe, fewer demonstrations in North America where regulators are reluctant

Criteria for Utilization

Adsorption of the contaminant from the gas is the first process that must occur and is described by the formula (Van't Riet and Tramper 1991):

$$F" = K_l a \, (C^* - C)$$

where $C^*=C_g/H$ and F" is the mass transfer rate (mol/m³-s), K_l is the overall interfacial mass transfer coefficient (m/s), a is the interfacial area per unit of volume (m²/m³), C^* is the concentration in the liquid phase at equilibrium with the gas phase (mol/m³), C is the concentration in the liquid phase (mol/m³), C_g is the concentration in the gas phase (mol/m³), and H is the Henry coefficient (dimensionless).

The Henry coefficient is the determining factor for adsorption. Values of higher than ten indicate that the contaminant is of low water solubility and, thus, is not a good candidate for treatment using biofiltration. Sulfur-containing compounds have high coefficients and are of low solubility. Waste gases from wastewater treatment, composting, manure, etc., contain high levels of these compounds, resulting in longer residence times and higher specific areas. Flows are typically in the order of 25 to 50 m³/m³-h (Heslinga 1994). For components of high Henry coefficient, the mass transfer rate at the interface between the air and the water film in the biofilter would be very low (Kok 1992). Other factors that influence the mass transfer rate are the pore area and particle size of the supporting material. The gas flow rate can also have an effect on the mass transfer coefficient.

The concentration profile of the substrate as it is transported through the various phases is schematically shown in Figure 3.2. Equilibrium concentrations are assumed at the liquid/gas interphase. Biodegradation and diffusion then occur in the liquid phase. The kinetics of biodegradation is first order at low concentrations and zero order at high VOC concentrations (Edwards and Nirmalakhandan 1996). Absorption into the support media may also occur in the case of activated carbon. It has been postulated that this phenomenon can increase the rate of biodegradation due to higher availability of the substrate to the bacteria, but this has been disputed (Govind et al. 1995).

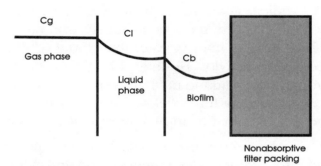

Figure 3.2 Transfer of Gaseous Contaminants Across the Various Phases During Biofiltration with a Nonabsorptive Packing. Cg Represents the Concentration of the contaminant in the Gaseous Phase; Cl Represents the Concentration in the Liquid Phase; and Cb the Concentration in the Biofilm

The Michaelis-Menten equation can be used to describe microbial degradation in the biofilter as follows (Hansen 1997):

$$r_s = V_{max} \times S \times \frac{B}{(K_m + S)}$$

where r_s is the rate of substrate utilization, V_{max} is the maximum rate of reaction, S is the substrate concentration, B is the population density, and K_m is the half saturation constant. The terms, $V_{max,}$ B, and K_m can be determined from pilot tests.

Operating Conditions

The microbial composition varies according to the composition of the contaminants and the environmental conditions (Cox et al. 1993). Rieneck (1992) determined that most of the microorganisms are bacteria and include coryneforms, bacilli, pseudomonads, and actinomycetes such as *Streptomyces* spp. Bacterial counts of 5 x 10^7 to 3 x 10^{10} per g of heather, the support medium, have been demonstrated (Pearson 1992). Fungi (Rieneck 1992) including *Mortierella, Rhizopus, Penicillium, Aspergillus, Fusarium, Alternari,* and *Botrytis*, although less common than bacteria (10^5 to 10^7 per g heather, soil or compost), are also present. They can be useful in breaking down more complex molecules due to the enzymes they produce.

Before optimal biotreatment occurs, adaptation of the microorganisms to the incoming contaminants in the air is required. Inoculation is usually necessary, even though the packing materials such as compost can contain substantial microbial counts. Activated sludge from wastewater treatment plants of material from currently operating biofilters can be added to the packing material.

Microorganisms capable of utilizing the contaminants as substrates will grow in the biofilter. Oxygen, required as an electron acceptor, is supplied by the gas being treated. Only at high loading rates (greater than 100 g C/m^3-h) could oxygen become limiting and cause a decrease in activity (Heslinga 1994). Other nutrients originate from the packing material. Growth will continue until an element such as carbon, hydrogen, or oxygen is limiting. This occurs usually a few weeks after inoculation. At this time, growth conditions are stationary and the microbial count will remain at a constant level.

The temperature should be maintained between 20 and 40°C with a pH between 6 and 9 for biological treatment (Leson and Winer 1991). Although biofilters have been used in cold climates such as Finland (Lehtomaki et al. 1992), insulation of the piping for the inlet gases and the biofilter is usually necessary to maintain the temperature. In the case of high temperature streams, cooling may also be required to obtain the desired temperature. However, operation at 50 to 65°C is possible with thermophilic bacteria if the temperature is constant at this level.

If mineral acids such as HCl are produced during the degradation of chlorinated hydrocarbons, nitric or sulfuric acids, or organic acids, pH control will then become necessary. Fungi, which grow at lower pH values than bacteria could adversely affect the performance of the biofilter (Cox et al. 1993). Other options to avoid problems include the prior addition of a buffer and the addition of $Ca(OH)_2$, unslaked lime (CaO), limestone ($CaCO_3$), dolomite ($CaMg(CO_3)_2$), or other materials for precipitation, washing or draining of the filter. Oude Luttighuis (1989) found that the effects of HCl accumulation could be neutralized by buffering and calcium carbonate with occasional washing and draining. The washing must not be too frequent, though, since this can deplete the nutrients. Neutralization of the acids with caustic soda is generally not a good strategy since this can cause an increase in the ionic strength of the aqueous biofilm. A decrease in pH to 2.5 to 3.5 is not a problem for H_2S biofilters, however, since the *thiobacilli* bacteria can tolerate this pH range (Schelchshorn and Vink 1989).

An extremely important factor for the optimal performance of biofilters is the moisture content. Poor moisture control accounts for 50 to 75% of biofiltration performance problems (Helsinga 1994). In general, 40 to 60% is a good level (Ottengraf 1986). The lower range is better for hydrophobic VOC treatment and compact, poorly drained support media while the upper range is better for hydrophilic VOC treatment and light, highly porous media (van Lith et al. 1997). Moisture content is defined as:

$$Moisture\ content = \frac{mass\ of\ water}{mass\ of\ water\ +\ mass\ of\ dry\ material}$$

The moisture can be maintained by spraying the surface of the bed with water and/or by humidifying the incoming gases to obtain relative humidity in the range of 98 to 100%. Relative humidity (%) is defined as the partial pressure of the vapor in the gas (Pa) divided by the vapor pressure of the water (Pa). This prehumidification step reduces drying out of the biofilter. The gas temperature is usually reduced by the humidification process. Higher contents lead to mass transfer problems, particularly for hydrophobic contaminants due to decreased gas/liquid surface areas and plugged pore spaces, and can cause the washout of small particles, slime formation, and the development of anaerobic zones. Low moisture contents allow fungus to prosper over bacteria and cause loss of bacterial activity due to an insufficient biofilm and degradation of the filter packing. Drying occurs, in particular, at loads high than 100 g/m^3-h (Heslinga 1994).

Since inhibition can occur at high concentrations, laboratory tests are useful to determine the maximum substrate concentration that can be effectively degraded. These studies can also indicate if any accumulation of intermediates of one compound is interfering with the degradation of another. For example, Engresser (1992) showed that although chlorobenzene and toluene are individually degradable, when combined toluene degradation was inhibited. Thus, the efficiency of the biofilter depends on the gas-liquid mass transfer rate, available biomass, the rate of biodegradation, and inhibition.

Before treatment in a biofilter, the waste-gas should be free of dust, fat, and aerosols to prevent plugging of the bed (Boyette 1998). This can be accomplished through means of a filter or scrubber. In general, 5 mg/m^3 can be taken as the maximum allowable particle level (Leson 1998). Ammonia levels may also have to be lowered by acid addition. If the levels of ammonia are not substantially elevated, water scrubbing may remove adequate amounts.

Design

Filter bed heights are mainly between 0.5 and 1.5 m to avoid channelling in too short beds and high flow resistance in too tall beds (van Groenstijn and Hesselink 1993). Most of the biodegradation occurs in the first 250 mm of the bed. When higher beds are used for higher contaminant concentrations, clogging can occur due to the increased microbial activity at the inlet. It is also generally better to enclose the biofilter, since open biofilters may be unstable or not perform as well due to weather conditions. Whether a system is open or closed can be determined depending on whether they can be operated under vacuum and pressure conditions. Open biofilters are constructed of concrete (precast or poured on-site), wood, and coated or stainless steel. Some open systems in the U.S. are in excavated sites without constructed walls (van Lith et al. 1997). Closed biofilters are constructed of steel with insulated steel walls or concrete (precast or poured on-site). Biofilters less than 60 m³ can be made of steel transport containers or fiberglass-reinforced epoxy resin. Closed systems are used for VOC and Hazardous Air Pollutant (HAP) treatment since maintenance, process control, and gas monitoring are also easier (van Groenstijn and Hesselink 1993). Floor space requirements can be substantial for biofilters greater than 10,000 m³/h and are usually in the order of 4 to 22 m² per 1000 m³/h of gas treated (Leson 1998).

Empty-bed residence time, contaminant loading rate, gas flux, and elimination capacity are the main parameters used to design the biofilter. The residence time, t, is the time a gas molecule would spend in the bed and is calculated as:

$$t = \frac{V}{Q}$$

where V is the volume of the bed without packing, and Q is the volumetric gas flow rate. Typical residence times are in the order of 0.3 to 12 minutes (Eweis et al. 1998). Longer residence times are required for the more difficult to degrade compounds. For example, 90% removal of alcohols can be achieved within 30 seconds, whereas 90% removal of trichloroethylene requires up to 150 minutes (Bohn 1992).

Gas flux or superficial velocity (m/h) can be determined from the residence time by dividing the height of the bed by the residence time. This is the average fluid velocity passing through an empty bed. Another way to calculate it is by dividing the volumetric flow rate of the gas by the cross-sectional area of the bed. The velocity of the gas through the pores is calculated by dividing the superficial velocity by the void fraction of the bed (dimensionless).

The other important factor is the contaminant loading rate, R_m (g/m³-h), which is determined from the inlet gas concentration, C_i (g/m³), by the equation:

$$R_m = \frac{QC_i}{V}$$

The removal efficiency (RE) is defined as:

$$RE(\%) = (1 - \frac{C_2}{C_1}) \times 100\%$$

where C_2 is the concentration at the outlet and C_1 is the concentration at the inlet. Removal efficiencies are usually low at start-up. However, if optimal conditions are maintained, removal efficiencies of 99% can be obtained once adaptation has taken place. The biofilters are usually resistant to shock loads, which are instantaneously high contaminant loads. The requirements for removal efficiencies are usually based on regulatory agencies which can require 90% removal of VOCs and 99% removal of odors. Examining data from pilot tests is the best method to design the size of a biofilter. Typically the removal efficiency is obtained as a function of the loading rate to determine the optimal loading rate. The volume of the biofilter can be calculated from the loading rate. Undersizing biofilters results in lower removal efficiencies than desired due to inaccurate estimates of the gas stream characteristics (composition or air flow) by pilot testing non-representative gas streams. Therefore, it is important to plan and run pilot tests in an optimal manner to design the full-scale unit properly.

Gas can be passed through the bed in the upflow or downflow direction. Open biofilters are usually operated in the upflow mode. Upflow also may be preferred for treatment of off-gases such as sulfur and nitrogen compounds and chlorinated VOCs, since flushing of the acidic components can be done (Yang and Alibeckoff 1995). Advantages of downflow operation are that the lower part of the bed does not dry out due to the sprinklers on top and that VOCs in the drainage water are less likely to evaporate into the atmosphere. Four types of commonly-used gas distribution systems include perforated pipes, pressure-chamber systems, sinter-block systems, and plenums (Eweis et al. 1998). Perforated pipe systems involve a network of perforated pipes under the biofilter in a gravel bed. Although these systems work well for odor control, clogging of the pores, drying of the beds, and uneven gas distribution often occur (Heslinga 1994). In another type of system, a large pressure chamber at the top or bottom of the bed is used to supply and distribute the air. Due to the instability of large pressure chambers and the weight of large filters, this type of system is restricted to smaller biofilters. For sinter-block systems, air is distributed through slotted concrete blocks which must be strong enough to provide access to the packing material by a front-end loader. The last type of gas distribution system involves the use of plenums of 150 to 300 mm in height. They are used in small biofilters that can be supported by the expanded metal grids.

Maintenance

Maintenance of biofilters includes supplementation of nutrients and addition of water to maintain bed moisture. Measurements of the pH, water content, pressure drop, gas flow, filter weight, and performance are required to determine the amount and frequency of additives. Moisture content is one of the most important factors influencing biofilter performance. Methods of measuring and controlling moisture are shown in Table 3.1. Spot measurements can be made by manual sampling, electrical conductivity, or capacity measurements. This approach is limited, since only a small sample can be determined from a large bed and the sensitivity is low. Load cells are sometimes used to automatically monitor and control moisture content of closed biofilters since it gives an overall indication of the bed moisture content (van Lith 1989). Corrections have to be made for media decomposition and loss. Nutrients can be added with the humidification system or by occasionally soaking the bed with a nutrient solution (Ergas et al. 1994).

The choice of support media is important since it serves as a support for the microorganisms and in some cases can assist in adsorption of the contaminants to increase the availability to the microorganisms. High permeability and a specific surface area, in addition to an excellent supply of nutrients, are selection criteria for the packing media. Plugging can occur as a result of excessive microbial growth or bed compaction due to settling. Increased pressures and flow channelling can result from plugging. The advantages and disadvantages of the various support media are shown in Table 3.2 (Edwards and Nirmalakhandan 1996).

Compost from manure, yard waste, and wastewater sludge is commonly used as support media. Variability, homogeneity, channelling, maldistribution of air, and stability can be problems, however. Tree bark and chopped wood can also be used, but their microbial activity is lower than compost. Aging decreases the fibrous structure and affects bed porosity, pressure drop, and bed height. For example, more mature composts typically have lower nitrogen contents than more active ones. Recent developments involve the formation of spheres of compost which reduces the aging effects and the separation of particles (Oude Luttinghuis 1997). Inoculation may be necessary at start-up of the biofilter to decrease the acclimation or lag time due to small microbial populations or to introduce a specific enzyme for contaminant degradation to take place after a change in the substrate. For example, initial experiments using a compost-based biofilter showed that the acclimation could be reduced from one year to three weeks for degradation of methyl tert-butyl ether if an inoculum from the previously used compost biofilter was added (Ergas et al. 1994).

Table 3.1 Methods of Controlling Moisture Content in Biofilters

Method	Description
Manual/Ad hoc	No spraying system. Moisture is determined several times a year and hose spraying is performed as necessary. Difficult and time-consuming. Suitable if moisture loss is minimal (less than 50 g/m^3-h).
Manual/Periodic	Bed spraying installed. Spraying based on periodic sampling. Suitable for moisture losses of 50 to 180 g/m^3-h.
Semi-automatic	Spraying frequency and duration according to timer which is adjusted according to sampling or automatic monitoring. Suitable for moisture losses of 180 to 400 g/m^3-h.
Automatic	Moisture content is measured automatically. Results that are too low are used to initiate spraying. Suitable for coarse, well-drained media; off-gases with high VOC loadings; high temperatures; insufficient prehumidification; beds with moisture losses of greater than 400 g/m^3-h.

Adapted from van Lith et al. 1997

Table 3.2 Advantages and Disadvantages of Specific Support Media for Biofiltration

Support media	Advantages	Disadvantages
Ceramic straight passages	Can handle high contaminant concentrations	Slow start-up
	Large channel sizes for long life	Weak adhesion of biomass
	Low pressure drop	Low capacity for adsorption
	High surface area per volume	High cost
Pelletized activated carbon	High adsorption and biodegradation capacity	Difficult to clean
	Good biomass adhesion	High cost
	Suitable for high contaminant concentrations (>200 mg/m³)	Eventual plugging
	Fast start-up	
Activated carbon and ceramic straight passages	High biodegradation capacity	High cost
	Advantages of both support media	
Pelletized ceramic	Ability to handle high contaminant concentrations (>200 mg/m³)	More expensive than soil or peat/compost
	Less expensive than activated carbon	Eventual plugging
	Fast start-up	
	Easy to clean and maintain	
Peat/Compost	Commercially available	Channelling and poor distribution frequently occur
	Low cost	Limited buffering capacity
	Suitable for low contaminant concentrations (less than 50 mg/m³)	Low capacity for biodegradation
Soil	Well-established technology	Limited nutrient supply
	Low cost	Channelling and poor distribution frequently occur
	Suitable for odor control or low contaminant concentrations (less than 50 mg/m³)	Replacement required
		Limited buffering and nutrient supply capacity
		Low adsorption and biodegradation capacity
		Replacement required

Adapted from Govind et al. 1995

In comparison to compost, soil is less permeable and the pores are smaller. Thus, larger areas are required. Sandy soils are preferred for highly degradable gases since they are more permeable. For less degradable gases, soils which have a higher sorption capability are preferred. Due to the base and mineral content of soil, the acidic products, SO_2, NH_3, H_2S, and NO_x, can be neutralized (Bohn 1992). In addition, lime can be added to soil more extensively than compost, which can compact if limed too much. However, too much salt can inactivate the microbial activity and drainage may be required. Soil can be more easily rewetted than compost because of its hydrophilic nature. The lifetime of the soil bed is extensive compared to compost, which must be replaced between one to five years due to its biodegradability. Its ability to neutralize acids (such as sulfuric and nitric) may limit the lifetime of the soil bed.

Synthetic packing includes granular activated carbon (GAC), plastic Pall rings, Raschig rings, expanded polystyrene spheres, sintered glass disks, diatomaceous earth pellets, and ceramic passages. Although head losses can increase up to 1 to 3 cm after several years of operation, GAC has low head losses, low biofouling rates, and large specific surface areas. The most common packing materials are tree bark, heather, peat/heather, compost/wood bark, compost/polystyrene, and peat/heather/pine branches. The addition of the structural materials such as bark, wood chips, mineral granulates, and heather helps to decrease the occurrence of increasing pressure drop and channelling. Many advances have been made recently concerning packing material (Oude Luttighuis 1997). For example, Envirogen (Lawrenceville, NJ) uses organic materials such as wood chips or peat for the microorganisms but adds a proprietary material to keep the biofilter warm and humid (Togna et al. 1997).

Applications

The degradability of various components is shown in Table 3.3. Designs for biofilters are based on the chemicals present, gas flow rate, and composition. Lower residence times are required for more easily degradable components. Alcohols are very biodegradable, followed by ketones, straight chain alkanes, and aromatics (Standefor 1996). For example, methyl ethyl ketone can be treated with a residence time of 25 seconds compared to four minutes for ethyl benzene (both with 50 mg/m^3 concentration at 85,000 m^3/h). In addition, volatile inorganic compounds such as ammonia, NO, and N_2O can also be treated by biofiltration. In ammonia treatment systems, pH control is necessary to maintain the pH above 6.5 for nitrifying bacteria. Denitrification has been demonstrated for NO and N_2O removal (du Plessis et al. 1996; Apel et al. 1995).

Due to the remediation of soil and groundwater via soil vapor extraction (SVE), the development of treatment technologies for off-gases containing hydrocarbons and benzene/toluene/ethylbenzene/xylene (BTEX) is necessary. Incineration and carbon adsorption are not very cost-effective. Biofiltration is a promising technique for this application (Togna et al. 1994). BTEX removal rates of between 70 and 95% have been obtained, depending on the residence time which varied from 0.5 to 6 min (Togna and Singh 1994). BTEX is usually treated more efficiently than the aliphatic less-soluble fraction of petroleum which is included in the total petroleum hydrocarbon (TPH) value. Thus, TPH removal levels by biofiltration are usually lower. In the United States, removal of gasoline vapors by biofiltration has become popular with over 50 units, each treating less than 425 m³/h of SVE off-gas.

Table 3.3 Biological Degradability of Various Components in Biofilters

Rapidly degradable VOCs	Rapidly degradable VICs	Good VOC degradability	Low VOC degradability
Alcohols	H_2S	ketones	*Halogenated*
methanol	NO_x	hexane	*hydrocarbons*
ethanol	SO_2	benzene	trichloroethylene
propanol	HCl	styrene	trichloroethane
butanol	PH_3	phenols	carbon tetrachloride
Aldehydes	SiH_4	ethyl acetate	pentachlorophenol
formaldehyde	HF	methyl acetate	dichloromethane
acetaldehyde	NH_3	methacrylate	perchlorethylene
butyraldehyde	CO	*Sulfur compounds*	*Ethers*
toluene		dimethyl sulfide	diethylether
xylene		thiocyanates	dioxane
esters		thiopene	Polyaromatic
organic acids		methyl mercaptan	hydrocarbons
amines		mercaptan	CS_2
trimethylamine		*Nitrogen compounds*	nitro compounds
tetrahydrofuran		pyridine	methane
acetone		acetonitrile	pentane
methyl ethyl ketone		amides	
		chlorophenols	
		thiocyanate	

Adapted from Bohn 1992 and Society of German Engineers (VDI 3477) 1991

For petroleum hydrocarbons, biofiltration could be used alone or in combination with other treatment technologies by acting as a pretreatment. This decreases the cost of carbon disposal by 65 to 85% (Togna et al. 1994). When the VOC concentrations are higher, such as when SVE is initiated, thermal technologies can be used. However, when the VOC concentrations decrease, it may be more cost efficient to use biofiltration due to the high operating costs for the required fuel for the thermal oxidation process.

It has been demonstrated recently that H_2S and CS_2 can be treated by biofiltration (Fucich et al. 1997b). Traditionally, absorption, adsorption, and thermal oxidation are used for the treatment of these compounds. These methods are expensive or inefficient for waste streams containing concentrations less than 10,000 mg/m³. *Thiobacillus* species are thought to be the main type of bacteria involved in oxidation of these two chemicals. The reactions are as follows (Yang and Alibeckoff 1995):

$$1 \qquad CS_2 + H_2O \rightarrow COS + H_2S$$

$$2 \qquad COS + H_2O \rightarrow CO_2 + H_2S$$

$$3 \qquad H_2S + 2O_2 \rightarrow H_2SO_4 + Energy$$

Although H_2S can degrade very quickly, CS_2 is much more difficult. Removal efficiencies of 95% of H_2S and CS_2 can be obtained at concentrations up to 4,000 and 1,000 mg/m³, respectively, after proper acclimation. Since sulfuric acid is produced, adequate pH control must be available.

In Ojai Valley (California) Sanitary District's wastewater treatment plant, a two-stage biofilter was installed (Devinny et al. 1998). The first stage is an enclosed system with porous stones called lava rock for removal of the acid H_2S gas. Lava rock was chosen since it works well at low pH. The second stage is a traditional open biofilter containing wood chips. Greater than 90% removal of the H_2S was obtained in the first filter.

There have been many recent industrial applications of biofilters. Many companies such as Kodak, Bush Boake Allen, Mercedes-Benz, and Coca-Cola are using biofiltration (Willingham 1996). Vapors from contaminated soil or groundwater can also be treated by biofilters. The Dutch Company Comprino B.V. (Biobox®) was one of the first companies in the U.S. to install a full-scale biofilter that was used to treat 60,000 m³/h of 1 g/m³ ethanol from a fermentation waste gas. Odor was reduced by 95% and VOC by 85%. Another Dutch company, Clairtech, which was recently sold to Monsanto Enviro-Chem Systems, has used the BIOTON® biofilter in over 100 applications (Figure 3.3). Gases up to 200,000 m³/h can be treated by biofilters up to 3,000 m³ (Huber 1992). This translates into volumetric loads (volume of gas treated per volume of reactor) of 100 to 200 m³/m³-h. Concentrations of the contaminants are usually below 1,000 mg/m³ with most applications in the range of 5 to 500 mg/m³. A summary of the industrial applications is shown in Table 3.4. Composting facilities, emissions from wastewater treatment facilities, food flavoring, corn processing, can coating, polystyrene, optic

Figure 3.3 An Enclosed BIOTON Biofilter System (courtesy of Enviro-Chem Systems)

lens coating, pesticide, tobacco, rubber production, breweries, foundries, fragrance, printing, plastics, paper, and petrochemical are some of the other industrial applications for biofilters, as shown in Table 3.5.

Table 3.4 Industrial Applications of Biofiltration

Compound	Industry	Efficiency (%)	Flow rate (m³/h)	Volumetric load (m³/m³-h)	Reference
Odor	Animal rendering	94–99	241,000	66	Huber (1992)
Odor	Tobacco	95	180,000	109	Kersting (1992)
Odor	Vegetable oil	97	39,000	120	Eitner (1992)
Odor	Gelatine works	70–93	35,000	146	Kirchner et al. (1996)
Odor	Cocoa roasting	>99	4,000	73	Hofmann (1989)
Odor	Wastewater treatment	80–90	5,000	7	Mildenberger (1992)
VOCs	Storage tanks	90	2,000	8	Mildenberger (1992)
VOCs	Wastewater treatment	70–90	65,000	31	Mildenberger (1992)
VOCs	Fish processing	95	6,300	105	Liebe (1989)
VOCs	Fish processing	85	10,300	184	Liebe (1989)
H₂S	Landfill gas	>99	300	17	Sabo (1989)
Alcohols	Foundry	>99	30,000	150	Maier (1989)
Benzene	Foundry	80	40,000	120	Maier (1989)
Aromatics	Surface coating	85–90	1,500	79	Demiriz (1992)
Styrene	Resins processing	65	pilot	100	Demiriz (1992)
Phenol	Phenol resins	97	pilot	200	Demiriz (1992)
Formaldehyde	Plywood production	80	1,450	426	Mackowiak (1992)
Odor	Flavor/fragrance	98	25,400		Ottengraf and Diks (1992)
Ethanol	Ceramics	98	30,000	200	Ottengraf and Diks (1992)
VOCs	Photo film	75	140,000		Ottengraf and Diks (1992)
VOCs	Pharmaceutical	80	75,000		Ottengraf and Diks (1992)
VOCs	Particle board	80	140,000		Standefor (1996)
CS₂ and H₂S	Cellulose sponge manufacturing	80 (CS₂) 90 (H₂S)	51,000		Fucich et al. (1997b)
H₂S	Wastewater treatment	>99	2x3,000		Hansen (1997)
Benzene	Sheet fabrication	86	48,000		Eitner (1998)
VOCs	Screen printing	85	27,000	50	Bhat et al. (1998)
VOCs including BTEX	Sewer system and interceptor sewer	85–95 (VOC) 98 (BTEX)	19,000	22.8	Marran and Laustsen (2000)
Formaldehyde	Panel board	>90	80,000–100,00	478	Ferranti and Conca (2000)

Table 3.5 Other Industrial Applications of Biofilters

Industrial and Commercial Sources	Food Processing and Production	Waste Management

Coating and Printing
  Automotive coating
  Can and drum coating
  Photographic film coating
  Piano manufacturing
  Plastics coating
  Wire coating
  Wood coating
 	
Off-set printing
Screen printing
 	
Manufacturing
Adhesive production
Brake pad and clutch
    production
Coatings production
Fiberglass application
Fragrance production
Furniture manufacturing
Ink production
Investment foundries
Plastics production
Polymer production
Rayon production
Sanitation products
Solvent production
Wood products manufacture
 	
Storage and transfer of organics
Chemical storage
Crude oil storage
Tanker cleaning
	Baking processes
	  Bread
	  Eggroll
	Broth production
	Fish flour production
	Flavor production
	Meat and fish frying
	Meat and fish rendering
	Bone processing
	Blood drying
	Smoke house
	Sugar production
	Yeast production
	
	Coffee processing
	Corn processing
	Hay drying
	Soybean processing
	Pet food production
	
	Feed lots
		Wastewater treatment
		  (residential and industrial)
		Waste transfer stations
		Composting operations
		Sludge and manure drying
		Hazardous waste storage
		Landfill gas extraction
		Plastics recycling
		Waste oil recycling

Adapted from Leson 1998

Costs and Other Considerations

The capital cost of a system depends on pretreatment requirements, concentration and flow rate of the air to be treated, removal efficiency requirement, duct work requirements, whether it is open or closed, and the degree of process control employed. Open systems with fewer gas distribution systems, little moisture, and less temperature control are the least expensive while enclosed, controlled, multi-level biofilters are the most costly. The same removal efficiencies, residence times, and media volumes are used for comparing the two systems. Installed costs of $2 per cubic meter per hour of gas have been reported (van Lith et al. 1997) for larger systems while costs are in the range of $5 to 25 per cubic meter per hour for smaller beds (less than 25,000 m³/h). Competitiveness with incineration systems is particularly noted as the residence time decreases below one minute. Accurate costing is obtained mainly from the results of pilot tests since commercial installations are still fairly limited in North America.

Operating costs include electricity, media replacement, water consumption, system monitoring, maintenance, and repairs. Energy requirements are based on the degree of transport of the gas required through the ducts, humidification system, and biofilter bed. Pumps and controls do not significantly add to the electricity cost. Single beds have one-fourth the pressure drop of a double bed system as a result of twice the gas velocity and twice the bed height. For industrial systems, pressure drops in the order of 50 to 3,000 Pa (0.2 to 12 inches of water) are typical. This corresponds to a power requirement for a 30,000 m³/h gas flow between 0.7 to 41 kW, depending on the selected media and biofilter design. Water requirements are usually less than $200 per year for a 1,000 m³ system, depending on the type of system and the weather for open systems (van Lith et al. 1997). Labor requirements depend on the type of moisture control system chosen. Requirements increase as the degree of automation decreases. While open filters with varying inlet VOC concentrations can require up to several person hours per day to maintain an appropriate moisture content, others used for odor control require as little as five to ten minutes per day. Inspections also have to be made for compaction.

Recently, Marran and Laustsen (2000) performed a cost benefit analysis of a full-scale system for treatment of VOCs including benzene, toluene, ethylbenzene, and toluene (BTEX). It uses a contact time of 2.6 minutes with filter material of soil, leaf, compost, wood chips, bark, or other wastes. Capital cost and annual operating and maintenance costs were compared with carbon adsorption and thermal oxidation systems as follows:

Table 3.6 Cost Benefit Analysis of a Full-Scale System

Treatment System	Capital Cost ($)	Rating	Annual Operating and Maintenance Cost ($)	Rating
Biofilter	430,000	+	94,000	+
Carbon Adsorption	625,000	++	1,950,000	+++
Thermal Oxidation	612,000	++	380,000	++

Notes: + denotes least expensive and +++ most expensive

In comparison to the other options, the capital and operating costs for the biofilter are lower. Operating costs are substantially lower due to the lower energy requirements. In addition, operator requirements are not extensive.

Frequency of replacement of the support media depends on the operating conditions (gas temperature, loading rates, and concentration) and the composition of the media. When the pressure drop increases to an unacceptable level, the media has to be replaced. Replacement frequency is in the order of two to seven years. Open filter systems can be periodically mixed with fresh media to extend media life. Lower cost media ($30 to $50/m³) have shorter life expectancy and lower performance (van Lith et al. 1997). Higher cost media ($100 to $300/m³) are designed for low pressure drops and are often used for closed systems in which the media are more difficult to replace. The capital costs increase for closed systems and with the complexity of the system and level of control.

Overall economic analyses have shown annual operating costs to be in the order of $1 to $8 per cubic meter per hour (Boyette 1998). Willingham (1996) showed that this was 35 to 40% less expensive than for recuperative catalytic oxidation. Biofilters have the additional benefit of operating at ambient temperatures. Thus, biofilters are safer to use than thermal technologies; there is less risk of fire or explosion since they are moist systems, and there is no generation of NO_x compounds like thermal or catalytic oxidizers.

A recent study (Adler 2001) compared various technologies for the treatment of an air stream (170,000 m³/h) containing 100 mg/m³ of methanol and an air stream containing 30 mg/m³ of toluene. In particular, the operating costs for biofiltration ($0.13 to $0.30/1000 m³) for the highly-degradable methanol were very competitive with thermal oxidation ($0.30 to $0.56/1000 m³). In contrast, the operating costs for biofiltration of toluene ($0.36/1,000 m³) are in the middle of the costs for the other systems ($0.20 to $0.55/1,000 m³).

Case Studies

An enclosed biofilter system with two beds in a series was installed in the Netherlands in the early 1990s (van Lith et al. 1997). A total of 1200 m³ of media was used to treat 140,000 m³/h containing solvents with an average concentration of 500 ppm of carbon. The residence time was 30 seconds for a 90% VOC removal efficiency. The gas velocity was 240 m/h. Moisture control was based on a load cell, process controllers, and overhead sprayers. The system was operated for six years with a 25% increase in the pressure drop to 500 Pa (2 inches of water). Some compaction was noted and remedied by raking the top 10 cm of the filter material which decreased the pressure drop. Raking is now done annually.

Another biofiltration system (an open system for odor control) has been operating since 1991 at the Solid Waste Authority of Palm Beach County, Florida (Goldstein 1996). The facility is an 18 dry tonne/day biosolids composting facility with three 1,100 square meter biofilters handling 100,000 m³/h of air. A perforated pipe is used for air distribution. This is covered with a 45 cm bed of washed stones. A permeable plastic cover is used on top of the stones. The filter material is 1.0 to 1.2 m in depth and consists of shredded pine bark and Southern pine wood chips (1:1 ratio). The wood chips are used to maintain the porosity of the bed. The top of the bed is covered by six inches of shredded pine bark. A sprinkler system operating on a timer basis is used to maintain the humidity of the bed. Humidification of the inlet gases was not used since the air was already at 100% humidity. Moisture content and pH is monitored weekly, and back pressure is measured monthly using a variable speed fan. The pH, which can drop due to sulfuric acid production, is maintained above the five to six pH range by flushing with seven to thirteen inches of water or waiting for heavy rains. The system operated without loss of performance over the 18-month evaluation period.

PPC Biofilter (Longview, Texas) installed a 240,000 m³/h BIOTON® biofilter in Georgia at a particle board manufacturing plant for treating the emissions from press vents (Standefor 1996). Formaldehyde, ketone, and alpha and beta pinenes were treated. The support media was a mixture of inorganic plastic beads, compost, and nutrients. VOC emissions were reduced by 80%.

Comparison to Other Methods

Overall, volatile biodegradable organic, and in some cases inorganic, compounds in air can be treated by biofiltration. They are becoming an acceptable and mature air pollution control technique. Typically 95% removal of VOC is required by U.S. regulatory agencies while odor control is not always regulated in North America. Economical systems can be designed based on experience and pilot testing. Very little secondary pollutants are generated. These processes are flexible despite variations in load. Maintenance is fairly simple. Due to the inexperience of regulatory authorities, lack of operating data, and bad design and maintenance of some earlier systems, the authorities may require continuous monitoring of the systems. Biofilters have been used for odor control with ammonia, thiols, hydrogen sulfide, food processing wastes, wastewater treatment, and composting facilities. VOC treatment has been demonstrated for propane, butane, styrene, phenols, methylene chloride, and methanol, among others. Although biofilters have been successfully used for many years, it is only recently that advances have been made into the understanding of the basics of biofiltration, which can lead to the design of high-performance biofilters, biotrickling filters, and bioscrubbers.

In comparison to other treatment technologies, these biological techniques have low installation and operating costs and primarily generate water and carbon dioxide (Table 3.7). When comparing chemical and biological scrubbers, the compounds to be treated must be soluble for both types. The difference is the type of treatment where either chemical or biological reactions take place. The supply of chemicals for the chemical scrubbers can become large. In addition, if the pH of the liquid effluent must be adjusted after chemical treatment, further expenses will be incurred. This will not be the case for the biofilter.

Activated carbon systems for air treatment require regeneration costs and replacement of the carbon. These systems are designed according to the inlet organic load. When the carbon cannot be regenerated, it has to be treated as chemical waste. Removal rates are high for carbon, but costs increase as the concentration increases. Biofilters may have higher capital costs, but operating costs can be lower at moderate to high concentrations.

Regarding thermal oxidation, the inlet concentration of contaminants is critical in determining if supplemental fuel is required. Non-catalytic systems require concentrations higher than 1,000 mg/m³ while catalytic systems can function in the 500 to 1,000 mg/m³ range without additional fuel requirements. For gas streams for concentrations below 200 mg/m³, additional fuel will always be required for thermal technologies; thus, their operational costs will become high (Monsanto 1998). The design of biofilters is based on the

Table 3.7 Comparison of Vapor Control Technologies

Technology	Secondary products	Pretreatment requirements	Feed concentration range (mg/m³)	Capacity (m³/h)	Removal Efficiency (%)	Limitations	Capital costs ($/m³/h)	Annual operating costs ($/m³/h)
Carbon adsorption	solid waste, collected organics	cooling, dehumidification, particulate removal	0 to 5,000	60 to 40,000	90 to 99	Limited to <50% humidity streams	10 to 70	6 to 20
Absorption	solid waste, wastewater	particulate removal	1,000 to 20,000	3,000 to 160,000	95 to 98	Special liquids may be required	10 to 50	20 to 80
Condensation	condensate	none	> 5,000	160 to 30,000	50 to 95	Not for compounds with boiling points <40°C	6 to 50	10 to 70
Thermal Oxidizer	CO_2, SO_x, NO_x	preheating	20 to 20,000	2,000 to 800,000	95 to 99+	Halogenated compounds may require further treatment	5 to 120	5 to 50
Catalytic Oxidizer	CO_2, SO_x, NO_x, solid waste	preheating, particulate removal	50 to 10,000	2000 to 200,000	90 to 95	Halogenated compounds may require further treatment	10 to 150	5 to 45
Biofilter	CO_2, solid waste	humidification	10 to 5,000	200 to 500,000	90 to 98	Restricted to biodegradable compounds	5 to 50	5 to 20

Source: Adapted from Siegall 1996; OECD 1994; Hansen 1997; Govind and Bishop 1998

amount and the type of the contaminant that must be treated, which is not the case for thermal oxidizers. For low contaminant concentrations, the biofilter can be fairly small and still achieve a required removal rate at low capital and operating costs.

In summary, the choice of the treatment process depends on the characteristics of the effluent to be treated (flow, temperature, composition, concentration, biodegradability, etc.), the treatment level required, space availability, secondary environmental impact (wastewater, NO_x generation, etc.) and capital and operating costs. Various technologies are compared in Figure 3.4. The figure indicates the range of contaminant

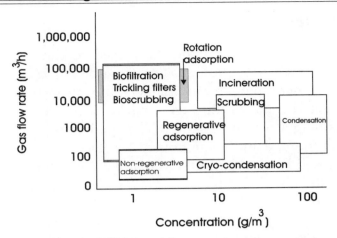

Figure 3.4 Optimal Range of Gas Flow Rates and Contaminant Concentrations for Various VOC Treatment Technologies. Adapted from KPMG (1994)

concentrations and flow rates where each technology performs in an optimal manner. Depending on the flow conditions, the best technology can then be chosen.

Biotrickling Filters

Box 3-2 Overview of Biotrickling Filters

Applications	Suitable for moderately contaminated air with aromatic compounds, aldehydes, alcohols, mercaptans, amides, hydrogen sulfide, and ammonia
Cost	Operating: $1.2 to $3.0 per 1,000 m^3, Capital costs $3 to $30 per m^3/h
Advantages	Ability to control pH and add nutrients
	Good retention of biomass
	Low capital and operating costs
Disadvantages	Start-up is complicated
	Operational costs higher than biofiltration
	Biomass can clog system
Concentrations treated	100 to 2,000 mg/m^3 with volumetric loads of 50 to 600 m^3/m^3-h
Other considerations:	Few installations in North America

Description

Biotrickling filters are considered an intermediate in the transition from biofiltration to bioscrubbing. As in biofilters, the gas to be treated is passed through a packed bed containing a biofilm (Figure 3.5). Liquid is sprayed on the bed and continuously circulated in a closed loop. Compost and many other organic materials cannot be used since water accumulation within the compost will cause a decrease in porosity, the development of anaerobic conditions, and an increase in pressure drops.

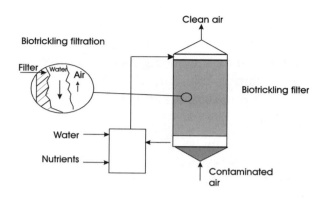

Figure 3.5 Schematic of a Biotrickling Filter. The Inset Shows the Three Phases within the Filter

Operating Conditions

The packing material must be able to accommodate the air and liquid flows; therefore, the dimensions of the pellets are larger. For example, diatomaceous earth pellets with 5 mm in diameter and 10 mm in length have given good gas treatment results but with biofouling (Smith et al. 1996). Polypropylene rings are also used. Loy et al. (1997) investigated the influence on elimination rates of various filter materials including polyurethane foam, coated polyurethane foam, synthetic polyamide cord textiles, and polyethylene plastic packing. Although coated polyurethane foam performed very well, a mixture of coated and pure polyurethane foam would be economical on an industrial scale.

Other modifications include adding makeup water to compensate for water losses and purging to avoid accumulation of salts in the recirculation loop, respectively. Since the specific surface area (100 to 300 m²/m³) is lower than for biofilters, the gas phase contaminants must be highly water-soluble (Ottengraf 1986). In general, the compounds should be at a concentration less than 0.5 g/m³ with Henry coefficients less than one (Kok 1992). Superficial gas velocities are usually 5 to 20 m/h. The spraying rates on the bed are usually about 1 m³/m²-day (Sorial et al. 1995).

Design

An analysis was performed by Deshusses (1999) in an attempt to correlate biomass growth, nutrient feed, and toluene removal. It was found that these parameters were not linearly correlated. A subsequent analysis of these results was then performed to determine VOC treatment costs in terms of operating, maintenance, and capital costs. This approach enabled an optimum reactor size to be designed, which could impact the design and operation of future full-scale biotrickling filters.

Maintenance

Since the biofilm is not removed by hydraulic shear, it can accumulate and cause plugging, channelling, and pressure drop problems. Clogging as a result of biomass growth has been a major problem, even at as low organic loading rates as 1.2 kg organic carbon/m³-day. The overgrowth can limit the surface area available for mass transfer and can cause plugging. Biofouling (Eweis et al. 1998) can be reduced by:

- Increasing the packing size to give larger pore sizes. Care must be taken, though, not to decrease the specific surface area such that the removal rates will decrease extensively;
- Backwashing of the packing periodically;
- Limiting the organic load;
- Evenly distributing the volumetric loading by using aerosol sprays for maintaining moisture and directional switching; and
- Controlling cell growth by limiting nutrient addition or addition of inhibitors such as NaCl.

Addition of protozoa (Cox and Deshusses 1997) seems to be a promising method to control biomass growth since bacteria are a food source for protozoa.

Applications

Laboratory and pilot tests (Diks and Ottengraf 1994) on units ranging from 0.1 to 100 L have shown that biotrickling filters can be successfully used for:

- chlorinated hydrocarbons (dichloromethane, 1,2- dichloroethene);
- aromatic compounds (styrene, toluene, nitrobenzene);
- alcohols (methanol, ethanol);
- aldehydes (propionaldehyde, acetone); and
- ammonia, hydrogen sulfides.

These concentrations are generally in the range of 100 to 2,000 mg/m³. Industrial applications are more difficult because of mixed and fluctuating compositions, temperatures, concentrations, and flow rates.

A two-stage biotrickling filter (Kraakman et al. 1997) has been tested at pilot scale to treat high concentrations of H_2S (Figure 3.6). Removal was 95% for concentrations of 70 to 95 mg H_2S/m³ from a wastewater treatment plant after a ten-second retention time through a synthetic packing. The gas loading was 350 m³/m³-h and contaminant loadings were 150 g H_2S/m³-h.

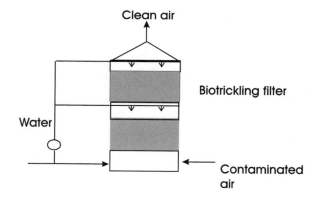

Figure 3.6 Schematic of a Two-Stage Biotrickling Filter. Adapted from Kraakman et al. (1997)

Another variation was developed to enable biotrickling filters to treat hydrophobic compounds such as alkanes. Van Groenestijn and Lake (1998) used a mixture of an organic solvent and water to trickle over a packed bed to capture hexane and pentane. The alkanes were absorbed in the oil phase (silicon oil), which were then transferred to the bacteria for degradation. A loading rate of

100 g hexane/m³ filter bed-h could be treated at a 90% efficiency. The influent gas contained 10 g hexane/m³. Clogging was not a problem over an eight-month period.

Costs

Tests at a wastewater treatment facility in Los Angeles demonstrated that 98% removal of H_2S was possible using a rock-based medium at loads under 39 g H_2S/m^3-h. The operational costs of a full-scale unit will be approximately $240 per million m³ of air treated compared to a chemical caustic scrubber, which would be $1,150 per million m³ (Morton and Caballero 1997). The air flow rate will be 2550 m³ per hour with a retention time of 15 seconds; thus, the corresponding reactor volume will be 10.6 m³.

Case Studies

Only a few installations have been done, but applications are increasing. A full-scale unit was used to treat phenol and ammonia at a Danish manufacturer of rock wool (Rydin et al. 1994). The air flow rate was 12,000 m³/h with concentrations of ammonia and phenol in the range of 50 to 125 mg/m³. The filter material was made of mineral wool slabs of 75 mm placed in a zig-zag manner (Figure 3.7). This material was chosen since it is light, stable with a large surface area, a good support for the microbes (mainly *Pseudomonas*), and because it gives a low pressure drop. The dimensions of the overall filter housing were 6 m x 3m x 2m (l x w x h) with each of the two filters in series having an area of 75 m². Degradation rates were 95 to 98% for phenol and greater than 98% for ammonia. The annual operating cost was approximately $4,090, which includes electricity, heat, acid, and labor.

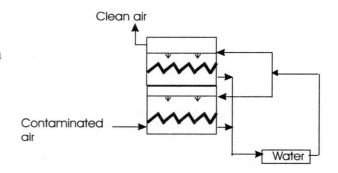

Figure 3.7 Diagram of a Biotrickling Filter Installed at a Danish Manufacturer of Rock Wool (adapted from Rydin et al. 1994)

Another set of tests (Wittorf et al. 1997) using the same type of unit was tested at a varnish and painting plant. The air flow rate was 1,200 to 3,000 m³/h with a gas load of 600 m³/m³-h and a solvent loading of 200 g of solvents/m³-h. The water flow rate over the packing was 2,000 to 5,000 L/h, the pH was 7 and the retention time was between six and 15 seconds. Solvent removal efficiency was 90 to 95%.

Another full-scale application of the trickling biofilter involves the treatment of N,N-dimethylacetamide (DMAc). The oxidation of this product gives ammonia. The concentration had to be lowered from 300 mg/m³ to 35 mg/m³ in a Dutch Lycra® factory. Gas flow rates were 2800 m³/h. Pilot tests had previously demonstrated that loads of 460 gC/m³-day could be treated with an efficiency of 95% (Waalewijn et al. 1994). The trickling filter used was cross-flow with a PVC packing. The pH was maintained at pH 7 by hydrochloric acid addition.

Recently a full-scale trickling biofilter was installed at a fiberglass bathtub manufacturer in Southern California (Webster and Deshusses 1997). Two reactors (1.5 m in diameter with a height of 2.4 m) were placed beside each other to treat between 320 and 1,500 m³/h of air containing styrene. The unit includes a blower, two water pumps, and two heating elements. Removal efficiencies were 70 to 95% under loading conditions of 20 to 40 g/m³-h.

Air Cure has developed a triple-stage, cross-flow trickling filter which has been installed for odor removal (amines, mercaptan, hydrogen sulfide) at concentrations of 50 mg/m³. The efficiency is 99% with a 60 mm water pressure drop for a unit treating 28,000 m³/h. In the winter, it is run as a chemical scrubber (using acid, bleach, and caustic addition), and in the summer it is run biologically.

SEUS (Germany) has installed a biotrickling filter at a foam production plant treating 3,000 m³/h of air containing 6 g/m³ of VOCs. The removal rate is 80% at a retention time of 40 to 60 seconds. The bed contains a packing of mineral expanded silica with a volume of 22.5 m³ and a low density of 280 kg/m³. This packing is disposable and very stable, which is desirable for a low air flow. Trickling filters have the potential to treat chlorinated compounds such as methyl chloride, dichloromethane, and 1,2-dichloroethane (Fucich et al. 1997a).

Envirogen has constructed the first of its full-scale biotrickling filters (Webster et al. 2000). The filter has dimensions of 3.1 m in height and 9.1 m in height. The air flow is downflow with a concurrent water phase. A programmable logic controller minimizes operator requirements. It is installed to treat 7 to 521 ppm$_v$ of VOCs and HAPs and odorous compounds (1 to 2 ppm$_v$) at a facility for oil recovery load equalization treatment. Operating since December 1999 at the North Island Naval Air Station, the reactor is constructed of fiberglass resin polymer with random inorganic packing. The air flow rate is 2970 m³/h into a 31 m³ bed volume with a gas residence time of 36 seconds. The removal rate is over 85%. The operating costs are $5,000 per year, compared to $36,000 per year for activated carbon.

Comparison to Other Methods

The major advantages of biotrickling filters over biofilters are their smaller size for the same volume of air and that only one unit is required. Higher contaminant concentrations can be treated in biotrickling filters than biofilters since control of the growth conditions (pH, nutrients, temperature, concentration of inhibiting compounds, etc.) is easier (Diks and Ottengraf 1994). Neutralization of acidic components is possible, such as sulfuric acid neutralized from hydrogen sulfides. The personnel for operation and maintenance of the unit, however, must be more highly skilled than biofilter operators.

Bioscrubbers

Box 3-3 Overview of Bioscrubbers

Applications	Suitable for high concentrations of VOCs, hydrogen sulfide (35 to 50 ppm) and ammonia
Cost	Operating: $0.6 to $5/1,000 m^3; Capital: $3 to $25 per m^3/h
Advantages	Ability to control pH and add nutrients
	High mass transfer
	High operational stability
Disadvantages	High capital and operating costs
	Water disposal required
	Complicated start-up
Concentrations treated	Up to 4 g/m^3 with volumetric loads of 50 to 1,500 m^3/m^3-h
Other considerations:	Few installations in North America

Description

Bioscrubbers are a combination of two well-known technologies: the water scrubber and activated sludge water treatment. They have been used for over 15 years in Holland and account for as much as 15% of the biological treatment processes in Germany (KPMG 1994).

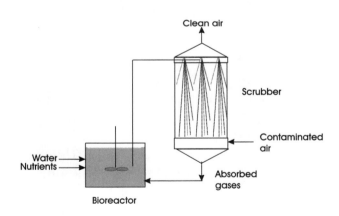

Figure 3.8 Flow Sheet of a Bioscrubber Including Scrubber and Bioreactor

The gas is passed into a tower where water is sprayed (Figure 3.8). The components are transferred to the water phase containing sludge (1 to 10 g per liter of water). Inert packing such as activated carbon may be added to increase the surface area. The water is then sent to the activated sludge bioreactor where the biodegradation step takes place. The treated aqueous phase containing microorganisms is then recirculated to the scrubber. This mobile phase makes pH and temperature control and nutrient addition much easier than in biofilters with no pressure drop problems. Variable air flows and contaminants can be accommodated in the bioscrubber. High boiling temperature components such as silicone fluids or phthalates can be added to increase the solubility of the contaminants in the scrubber and decrease the time of biodegradation (Schippert 1989 a, b).

A purge is used to avoid accumulation of salts and end-products. Fresh water is added as make-up. Like trickling filters, bioscrubbers are smaller in size than biofilters, but they are more complicated and contain two units compared to the one for the tricking filters. The design of the bioscrubbers can vary substantially. The bioreactor can be under or beside the scrubber. The bioreactor volume is typically larger than the scrubber. For example, the reactor and scrubber volumes are 28 and 15 m^3 (Hansen 1997) and 270 and 180 m^3 (Schippert 1989 a, b), respectively.

Criteria for Utilization

Since the gas/liquid surface area is lower than biofilters, this technique is restricted to compounds with Henry coefficients less than the five to ten range (Schippert 1989a, b) or less than 0.01 if high spray column and high water flow rates are to be avoided, since this increases energy consumption (Kok 1992). Compounds that can be treated include aromatic hydrocarbon, aldehydes, alcohols, mercaptans, amides, hydrogen sulfide, and ammonia. Applications include odor control at rendering, sewage treatment, food processing plants, and composting facilities, and for control of VOCs from paint, lacquering, and printing operations (Table 3.8). Bioscrubbers have also been used to control odor and ammonia at piggeries (Schirz 1994).

Table 3.8 Industrial Applications of Bioscrubbers

Compound	Industry	Removal efficiency (%)	Flow rate (m^3/h)	Volumetric flow rate (m^3/m^3-h)	Reference
Odor	Fish product	95	25,000	581	Hansen and Rindel (1992)
Odor	Fish feed	95	40,000		Hansen et al. (1994)
VOCs	Surface coating	>99	26,000	58	Schippert (1989a)
VOCs	Flexography	80–95	150,000		
VOCs		>94	14,000		
Phenol	Foundry	>50	36,000	1428	Büren (1989)
Ammonia	Foundry	>96	36,000		Büren (1989)
Methanol		>99	40,000		Wolff (1989)
Isopropyl alcohol	Contact lens	>99	8,500		Whaley et al. (1998)
H_2S	Wastewater treatment	>95	6,000		Hansen (1997)
VOCs and odor control	Fish feed	>95 for amines	63,500	827	Vaidila and Welch (2000)

Operating Conditions and Design

When the absorbing liquid reacts with the activated sludge that degrades the component quickly, the vapor phase concentration approaches zero and the efficiency of removal (E) for the scrubber can be expressed as (Hansen 1997):

$$\ln(\frac{100}{100-E}) = K_g a \times A \times Z \times \frac{P}{G}$$

where $K_g a$ is the mass transfer coefficient, P is the total pressure, G is the gas flow rate, A is the cross-sectional area, and Z is the packing height. $K_g a$ can be calculated by plotting 1/G versus ln (100/(100-E)) for various values of the liquid rate L. Since the mass transfer of the contaminant is liquid controlled, the equation $K_g a = cL^b$ where the constants, c and b, can be determined by the log plots of $K_g a$ versus L.

Applications

The results of a full-scale bioscrubber are shown for a system which was run under nitrification conditions (aerobic) in the scrubber and denitrification (anoxic) conditions in the bioreactor for odorous nitrogen components from a fish feed plant (Rasmussen et al. 1994; Hansen et al. 1995). This mode allows the oxidation of ammonia nitrogen to nitrate and then denitrification to produce nitrogen. If the system were run totally in aerobic mode, all the ammonia nitrogen would be shifted to nitrate and twice as much of the chemicals would be required to maintain pH. Nitrate buildup is avoided in the mixed mode, and water drainage is not required.

Another similar system included a 2 m^3 scrubber, a 4 m^3 trickling filter, and a 1 m^3 denitrification reactor (van Groenestijn et al. 1997) to treat a 8,000 m^3/h gas stream from a pig stable containing 15 mg NH_3/m^3. The following reactions occurred in each of the process units.

Absorption: $NH_3 + H_2O \rightarrow NH_4^+ + OH^-$

Nitrification 1: $2NH_4^+ + 3O_2 \rightarrow 2\,NO_2^- + 4H^+ + 2H_2O$

Nitrification 2: $2NO_2 + O_2 \rightarrow 2NH_3^-$

Denitrification: $5CH_3OH + 6NO_3^- \rightarrow 5\,CO_2 + 7H_2O + 3N_2 + 6\,OH^-$

Overall: $5CH_3OH + 6NH_3 + 12O_2 \rightarrow 5\,CO_2 + 19H_2O + 3N_2$

The overall removal rate for the ammonia was 93%.

Costs

Operating costs of the biological processes are $0.12 to $0.18 per 1,000 m^3 for a biofilter, $0.6 to $1.2 per 1,000 m^3 for a bioscrubber, and $1.2 to $3.0 per 1,000 m^3 for trickling filters (Herrygers et al. 2000). Capital for biofilters and bioscrubbers are $1.2 to $3.0 per

m^3/h and \$3 to \$12 per m^3/h, respectively. Biofilters are clearly the least expensive to build and operate.

Costs were also estimated by Whaley (1999) for a bioscrubber at a contact lens facility in Puerto Rico. Capital costs for the system were \$24 per cubic meter per hour and annual operating costs were \$5 per cubic meter per hour. Electrical requirements were approximately 390 kW h per year and 150 hours of labor were required. For this application, a catalytic thermal oxidizer would have capital costs in the range of \$35 to \$41 cubic meter per hour and annual operating costs in the range of \$11 to \$12 cubic meter per hour. Solvent recovery systems would have initial costs in the range of \$66 to \$280 per cubic meter per hour. The operation of the bioscrubber has been reliable without problems from sludge or odor production or nutrient consumption. The system has been proven to be economical in comparison to catalytic thermal oxidation (Whaley 1999).

Case Studies

SEUS Systemtechnik of Germany was one of the first to market bioscrubber systems for film printing plants and for the food industry for the treatment of ethyl acetate, ethanol, methanol, isopropanol, toluene, acetone, ammonia, and cyclohexane at concentrations up to 4 g/m^3. The mass flow rate was 160 kg/h of solvents, and the removal efficiency was around 97%. The air flow rate was 40,000 m^3/h. Electricity requirements were 150 kW and the decomposition time in the reactors was 80 minutes. The scrubber works in the counter-current direction. Once the air passes through the scrubber, the water enters a series of two bubble reactors containing activated carbon pellets or cross-linked plastic pellets containing immobilized and free microorganisms. Fresh air for the aerobic bacteria is added via a compressor. Once the water has passed through the two reactors, it is now contaminant free and is sent back to the scrubber; thus, no wastewater is produced.

Paques has developed a THIOPAQ® Bioscrubber for the treatment of air containing up to 300 mg/m^3 of H_2S with 99.9% removal (Paques 1998). The process consists of a scrubber with packing material containing sulfide-oxidizing biomass (Buisman and Prins 1994). Air is passed through the column in the counter-current direction to the scrubber water containing sodium hydroxide (pH 8 to 9) according to the reaction:

$$H_2S_{(gas)} + OH^- \rightarrow HS^-_{(liquid)} + H_2O$$

The sulfide is then oxidized in the aerobic reactor to produce sulfur:

$$HS^-_{(liquid)} + \frac{1}{2}O_2 \rightarrow S^\circ + OH^-$$

The caustic is recovered for recirculation to the scrubber. The last step is the recovery of the high quality sulfur by pre-concentration in a settler and then de-watering by drum filtration. The sulfide is converted to sulfur by the bacteria, which is removed by settling. The water is then recycled to the scrubber (Figure 3.9). Several full-scale installations are in operation, one in the Netherlands treating 450 m³/h of 1 to 2% H₂S, and the other in Germany at a papermill where the hydrogen sulfide concentration is reduced to 10 ppm. There is a 90% savings of caustic at these plants where consumption is less than 0.95 kg NaOH per kg of sulfur removed (Janssen et al. 2000). A unit at BP Chemicals in the United Kingdom treats purged air with concentrations of 10 to 500 ppm of H₂S. Despite peaks of 2,000 to 20,000 ppm, complete H₂S removal is obtained. Two other full-scale units have been built in India and the U.S.

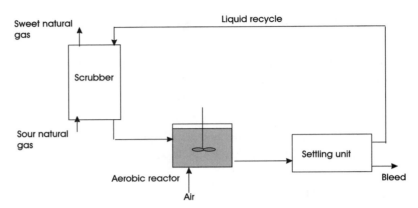

Figure 3.9 Bioscrubber for H₂S Removal from Sour Natural Gas (Paques 1998)

Sulfur dioxide removal for gas streams has been pilot tested by Paques at the Electriciteits Produki Maatschappij Zuid-Nederland (EPZ, Geetruidenberg, The Netherlands) (Janssen et al. 2000). A four-part process is used including an absorber, an anaerobic reactor, an aerobic reactor, and a sulfur recovery step. The absorber is a vertical spray tower in which the SO_2 is absorbed and forms HSO_3^- in the presence of water. Removal of 98% of the sulfur dioxide can be achieved. In the pilot plant, 6,000 m³/h of 120°C flue gas containing 2,000 mg/m³ of SO_2 is treated. In the anaerobic reactor, the following reaction takes place:

$$HSO_3^- + 3H_2 \rightarrow HS^- + 3H_2O$$

A gaslift reactor is used with recirculation of hydrogen. It is possible to use ethanol or methanol instead of hydrogen gas. A three-phase settler at the top of the reactor is required to retain the biomass and to separate the gas and liquid effluents. This liquid effluent is then sent by gravity to the aerobic reactor where *Thiobacillus* convert the sulfide into elemental sulfur. The sulfur cake (60% solids) has a purity of 95% which can be used for sulfuric acid production. If sulfur is used in the liquid form, it can be sold commercially. Other waste streams that contain sulfur dioxide and that potentially can be treated by this method include effluents produced by sulfuric acid plants, power stations, chemical plants, and municipal sludge or waste incineration.

Biopaq has also pilot tested a bioscrubber contained in one unit (Figure 3.10). Tests were performed on an air stream of 160 m³/h containing 70 to 500 mg/m³ of H₂S (Janssen et al. 1997). This integrated scrubber was able to remove over 99.9% of the hydrogen sulfide. Biothane also has a similar process called Biopuric for the biological removal of hydrogen sulfide. The biogas enters the reactor and is treated by sulfur oxidizing bacteria on the packing. The products, in this case are sulfate and sulfur.

Another variation developed by SNC Research Corp. (Mulligan et al. 1997) has not been applied at full scale but has been pilot tested on emissions from a flexography plant with total concentrations up to 1.6 g/m³. The process is comprised of either a water scrubber or activated carbon column, a buffer tank, and multiplate anaerobic reactor (Figure 3.11). The water scrubber can be used in the case of water soluble components such as alcohols (ethanol, methanol, propanol, etc.), and the activated carbon is used for solvents of low solubility such as ethyl acetate, which is difficult to capture by water scrubbing. Operating costs for this system would be lower than for aerobic systems since aeration is not necessary and less sludge is produced. In addition, biogas is generated which can be used as a source of energy due to its high methane content (greater than 80%).

Another configuration of the bioscrubber that is under development to treat components of low solubility involves the combination of absorption and biological degradation in the same unit (Laurenzis and Werner 1997). This involves a three-phase air lift reactor with an

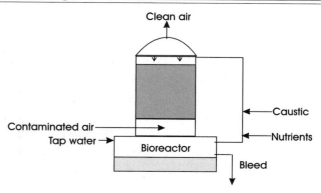

Figure 3.10 An Integrated Pilot Bioscrubber for H_2S Removal. Adapted from Janseen et al. (1997)

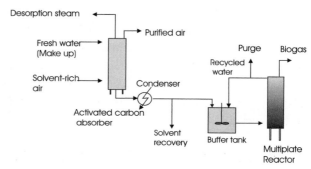

Figure 3.11 Schematic Flow Sheet of an Activated Carbon-anaerobic Reactor System for the Treatment of VOCs

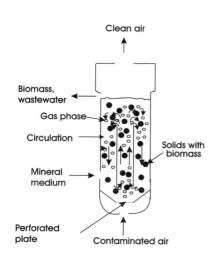

Figure 3.12 Three-phase Bioscrubber with Concentric Draught Tube (adapted from Laurenzis and Werner 1997)

internal loop (Figure 3.12). The contaminated air stream enters the scrubber at the bottom and rises in the middle tube in the liquid phase. The liquid then falls in the downcomer and the air exits out the top. Solid particles (up to 10%) are added to hold the biomass in the reactor. The particles are fluidized as the critical gas velocity is surpassed. The particles of polyamide spheres and polyurethane cubes were tested, and it was found that microorganisms were able to grow on the outside of the spheres but could grow on the outside and inside pores of the cubes. Degradation of 600 g toluene/m³-h was achieved. Energy is saved since pumping is lower.

Recently Allergan (Whaley et al. 1998) started up a bioscrubber for the treatment of isopropyl alcohol, which is emitted from lens manufacturing operations. A full-scale unit for the treatment of 8,500 m³/h was installed at their Puerto Rico plant. The inlet conditions ranged from 200 to

500 mg/m³ of isopropyl alcohol with some acetone and heptane. The system consists of two scrubbers (Figure 3.13). Water is recirculated at a rate of 56 to 76 liters per minute from the bioreactor. The water from the scrubber enters the fixed biological film bioreactor where the retention time is six hours. Additional air is sparged into the reactor by a blower to maintain the dissolved oxygen above 3.0 mg/L. A place at the end of the reactor is used for sludge settling which is either removed or recirculated to the bioreactor. After the initial start-up, two weeks were required to reach steady state. Nutrients in the form of ammonia sulfate,urea and sodium phosphate, and bacteria were added initially.

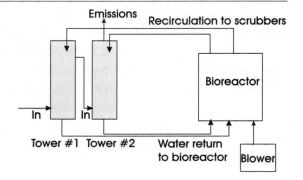

Figure 3.13 Schematic of a Two-stage Bioscrubber at a Pharmeceutical Plant in Puerto Rico (adapted from Whaley et al. 1998).

However, since the ammonia sulfate caused a decrease in the pH to 5, after a couple of months only urea was added. This was done on a weekly basis. Packing was installed in the scrubber to enhance the mass transfer rate. However, the packing was eliminated due to increased pressure drop from the plugging of the packing because of the bacterial growth. The removal efficiency was 98% without the packing compared to 99% with the packing. Sludge production was not a problem. Overall removal of 99% of the emissions was achieved.

Comparison to Other Methods

The three types of biological reactors can be differentiated by the whether the liquid phase is mobile (bioscrubber or trickling filter) or dispersed (biofilter) or whether the microbes are dispersed (bioscrubber) or immobilized (trickling filter or biofilter). Bioscrubbers or biotrickling filters would be more appropriate for the treatment of acid generating components such as chlorinated solvents. Comparisons between the three main biological treatment technologies are shown in Table 3.9.

Table 3.9 Advantages and Disadvantages of Biofiltration, Biotrickling Filters, and Bioscrubbing

Reactor type	Advantages	Disadvantages	Operating Costs	Capital Costs
Biofilters	simple to start-up and operate low capital costs low operating costs suitable for odorous compounds, aromatic compounds, aldehydes, alcohols, mercaptans, amides, hydrogen sulfide, ammonia and other water soluble, biodegradable, compounds of low concentration high gas/liquid surface area	poor control of conditions slow to adapt to fluctuating conc. large area requirement channelling is normal limited bed life excess biomass not disposable limited process control low surface area for mass transfer disposal of sludge difficult	+	+
Biotrickling filter	simple to operate low capital costs low operating costs suitable for moderately contaminated air with aromatic compounds, aldehydes, alcohols, mercaptans, amides, hydrogen sulfide and ammonia ability to control pH ability to add nutrients good retention of biomass	start-up is complicated operational costs higher than biofiltration few industrial installations biomass can clog system	++	++
Bioscrubbers	good control of pH and nutrients compact high mass transfer suitable for higher concentrations (35 to 50 ppm) of hydrogen sulfide and ammonia capability of avoiding product accumulation high operational stability	higher capital and operating costs water disposal required plugging possible in adsorption step complicated start-up limited buffering capacity	+++	+++

Adapted from Kok 1992; Wittorf et al. 1993. Notes: Rating system: +least expensive, +++ most expensive

References

Adler, S. F. 2001. Biofiltration: A primer. *CEP*, 97(4):33–41.

Apel, W. A., J. M. Barnes, and K. B. Barrett. 1995. Biofiltration of nitrogen oxides from fuel combustion gas streams, *Proceedings of the 88th Annual Meeting and Exhibition, Air and Waste Management Association*. San Antonio, TX. 18–23 June.

Bhat, S., J. T. Ravn, R. Willingham, and S. Standefor. 1998. Biofilter as a VOC control device for an industrial screen printing operation. Presented at the *Air & Waste Management Association's 91st Annual Meeting & Exhibition*, Paper 98-MP20A.05. San Diego, CA. 14–18 June.

Bohn, H. 1992. Consider biofiltration for decontaminating gases. *Chemical Engineering Progress*, 88(4):34–40.

Buisman, C. J. M., and Prins, W. L. 1994. New process for biological (flue) gas desulfurization. *VDI Berichte 1104: Biologische Abgasreinigung, Praktische Erfahrungen und Neue Entwicklungen*. Düsseldorf: VDI Verlag. 95–102.

Büren, E. 1989. Biocatalytische Abgasreinigung in einer kernmacheri. *VDI Berichte 735: Biologische Abgasreinigung, Praktische Erfahrungen und Neue Entwicklungen*. Düsseldorf: VDI Verlag. 89–117.

Boyette, R. A. 1998. Getting down to (biofilter) basics. *Biocycle*, 39(5):58–62.

Cox, H. H. J., J. H. M. Houtman, H. J. Doddema, and W. Harder. 1993. Enrichment of fungi and degradation of styrene in biofilters. *Biotechnology Letters*, 15:737–742.

Cox, H. H. J., and M. A. Deshusses. 1997. Increasing the stability of biotrickling filters by using protozoa. *Biological Gas Cleaning* Maastricht, The Netherlands, 28–29 April. W. L. Prins and J. van Ham, eds. Düsseldorf: VDI Verlag. 233–240.

Demiriz, A. M. 1992. Neue einsatzgebiete biologischer filteranlagen: Gießerei-bereich und lösemittelalabscheidung. *Biotechniques for Air Pollution Abatement and Odour Control Policies, Proceedings of an International Symposium*. A. J. Dragt and J. van Ham, eds. Maastricht, The Netherlands, 27–29 October, 1991. Amsterdam: Elsevier. 293–296.

Deshusses, M. A. 1999. A cost benefit approach to reactor sizing and nutrient supply for trickling filters for air pollution control. Presented at *Air & Waste Management Association's 92nd Annual Meeting & Exhibition*, Paper F8-242. St. Louis, MO. 20–24 June.

Devinny, J. S., D. E. Chitwood, and F. E. Reynolds, Jr. 1998. Two-stage biofiltration for wastewater treatment of off-gases. Presented at *Air & Waste Management Association's 91st Annual Meeting & Exhibition*, Paper 98-MP20A.06. San Diego, CA. 14–18 June.

Diks, R. M., and S. P. P. Ottengraf. 1994. Technology of trickling filters. *Biologische Abgasreinigung, Praktische Erfahrungen und Neue Entwicklungen.* VDI Berichte 104 Düsseldorf: VDI Verlag. 19–37.

du Plessis, C. A., K. A. Kinney, E. D. Schroeder, D. P. Y. Chang, and K. M. Scow. 1996. Anaerobic activity in an aerobic, aerosol-fed biofilter. In *Proceedings of Conference on Biofiltration,* University of Southern California. 24–25 Oct. 158–163.

Edwards, F. G. and N. Nirmalakhandan. 1996. Biological treatment of airstreams contaminated with VOCs: An overview. *Water Science and Technology,* 34:565–571.

Eitner, D. 1992. Emmissionminderung in Ölmühlen durc biofilter-erfahrungsbericht. *Biotechniques for Air Pollution Abatement and Odour Control Policies, Proceedings of an International Symposium,* A. J. Dragt and J. van Ham, eds. Maastricht, The Netherlands, 27–29 October. 1991. Amsterdam: Elsevier. 197–205.

———. 1998. The degradation of organic solvents in biological filtering systems shown by examples from varnish, drying and printing industry. Presented at the *Air & Waste Management Association's 91st Annual Meeting & Exhibition,* Paper 98-MP20A.03. San Diego, CA. 14–18 June.

Engresser, K. H. 1992. Mikrobiologische aspekte der biologischen abluftreinigung. *Biotechniques for Air Pollution Abatement and Odour Control Policies, Proceedings of an International Symposium,* A. J. Dragt and J. van Ham, eds. Maastricht, The Netherlands, 27–29 October 1991. Amsterdam: Elsevier. 33–40.

Ergas, S. J., K. A. Kinney, M. E. Fuller, and K. M. Scow. 1994. Characterization of a compost biofiltration system degrading dichloromethane. *Biotechnology and Bioengineering,* 44(9):1048–1054.

Eweis, J. B., S. J. Ergas, D. P. Y. Chang, and E. D. Schroeder. 1998. *Bioremediation Principles.* Boston: WCB McGraw-Hill.

Ferranti, M. M., and A. Conca. 2000. Formaldehyde biological removal from exhaust air in the composite panel board industry from pilot tests to industrial plant. Presented at the *Proceedings of Air & Waste Management Association's 93rd Annual Meeting & Exhibition.* Salt Lake City, Utah. 14–18 June.

Fucich, W. J., A. P. Togna, and M. P. Pitre. 1997a. The treatment of methyl chloride by a biotrickling filter. *1997 AIChE Annual Meeting.* Session 18, Paper 28h. November.

Fucich, W. J., Y. Yang, and A. P. Togna. 1997b. Biofiltration for control of carbon bisulfide and hydrogen sulfide vapors. Presented at the *Air & Waste Management Association's 90th Annual Meeting & Exhibition.* Paper 97-RP114B.04. Toronto, Ont. 8–13 June.

Goldstein, N. 1996. Odor control experiences: Lessons from the biofilter. *Biocycle,* 37(4):70–75.

Govind, R., and D. F. Bishop. 1998. Biofiltration for treatment of volatile organic compounds (VOCs) in air. *Bioremediation: Principles and Practice, Vol. II,* S. K. Sikdar and R. L. Irvine, eds. Lancaster: Technomic Publishing Company. 403–459.

Govind, R., W. Zhoa, and B. F. Bishop. 1995. Biofiltration for treatment of volatile organic compounds (VOCs). Presented at the *Third* In Situ *and On-Site Bioreclamation Symposium,* San Diego, CA. April.

Hansen, N. G. 1997. Bioscrubber and biofilter for air purification at wastewater treatment plants. *Biological Gas Cleaning,* Maastricht, The Netherlands. 28–29 April. W. L. Prins and J. van Ham, eds. Düsseldorf: VDI Verlag. 397–412.

Hansen, N. G., and L. Rindel. 1992. Recent experience with biological scrubbers for air pollution control in Denmark. In *Proceedings of an International Symposium: Biotechniques for Air Pollution Abatement and Odour Control Policies.* A. J. Dragt and J. van Ham, eds. Maastricht, The Netherlands. 27–29 October 1991. Amsterdam: Elsevier. 143–154.

Hansen, N. G., H. H. Rasmussen, and K. Rindel. 1994. Biological air cleaning processes exemplified by applications in wastewater treatment and fish industry. In *Proceedings: Odor and Volatile Organic Compound Emission Control for Municipal and Industrial Treatment Facilities.* Jacksonville, Florida. 24–27 April. Florida Water Environment Association. 2–23.

Herrygers, V., H. van Langenhove, and E. Smet. 2000. Biological treatment of gases polluted by volatile sulfur compounds. *Environmental Technologies for the Treatment of Sulfur Pollution.* P. Lens and L. Hulshoff Pol, eds. London: IWA Publishing. 281–304.

Heslinga, D. C. 1994. Biofiltration technology. *VDI Berichte 1104; Biologische Abgasreinigung, Praktische Erfahrungen und Neue Entwicklungen.* Düsseldorf: VDI Verlag. 11–18.

Hofmann, W. 1989. Biofilter nach kakaoröstereien. *VDI Berichte 735: Biologische Abgasreinigung, Praktische Erfahrungen und Neue Entwicklungen.* Düsseldorf: VDI Verlag. 233–242.

Huber, J. 1992. Planung, durchführung und erste erfahrungen zum biofilter tierkörperbeseitigungsanklage platting. *Biotechniques for Air Pollution Abatement and Odour Control Policies, Proceedings of an International Symposium,* A. J. Dragt, and J. van Ham, eds. Maastricht, The Netherlands. 27–29 October 1991. Amsterdam: Elsevier. 161–166.

Janssen, A. J. H., K. de Hoop, and C. J. N. Buisman. 1997. The removal of H_2S from air at a petrochemical plant. *Biological Gas Cleaning.* Maastricht, The Netherlands. 28–29 April. W. L. Prins and J. van Ham, eds. Düsseldorf: VDI Verlag. 359–364.

Janssen, A. J. H., H. Dijkman, and G. Janssen. 2000. Novel biological processes for the removal of H_2S and SO_2 from gas streams. *Environmental Technologies for the*

Treatment of Sulfur Pollution, P. Lens and L. Hulshoff Pol, eds. London: IWA Publishing. 265–280.

Kersting, U. 1992. Behandlung großvolumiger abluftströme durch biofilter, vorgestellt an beispielen der tabindustrie. *Biotechniques for Air Pollution Abatement and Odour Control Policies, Proceedings of an International Symposium*, A. J. Dragt, and J. van Ham, eds. Maastricht, The Netherlands. 27–29 October 1991. Amsterdam: Elsevier. 155–160.

Kirchner, K., S. Wagner, H. J. Rehm. 1996. Removal of organic air pollutants from exhaust gases in the trickle-bed bioreactor. Effect of oxygen. *Applied Microbiology and Biotechnology*. 45(3): 415.

Kok, H. J. G. 1992. Bioscrubbing of air contaminated with high concentrations of hydrocarbons. *Biotechniques for Air Pollution Abatement and Odour Control Policies, Proceedings of an International Symposium*. A. J. Dragt and J. van Ham, eds. Maastricht, The Netherlands. 27–29 October 1991. Amsterdam: Elsevier. 77–82.

KPMG Management Consulting. 1994. *Technology Advancement Studies, Biological Gas Cleaning, Stakeholder Workshop #1*. Environment Canada, Clean Air Technologies Division. 7 December.

Kraakman, N. J. R., J. W. van Groenstijn, B. Koers, and D. C. Heslinga. 1997. Styrene removal using a new type of bioreactor with fungi. *Biological Gas Cleaning*. Maastricht, The Netherlands. 28–29 April. W. L. Prins and J. van Ham, eds. Düsseldorf: VDI Verlag. 225–232.

Laurenzis, A., and U. Werner, U. 1997. Exhaust gas purification in a three-phase bioscrubber. *Biological Gas Cleaning*. Maastricht, The Netherlands. 28–29 April. W. L. Prins and J. van Ham, eds. Düsseldorf: VDI Verlag. 115-122.

Lehtomaki, J., M. Toronen, and A. Laukkarinen. 1992. A feasibility study of biological waste-air purification in a cold climate. *Biotechniques for Air Pollution Abatement and Odour Control Policies*. A. J. Dragt and J. van Ham, eds. Maastricht, The Netherlands. 27–29 October 1991. Amsterdam: Elsevier.

Leson, G. 1998. Biofilters in practice. *Bioremediation Principles and Practice, Vol. III*. S. K. Sikdar and R. L. Irvine, eds. Lancaster: Technomic Publishing Company. 523–556.

Leson, G., and A. M. Winer. 1991. Biofiltration: an innovative air pollution control technology for VOC emissions. *Journal of the Air Waste Management Association*. 41:1045–1054.

Liebe, H. G. 1989. Einsatz von biofiltern zur minderung der emissionen von anlagen zum verarbeiten von fleish und fisch für die menschliche ernährung. *VDI Berichte 735: Biologische Abgasreinigung, Praktische Erfahrungen und Neue Entwicklungen*. Düsseldorf: VDI Verlag. 215–231.

Loy, J., K. Heinreich, and B. Egerer. 1997. Influence of filter material on the elimination rate in a biotrickling filter bed. Presented at the *Air & Waste Management Association's 90th Annual Meeting & Exhibition*, Paper 97-RA71C.01. Toronto, Ont. 8–13 June.

Mackowiak, J. 1992. Abscheidung von formaldehyd aus der abluft im biofilter. *Biotechniques for Air Pollution Abatement and Odour Control Policies, Proceedings of an International Symposium*. A. J. Dragt and J. van Ham, eds. Maastricht, The Netherlands. 27–29 October 1991. Amsterdam: Elsevier. 273–278.

Maier, G. 1989. Biofiltration von Gießereiabgassen. *VDI Berichte 735: Biologische Abgasreinigung, Praktische Erfahrungen und Neue Entwicklungen*. Düsseldorf: VDI Verlag. 285–292.

Marran, K. S., and T. A. Laustsen. 2000. Full-scale biofilter treats volatile organic compounds from a combined sewer at 12000 cubic feet per minute. Presented at the *Air & Waste Management Association's 93rd Annual Meeting & Exhibition*. Salt Lake City Utah. 14–18 June.

Mildenberger, H. J. 1992. Biofiltersysteme zur geruchsbeseitigung und zur reduzierung von organikaemissionen auf kläranlagen und in der chemischen industrie. *Biotechniques for Air Pollution Abatement and Odour Control Policies, Proceedings of an International Symposium*, A. J. Dragt and J. van Ham, eds. Maastricht, The Netherlands. 27–29 October 1991. Amsterdam: Elsevier. 187–196.

Monsanto Enviro-Chem Systems. 1998. Biological oxidation can solve odour problems. *Environmental Science & Engineering*. 11(3):34–35, 49.

Morton, R. L., and R. C. Caballero. 1997. Removing hydrogen sulfide from wastewater treatment facilities' air process streams with a biotrickling filter. Presented at the *Air & Waste Management Association's 90th Annual Meeting & Exhibition*, Paper 97-RA71C.06. Toronto, Ont. 8–13 June.

Mulligan, C. N., J. Chebib, and B. Safi. 1997. Anaerobic treatment of VOCs of low water solubility using the SNC-Lavalin Multiplate Reactor. Presented at the *Air & Waste Management Association's 90th Annual Meeting & Exhibition*, Paper No. 97-TA4B.01. Toronto, Ont. 8–13 June.

OECD. 1994. *Reducing environmental pollution: Looking back, thinking ahead. An examination of OECD member country progress, trends and opportunities for accelerated pollution reductions in the 1990s and beyond*. Paris: OECD.

Ottengraf, S. P. P. 1986. Exhaust gas purification. *Biotechnology*. H. J. Rehm and G. Reed, eds. Weinheim: VCH Verlags-gesellschafter. 8:425–452.

Ottengraf, S. P. P., and R. M. M. Diks. 1992. Process technology of biotechniques. *Biotechniques for Air Pollution Abatement and Odour Control Policies*. A. J. Dragt and J. van Ham, eds. Maastricht, The Netherlands, 27–29 October. 1991. Amsterdam: Elsevier. 17–31.

Oude Luttighuis, H. 1989. Möglichkeiten für biofiltration in der pharmaceutischen industrie. *VDI Berichte 735: Biologische Abgasreinigun, Praktische Erfahrungenund Neue Entwicklungen.* Düsseldorf: VDI Verlag. 341–348.

———. 1997. A new generation packing material for biofilters. *Biological Gas Cleaning.* Maastricht, The Netherlands, 28–29 April. W. L. Prins and J. van Ham, eds. Düsseldorf: VDI Verlag. 141–148.

Paques. 1998. *H₂S removal.* Brochure from Pacques Bio Systems BV, the Netherlands. Also available online: http://www.paques.nl.

Pearson, C. C., V. R. Philips, G. Green, and I. M. Scotford. 1992. A minimum-cost biofilter for reducing aerial emissions from a broiler chicken house. In *Biotechniques for Air Pollution Abatement and Odour Control Policies, Proceedings of an International Symposium.* A. J. Dragt and J. van Ham, eds. Maastricht, The Netherlands, 27–29 October. Amsterdam: Elsevier. 245–254.

Rasmussen, H. H., N. G. Hansen, and K. Rindel.1994. Treatment of odorous nitrogen compounds in a bioscrubber comprising simultaneous nitrification and denitrification. *VDI Berichte 1104: Biologische Abgasreinigung, Praktische Erfahrungen und Neue Entwicklungen.* Düsseldorf: VDI Verlag. 491–497.

Rieneck, M. G. 1992. Mikrobiologische methoden zur charakterisierung von biofiltermaterialien. In *Biotechniques for Air Pollution Abatement and Odour Control Policies, Proceedings of an International Sympsium.* A. J. Dragt and J. van Ham, eds. Maastricht, The Netherlands, 27–29 October 1991. Amsterdam: Elsevier. 85–95.

Rydin, S., P. Dalberg, and J. Bodker. 1994. Biological waste gas treatment of air containing phenol and ammonia using a new type of trickling filter. VDI Berichte 1104: Biologische Abgasreinigung, Praktische Erfahrungen und Neue Entwicklungen. Düsseldorf: VDI Verlag. 231–237.

Sabo, F. 1989. Praktische erfahrungen mit biofiltern zur reinigung geruchsintensiver deponiegase. *VDI Berichte 735: Biologische Abgasreinigung, Praktische Erfahrungen und Neue Entwicklungen.* Düsseldorf: VDI Verlag. 293–312.

Schelchshorn, J. and A. Vinke. 1989. Erfahrungsbericht über die mickrobielle entschwefelung von biogasen auf klärenlagen nach dem biopulC-verfahren. *VDI Berichte 735: Biologische Abgasreinigun, Praktische Erfahrungenund Neue Entwicklungen.* Düsseldorf: VDI Verlag. 129–138.

Schippert, E. 1989a. Biowäsher nach einer dosenlackieranlage. *VDI Berichte 735: Biologische Abgasreinigung, Praktische Erfahrungen und Neue Entwicklungen.* Düsseldorf: VDI Verlag. 77–89.

———. 1989b. Das biosolv-verfahren von keramschemies zur absorption bon schwer wasserlöslichen lösemitteln. *VDI Berichte 735: Biologische Abgasreinigung, Praktische Erfahrungen und Neue Entwicklungen.* Düsseldorf: VDI Verlag. 161–177.

Schirz, S. 1994. Minderung von ammoniak-emissionen aus der geflugelhaltung. *VDI Berichte-Verein Deutscher Ingenieure*, N. 1104. 533–538.

Siegall, J. H. 1996. Exploring VOC control options. *Chemical Engineering*, 103(6):92–96.

Smith, F. L., G. A. Sorial, M. T. Suidan, A. W. Breen, and P. Biswas. 1996. Development of two biomass control strategies for extended stable operation of highly efficient biofilters with high toluene loadings. *Environmental Science Technology*, 30(5):1744–1750.

Society of German Engineers (VDI 3477) 1991. Biologische Abluft-/Abgasreinigung Biofilter. Dusseldorf. 12.

Sorial, G. A., F. L. Smith, M. T. Suidan, P. Biswas, and R. C. Brenner. 1995. Evaluation of a trickle bed biofilter for toluene removal. *Journal of Air & Waste Management Association,* 45(12):801–810.

Standefor, S. 1996. Evaluating biofiltration. *Environmental Technology*, 6(4) July/August.

Togna, A. P., W. J. Fucich, R. E. Loudon, M. Del Vecchi, D. W. Barshter, and A. J. Nadeau. 1997. Treatment of odorous toxic air pollutants from a hardwood panel board manufacturing facility using biolfiltration. Paper 97-WA71A.01. *Proceedings of the Air & Waste Management Association's 90th Annual Meeting & Exhibition*. 8–13 June. Toronto, Ontario, Canada.

Togna, A. P., and M. Singh.1994. Biological vapor-phase treatment using biofilter and biotrickling filter reactors: Practical operating regimes. *Environmental Progress*, 13(2):94.

Togna, A. P., G. J. Skladany, and J. M. Caratura. 1994. Treatment of BTEX and petroleum hydrocarbon vapors using a field-pilot biofilter. In *49th Purdue Industrial Waste Conference Proceedings,* Chelsea: Lewis Publishers. 437–448.

Vaidila, A. K., and C. Welch. 2000. BIOSCRUB: An innovative process for VOC and odor control. *Proceedings of the Air & Waste Management Association's 93rd Annual Meeting & Exhibition*. Salt Lake City, Utah. 18–22 June.

van Groenestijn, J. W., M. P. Harkes, and R. F. W. Baartmans. 1997. A novel bioscrubber for the removal of ammonia from off-gases. *Biological Gas Cleaning*, Maastricht, The Netherlands, 28–29 April. W. L. Prins and J. van Ham, eds. Düsseldorf: VDI Verlag. 305–312.

van Groenestijn, J. W. and P. G. M. Hesselink. 1993. Biotechniques for air pollution control. *Biodegradation,* 4:283-301.

van Groenestijn, J. W., and M. E. Lake. 1998. Elimination of alkanes from off-gases using biotrickling filters containing two liquid phases. Presented at the *Air & Waste Management Association's 91st Annual Meeting & Exhibition*, Paper 98-TA20B.02. San Diego, 14–18 June.

van Lith, C. 1989. Design criteria for biofilters. *Proceedings of the 82nd Meeting of the Air and Waste Management Association*, Anaheim, CA, 1989.

van Lith, C., G. Leson, and T. Michelsen. 1997. Evaluating design options for biofilters. *Journal of Air & Waste Management Association*, 47(1):37–45.

van't Riet, K. and J. Tramper. 1991. *Basic Bioreactor Design*. New York: Marcel Dekker.

Waalewijn, E., C. D. Meijer, F. J. Weber, U. C. Duursma, and P. van Rijs. 1994. Practice runs of a biotrickling filter in the exhaust gases from a fiber factory. VDI Berichte 1104: *BiologischeAbgasreinigung, Praktische Erfahrungen und Neue Entwicklungen.* Düsseldorf: VDI Verlag. 545–553.

Webster, T. S., and M. A. Deshusses. 1997. Optimization of a full-scale biotrickling filter reactor in the treatment of industrial styrene off-gas emissions. *1997 AIChE Annual Meeting,* Session 220, Paper 220a. 18 November.

Webster, T. S., A. P. Togna, W. J. Guarini, B. Hooker, and H. Tran. 2000. Treatment of vapor emissions generated from an industrial wastewater treatment plant using a full-scale biotrickling filter reactor. *Proceedings of the Air & Waste Management Association's 93rd Annual Conference & Exhibition,* Salt Lake City, Utah. 18–22 June.

Whaley, M. B., P. Monroig, and E. Villarubia. 1998. Allergan: Isopropyl alcohol emissions removal using a unique biological system. *Air & Waste Management Association VOC/ Air Toxics Controls Specialty Conference.* Clearwater, FL. February.

Whaley, M. B. 1999. Update on Allergen's biological system for control of VOC emissions. *Proceedings of the Air & Waste Management Association's 92nd Annual Meeting & Exhibition*, Paper F6-448. St. Louis, MO. 20–24 June.

Willingham, R. 1996. Biofiltration: An economic alternative for controlling VOCs. *FLEXO,* 21(11):12–13.

Wittorf, F., S. Knauf, and H. E. Windberg. 1997. Biotrickling-reactor: A new design for the efficient purification of waste gases. *Biological Gas Cleaning*, Maastricht, The Netherlands, 28–29 April. W. L. Prins and J. van Ham, eds. Düsseldorf:VDI Verlag. 329–335.

Wolff, F. 1989. Biologische abluftreinigung mit einem neuen biowäscherkonzept. *VDI Berichte 735: Biologische Abgasreinigung, Praktische Erfahrungen und Neue Entwicklungen.* Düsseldorf: VDI Verlag. 99–107.

Yang, Y. and D. Alibeckoff. 1995. Biofiltration for control of CS_2 and H_2S vapors. *Proceedings of the 1995 University of Southern California and The Reynolds Group Conference on Biofiltration.* Los Angeles, CA. 5–6 October.

Water and Wastewater Treatment

Introduction

Biological treatment of wastewater has been employed successfully for many types of industries. It is important, however, to understand these processes fully to obtain optimal performance. Gathering information concerning wastewater characteristics is the first step in determining what type of process is best suited to obtain the desired results.

Total organic carbon (TOC), biological oxygen demand (BOD), oil and grease (O & G), and total petroleum hydrocarbons (TPH) are used as indicators of the overall amount of organic matter in the water. BOD indicates the compounds that can be biologically degraded whereas the chemical oxygen demand (COD) indicates those compounds that are chemically oxidizable, and the TOC shows the amount of organic carbon. In general, the higher the ratio of BOD to COD, the more likely the organics can be biologically degraded. A low ratio of TOC to COD would also indicate that the inorganic carbon, which is not usually biodegradable, is high. O & G and TPH are good indicators of the oil or petroleum content.

Some examples of wastewater streams include municipal wastewaters, industrial wastewaters, agricultural wastewaters, and stormwaters. Municipal wastewaters include domestic wastewater from residences, offices, and public areas, with some contributions from commercial and industrial sources. Metcalf and Eddy (1991) estimated that domestic wastewaters contain soaps and soil solids (33%), urine (20%), food residues (18%), feces (16%), paper (7%), and miscellaneous substances (5%). There are many sources of industrial wastewaters. They include pulp and paper producers, food processing plants, slaughtering and rendering houses, chemical plants, petroleum refineries, and landfill leachates. Typical pollutants include BOD, COD, total suspended solids (TSS), total nitrogen and grease, metals, organics (may or may not be toxic), and salts. These range from diluted to very concentrated. Characteristics of these pollutants are shown in Table 4.1. Dairies, feedlots, swine houses, and aquaculture facilities generate agricultural wastewater. The characteristics of these wastewaters are generally similar to domestic wastewaters. Stormwaters are the runoff from storm events which can originate from various sources, including agricultural, residential, urban and industrial runoff, land drainage, and streams or drains. The flows and pollutant concentrations of the various types of runoff are highly variable. For example, BOD levels can vary from 1.45 mg/L for runoff from undeveloped land to up to 150 mg/L for highly developed urban areas. Approximately 20 to 40% of the total suspended solids (TSS) from urban runoff is organic, but it is difficult to degrade since it contains wood, rubber, and road material. Typical pollutants include mineral solids, nutrients, pesticides, oil and grease, priority pollutants, and metals from vehicle emissions.

The major biological treatment processes for wastewaters include activated sludge processes, aerated lagoons or stabilization ponds, trickling filters or fixed-film reactors, and anaerobic processes. Biological treatment is commonly used as a secondary treatment. The major groups of biological processes are aerobic, anaerobic, and a combination of both. The systems are divided into suspended or attached growth processes for the removal of BOD, nitrification, denitrification, stabilization, and phosphorus removal. Aerobic processes, including activated sludge, trickling filters, aerated lagoons, and rotating biological contactors, have been used extensively. These developments, among others, are described in the following sections of this chapter.

Table 4.1 Characteristics of Various Wastewaters (adapted from Kadlec and Knight 1996)

Parameter (mg/L)[a]	Pulp and paper	Landfill leachate	Petroleum refinery	Textile mills	Starch production	Breweries	Municipal
BOD$_5$	100–500	42–10,900	10–800	75–6,300	1,500–8,000	1,500–3,000	110–400
COD	600–1,000	40–90,000	50–600	220–31,300	1,500–10,000	800–1,400	250–1,000
TSS	500–1,200	100–700	10–300	25–24,500	100–600	100–500	100–350
VSS	100–250	60–280	—	100–400		50–500	80–275
NH$_4$ - N		0.01–1,000	0.05–300		10–100		12–50
Total N		70–1,900	—	10–30	150–600	25–45	20–85
Total P		0.01–2.7	1–10				6–20
pH	6–8	3–7.9	8.5–9.5	6–12	3.5–8	5–7	
Sulfur[b]		10–260	ND–400				

[a] All parameters are in mg/L with the exception of pH, which has no units
[b] Sulfur is in the form of sulfate or sulfide

Selection of a Wastewater Treatment Process

As the principles of the various processes are understood and the wastewater has been characterized, the most appropriate process can be selected (Table 4.2). The decision of which process to chose can be based on the diagram in Figure 4.1. Once the decision to use a biological process has been made, then a specific biological process must be selected. The degree of treatment depends on whether the water will be recovered for reuse or disposed to a municipal sewer system or receiving body of water including streams, estuaries, and large water bodies. Performing bench-scale, pilot, treatability, and feasibility studies is useful to confirm the treatment feasibility, to obtain sufficient data needed to determine the treatment efficiencies, and to design and cost-estimate a full-scale system. Capital costs for biological treatment process can range from $10 to $130 per million liters treated for wastewaters of less than 1,000 mg/L BOD and around $260 per million liters treated for waste streams of BOD greater than 5,000 mg/L (Zinkus et al. 1998).

Since municipal and industrial wastewaters are high in organic content, most of the treatment processes have been developed to treat organic materials. In most plants, a combination of physical, chemical, and biological processes are used for pretreatment, primary, secondary, or tertiary treatment. Pretreatment is usually in the form of a physical or chemical process such as coarse and/or fine screening, grit removal, air stripping, flotation, oxidation/ reduction, and neutralization. Primary treatment is often in the form of sedimentation.

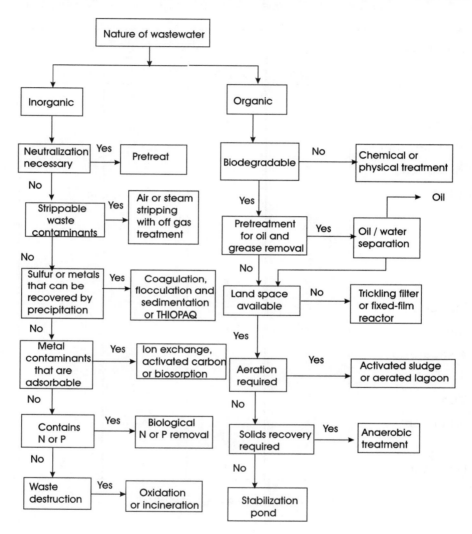

Figure 4.1 Decision Tree for Selection of Wastewater Treatment Processes

As discussed earlier, biological processes—aerobic, anaerobic, or a combination of both—are frequently selected as a secondary treatment. Tertiary treatment is often in the form of sand filtration, adsorption, chemical oxidation, and ozonation. Nutrient removal can be achieved by a biological process or by a combination of biological and chemical processes. Although biological nutrient removal (BNR) had been perceived as emerging and costly, these processes are now efficient and cost effective.

Selection of the appropriate process will depend on the effluent quality limitations, lowest wastewater availability of land, and costs (capital, operating, and maintenance). Other processes such as biosorption are under development and will be able to compete with ion exchange processes once the new processes are further developed. If space is available, wetlands can be used to remove BOD, TSS, nutrients, and heavy metals.

Table 4.2 Applicability of Treatment Technologies for Various Contaminants

Technology	Contaminant Type				
	Inorganic	Organic	Dissolved	Suspended	Biological
Biological	X	X	X	X	X
Carbon adsorption	X	X	X		
Centrifugal separation				X	
Chemical oxidation	X	X	X		
Crystallization	X	X	X		
Electrolysis	X		X		
Evaporation	X	X	X		
Filtration	X	X		X	X
Flotation	X	X		X	
Gravity separation	X	X		X	X
Ion exchange	X	X	X		
Membrane separation	X	X	X	X	X
Precipitation	X	X	X	X	
Solidification	X	X	X	X	
Solvent extraction	X	X	X		
Stripping	X	X	X		
Thermal treatment		X	X	X	X

Source Zinkus et al. (1998)

Aerobic Biological Treatment

Aerobic Biological Treatment: Activated Sludge Processes
Box 4-1 Overview of Activated Sludge Processes

Applications	Suitable for treatment of organic chemical, textile, municipal sewage, petrochemical, steel making, and pulp and paper industries for dilute compounds (less than 1%) of low toxicity
Cost	Operating: $0.16 to $0.8/m^3, Capital: $500 to $1,900/m^3 per day for 100 to 1,000 m^3/d
Advantages	Simple to start-up and operate
	Low capital and operating costs
	Numerous system variations available
Disadvantages	High production of sludge
	Slow to adapt to fluctuating concentrations
	Dependent on settling requirements
Organic loading	0.5 to 2.0 kg BOD/m^3-day
Hydraulic retention time	4 to 8 h (municipal wastewater)
BOD removal	85 to 95% (municipal wastewater)
Other considerations:	Highly chlorinated organics, aliphatics, amines, and aromatic compounds are not efficiently treated by this process. Alcohols and ketones can form filamentous organisms and can, therefore, be difficult to settle. Hydrogen sulfide may also be produced

Description

Activated sludge processes were first used in the early 1900s as fill and draw reactors (Arden and Lockett 1914). Continuous reactors were developed soon after and are now regularly used. Activated sludge processes involve the use of suspended aerobic microorganisms including bacteria, fungus, protozoa, and rotifers for the degradation of organic compounds. The sludge and the wastewater together are known as the mixed liquor. The process can either be aerated or be completely mixed (Figure 4.2). The mixing must be sufficient to prevent sedimentation of the microorganisms. Nutrients and oxygen are supplied to the activated sludge (living and dead microorganisms consisting of 70 to 90% organic matter). A general guideline for nutrient requirements is the ratio of BOD or COD to nitrogen to phosphorus of COD:N:P = 100:5:1 (mass basis). Nutrient supplementation would be necessary for wastewaters with nitrogen and phosphorus below this ratio. Supplying oxygen consumes the majority of the energy for the entire process and can be continuous or semi-continuous (Owen 1982). Complete oxidation of the substrate produces carbon dioxide, water, and inorganic substances such as nitrate and sulfate ions. Part of the substrate is used for biomass production. The supply of oxygen can cause stripping of volatile organic carbon compounds (VOC) into the air.

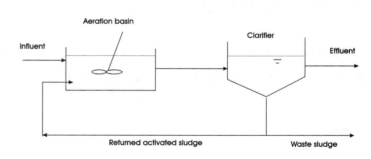

Figure 4.2 Schema of a Typical Activated Sludge Process

After the degradation step, the suspension is sent to the clarifier or settler where the microbial sludge is removed. Further treatment may be needed if the quality is not adequate; however, if the the quality is adequate, the treated water is then discharged. The sludge can either be recycled to the aeration tank or disposed of after sludge digestion or sludge thickening. Incineration or wet oxidation can also be used for sludge disposal. Sludge digestion can be performed aerobically with mixing or anaerobically without mixing. Due to a lack of carbon, the microorganisms will self-degrade which reduces the material for disposal and stabilizes the waste by decreasing sludge activity. Sludge thickening involves passing the sludge through belt or filter presses to remove the excess water and reduce the sludge volume. Concentrated streams produce much more sludge than dilute ones; therefore, the frequency of sludge disposal will depend on the organic concentration of the stream. When determining the most appropriate disposal method, consider that the sludge may contain other organic or inorganic compounds (e.g., metals) adsorbed onto the bacteria which may or may not be biodegradable. Sludge can be used as a soil conditioner if the levels of metals are not too high.

Recycling of a portion of the biomass (typically about 20% of the total amount) improves the degradation process in several ways. The sludge is called returned activated sludge (RAS) and contains a suspended solids content of approximately 10,000 to 20,000 mg/L (Kiely 1997). The recycle rate is defined as the ratio between the returning sludge and the influent wastewater stream. Since the sludge age increases as the number of recycles increases, the microorganisms become more acclimatized to the components in the waste stream, thus increasing their degradation ability. In addition, more types of microorganisms can grow. Slower growing microorganisms are washed out if they are not recycled. The sludge becomes very stable once it is well developed and provided with sufficient oxygen and nutrients including nitrogen and phosphorus. The suspended solid concentration of the mixed liquor (the activated sludge suspension) should be maintained at a minimum of 2,000 mg/L for a healthy population.

Operating Conditions

The process conditions in the plant can have a significant effect on the efficiency of the biological treatment process. The overall efficiency is determined by calculating the contaminant concentration (BOD, COD, etc.) in the stream from the settling tank and comparing this to the initial concentration (Table 4.3). Climatic changes, shock loadings, toxins, and other upsets in the process can influence effluent wastewater characteristics. Often an equalization tank is added as a pretreatment to decrease the effect of shock loadings. Since microbial activity is optimal at pH 6 to 8, pH adjustment may be necessary. Free oil may also have to be removed from the wastewater by oil/water separators or dissolved air flotation units. Excess oil can coat the sludge reducing its ability to get oxygen. Pretreatment may also be necessary to remove toxic metals which accumulate in the biomass and inhibit microbial activity.

Table 4.3 Typical Operating Parameters for an Activated Sludge Plant

Parameter (units)	Quantity
Treatment efficiency in BOD, COD or TSS (%)	80–95
Recycle ratio (%)	50–100
Sludge index (mL/g)	50–150
Oxygen respiration of activated sludge at 20°C (g O_2/kg VSS-h)	5–40
Minimum temperature of sludge and wastewater (°C)	6–10
Maximum temperature of sludge and wastewater (°C)	18–22

Adapted from Henze (1997)

The two major parameters to consider in operating activated sludge systems are the food:microorganism (F/M) ratio and the sludge age. The F/M ratio (U) is the ratio of BOD or COD in the influent to the amount of volatile suspended solids (VSS) in the aeration tank and is calculated as:

$$U = \frac{Q(S_o - S_e)}{VX_v} = \frac{(S_o - S_e)}{X_v \vartheta_d}$$

where Q is the water flow rate (m³/day), S_o the inlet substrate concentration (kg/m³), S_e the outlet substrate concentration (kg/m³), V is the reactor volume (m³), X_v is the biomass concentration (VSS in kg/m³), and ϑ_d is the hydraulic retention time (HRT in days). The F/M ratio is the degree of starvation and is useful for designing an activated sludge process. This ratio indicates a balance between the amount of substrate and the amount of organisms. The effluent substrate concentration is often removed since the substrate should be completely degraded.

The volumetric loading rate (B_v) is defined as:

$$B_v = \frac{Q \times S_o}{V}$$

It is often expressed as kg BOD/m^3-day and is used to design activated sludge plants. The sludge loading rate (B$_x$) is the amount of organic matter per amount of sludge per day and is defined as:

$$B_x = \frac{Q \times S_o}{V \times X_v}$$

and is expressed as the kg of BOD per kg of VSS per day and also is used to design the activated sludge unit for most processes, with the exception of phosphorus removal, nitrification, and denitrification. The volume of the aeration tank can be determined by rearranging the above equation to obtain:

$$V = \frac{Q \times S_o}{X_v \times B_x}$$

The type of activated sludge process is dependent on the F/M ratio. Extended aeration systems have F/M ratios of between 0.03 and 0.8, and conventional systems have F/M ratios between 0.8 and 2.0 while high-rate systems have ratios greater than 2.0. High-rate systems, with an excess of substrate, utilize bacteria that are in the accelerated growth rate phase. Conventional activated sludge systems operate with bacteria in the declining growth phase. However, the lower F/M ratios (less than 0.4) indicate that the bacteria are in the death or endogenous phase which is typical of plug flow (F/M of 0.2) and complete mix (F/M of 0.1) systems.

Sludge age is the residence time of the sludge in the system. The sludge requires a specific period of time to utilize the substrate and reproduce. If this time is not allowed, the sludge is washed out. Older sludges will decay more, which affects settleability and the sludge production rate. Since the sludge undergoes anaerobic conditions in the clarifier, it becomes dormant and its age does not increase. Therefore, the sludge age or retention time (SRT) or ϑ_x is only dependent on the time it spent in the aerated basin. ϑ_x = mass of solids in the aeration basin/solids removal rate in the system (wasted and lost) = VX_v/solids removal rate from the system.

The parameter, sludge age, is used to design plants for nitrification of wastewaters that contain contaminants such as phenol and cyanide, which slow bacterial growth. The volume of the aerated basin can thus be determined by the equation:

$$V = \vartheta_x \frac{\text{solids removal rate}}{X_v}$$

The F/M ratio and the SRT influence the settleability and compactability of the sludge. When the biomass is decaying, it forms polymers that enhance natural flocculation. Newer sludges are more filamentous and do not settle well. Under starved conditions or high SRT, the sludge forms a fine pinpoint-like floc that also does not settle well.

The sludge volume index (SVI) is defined as the volume of sludge in milliliters as measured in a 1 or 2 L cylinder with 1 g of sludge (with a known TSS content) after settling for a specified time period (30 min to 2 h). Therefore, the equation is:

$$SVI(\frac{mL}{g}) = \frac{y}{X_T(1000\frac{mg}{g})}$$

where y is the volume in mL at the end of the settling time and X_T is the TSS content of the sludge. Low SVI values of 80 to 120 indicate good settling for a sludge with a TSS of 2,000 to 3,500 mg/L.

Design

Complete Mix Reactors

As indicated by the name, complete mix reactors have uniform properties throughout the circular or square reactor. Either submerged bubble diffusers or surface aerators are used to maintain dissolved oxygen (DO) levels of 1 to 2 mg/L. Since the effluent substrate concentration is the same as the reactor concentration, the F/M ratio is low (typically 0.04 to 0.07). Because of this low F/M ratio, the system can withstand shock loads and a range of loads. The organic loading rates are usually less than 1 kg BOD/m³-day, and the dissolved oxygen levels should be above 2 mg/L. It is preferable to add the returned activated sludge (RAS) directly into the aerated reactor instead of the influent stream.

A mass balance can be made for biomass production including the aeration tank and the clarifier as shown in Figure 4.3.

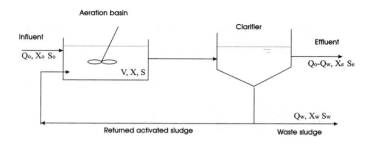

Figure 4.3 Mass Balances for a Complete Mix Activated Sludge Process

Influent biomass + biomass production = effluent biomass + sludge wasted biomass

$$Q_o X_o + V\frac{dx}{dt} = (Q_o - Q_w)X_e + Q_w X_w$$

where Q is the flow rate in m^3/day, X is the biomass concentration in mg/L, V is the aeration tank volume in m^3, and the subscripts o, e, and w represent the influent, effluent, and waste streams, respectively.

Since growth is in the endogenous phase with a die off rate, k_d, the Monod growth rate equation with die-off can be substituted into the above equation for dx/dt to give:

$$Q_oX_o + V(\frac{\mu_mSX}{(K_s+S)-k_dX}) = (Q_o-Q_w)X_e + Q_wX_w$$

where μ_m is the growth rate constant and S is the substrate concentration (kg/m^3 or mg/L). Assuming that X_o and X_e are equal to zero and since there is only a small amount of biomass in the influent and effluent, the equation can be simplified to the following:

$$\frac{\mu_mS}{K_s+S} = \frac{Q_wX_w}{(VX)} + k_d$$

A balance on the substrate can also be performed for the aeration tank and clarifier. Influent substrate + substrate utilized = effluent substrate + wasted substrate:

$$Q_oS_o + V\frac{dS}{dt} = (Q_o-Q_w)S_e + Q_wS_w$$

Substituting the Monod equation for substrate utilization where Y is the amount of substrate converted to biomass (mg/L of biomass per mg/L of substrate) gives the following equation:

$$Q_oS_o + \frac{V}{Y}(\frac{\mu_mSX}{K_s+S}) = (Q_o-Q_w)S_e + Q_wS_w$$

The equation can be simplified by assuming that the substrate concentration in the clarifier, S_e, and the effluent, S_w, is the same and equal to S and then combining it with the equation derived for biomass.

$$\frac{\mu_mS}{K_s+S} = \frac{Q_oY}{VX(S_o-S)} = \frac{Q_wX_w}{(VX)} + k_d$$

The hydraulic retention time (HRT), or ϑ, is then defined as V/Q_o and has units of time, and the solids retention time (SRT) or sludge age, ϑ_x, is defined as $VX/(Q_wX_w)$ in units of time. Since only a fraction of the sludge is recycled to the aeration basin, then SRT > HRT and the above equation can be shown as:

$$\frac{1}{\vartheta_x} + k_d = \frac{Y}{\vartheta X(S_o-S)}$$

The biomass in the aeration tank can then be calculated by rearranging the equation to obtain the following:

$$X = \frac{\vartheta_x}{\vartheta} Y \frac{S_o - S}{1 + k_d \vartheta_x}$$

The F/M ratio can be calculated for the complete mix systems from the following equation:

$$F/M(\frac{mg\ BOD\ applied}{mgVSS/day}) = \frac{S_o}{(\vartheta X)} = \frac{Q_o S_o}{VX}$$

Contact Stabilization

Contact stabilization (Figure 4.4) has two aeration phases. In the first phase with a retention time of 30 to 60 minutes, a contact tank is used for absorption of microbial biomass onto suspended organic matter. Here, the VSS is approximately 2,000 mg/L. In the second tank, the solids (20,000 mg VSS/L) which were removed from a settlement zone are stabilized by aeration in the stabilization tank with a retention time of 2 to 6 h. The volumes required for this type of system are generally 50 to 60% less than for conventional plug flow, enabling it to be used where space availability is limited. BOD removal rates are typically high (80 to 90%). Cell residence times are in the order of five to 15 days. Applications include expansion of existing systems and in package plants (Metcalf and Eddy 1991).

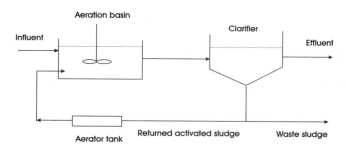

Figure 4.4 Contact Stabilization

Applications

Activated sludge processes are usually used for the organic chemical, textile, municipal sewage, petrochemical, steel making, and pulp and paper industries for dilute compounds (less than 1%) of low toxicity. Highly chlorinated organics, aliphatics, amines, and aromatic compounds are not efficiently treated by this process. Alcohols and ketones can form filamentous organisms and can, therefore, be difficult to settle. Hydrogen sulfide may also be produced.

Costs

Costs for traditional activated sludge systems are difficult to obtain. Estimates have been made for operating costs at $0.16 to $0.8/m³ and for capital costs at $500 to $1,900/m³ per day for 100 to 1,000 m³/d.

Case Studies

Variations have been made to the traditional activated sludge process. For example, the HCR process by Kvaerner Water Systems a.s. (Norway) incorporates a loop reactor with a patented two-phase jet aeration system with a standard clarifier. This design increases mass transfer to the biomass enabling more efficient treatment of contaminants, particularly toxic ones such as formaldehyde, phenol, furfural, and resin acids. High volumetric loads (10 to 70 kg COD/m^3-day) can be achieved. Conventional activated sludge systems generally handle 0.5 to 2.0 kg COD/m^3-day. Reactors of up to 1,200 m^3 have been installed in Norway, Canada, China, and France for the treatment of food processing, brewery, pulp and paper, landfill leachate, newsprint, and municipal wastewater.

Pure oxygen generated on-site is injected into the completely mixed basins instead of air for the treatment of highly-concentrated wastewater. The use of pure oxygen increases the oxygen saturation by up to four times the normal concentration (Reynolds and Richards 1996). Covering the basin decreases oxygen losses. Mechanical mixers are used to maintain complete mixing since the oxygen is only sufficient to achieve the oxygen demand. Concentrations of the mixed liquor can be higher (3,000 to 8,000 mg/L). However, the sludge is more highly stabilized. According to Metcalf and Eddy (1991), hydraulic retention times are reduced to between 1 and 3 h, solids retention time is from eight to 20 days, and loading rates are from 1.6 to 3.2 kg BOD/m^3-day for municipal wastewater. F/M ratios are 0.25 to 1.

Aerobic Biological Treatment: Oxidation Ditch

Box 4-2 Overview of Oxidation Ditch Processes

Applications	Suitable for treatment of organic chemical, textile, municipal sewage, petrochemical, steel making, and pulp and paper industries for dilute compounds (less than 1%) of low toxicity for small communities and small installations
Cost	Operating: $0.10 to $0.50/m^3, Capital: $660 to $1,100/m^3 per day
Advantages	Simple to start-up and operate
	Low energy costs
	Less sludge than activated sludge
Disadvantages	High levels of solids in effluents of sludge
	Requires large area
Organic loading	0.16 to 0.40 kg BOD/m^3-day
Hydraulic retention time	18 to 36 h (municipal wastewater)
Other considerations:	Can also be used for nitrogen removal

Description

One type of plug flow reactor is the oxidation ditch which consists of an oval with aerators at one or several positions (Figure 4.5). The flow moves from higher to lower oxygen concentration zones with the effluent exiting the reactor downstream. HRTs are usually about 24 h, whereas the SRTs are from 20 to 30 days. Suspended solids concentrations are in the range of 3,000 to 6,000 mg/L (Metcalf and Eddy 1991).

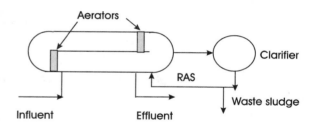

Figure 4.5 Oxidation Ditch Reactor

Operating Conditions

The amount of sludge in the process is usually higher than most processes; therefore, removal rates are very good for this type of reactor (75 to 95% BOD removal). However, the sludge can be difficult to settle. Reactor volumes can be determined from a known mixed liquor suspended solids (MLSS) value or by assuming a retention time and a recycle ratio and then calculating mass and reactor volume. Although this type of reactor is designed on the basis that no net sludge is produced, period wasting is necessary to avoid accumulation of cells with low viability. Velocities in the reactor must also be maintained at 0.24 to 0.37 mps to maintain the cells in suspension. A significant number of cell fragments are found in the reactor effluent due to the long aeration time and SRT.

Design

In plug flow reactors, substrate concentration or BOD varies along the reactor length with little mixing in elements before or after a specific section, as shown in Figure 4.6. At the

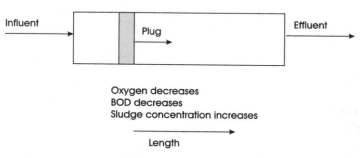

Figure 4.6 Plug Flow Reactor

inlet, BOD is high, F/M ratio is high, and growth is high in the log growth phase. At the end of the reactor, the substrate concentration is decreased substantially; therefore, microbial growth is in the endogenous phase. These reactors can be quite low to enable sufficient retention time. Length to width ratios are typically 10:1 with depths of 2 to 4 m.

Mass balances can be used to design the reactor. However, since they are more complicated than for the complete mix systems, an assumption is made such that the biomass concentration at the influent and effluent is the same (X).

Therefore,

$$X = \frac{\vartheta_x}{\vartheta} Y \frac{S_o - S}{1 + k_d \vartheta_x}$$

Solving for $1/\vartheta_x$ and using a as the recycle ratio and $S_i = (S_o + \alpha S)/(1 + \alpha)$ gives the equation:

$$\frac{1}{\vartheta_x} = \mu_m \frac{S_o - S}{(S_o - S) + (1 - \alpha)\left[K_s \ln\left(\frac{S_i}{S}\right)\right]}$$

Case Study

Improvements are constantly being made. For example, Purac-Kier Consortium (United Kingdom) built an oxidation ditch in the East of Scotland Water, which has a population of 16,000, for a cost of $1.8 million U.S. Usually oxidation ditches have surface brushes to aerate and circulate the wastewater and sludge. However, Purac used the Rosewater IBA™ system that separates aeration and mixing. Two Flygt flow inducers were used to prevent the sludge from settling while two 55 kW blowers with five aeration frames containing 25 tubular membrane diffusers on the floor of the ditch were employed. This configuration allows the aeration to be turned off when necessary, thus saving energy while enhancing oxygen transfer. The agitation can be left on to maintain the sludge in suspension.

Aerobic Biological Treatment: Sequencing Batch Reactor (SBR)

Description

The sequencing batch reactor (SBR), a variation of the activated sludge process, has recently been developed (Figure 4.7). In a single tank, the wastewater is added to the sludge, air is added by aerators or mixers for a specific time period, and then the air or mixers are shut off. Control of the aerators, mixers, pumps, and decanters during the cycle by the timing and sequencing is dependent on the wastewater characteristics and the desired removal efficiencies. The sludge is allowed to settle to the bottom. The treated water is removed from the top of the tank and the sludge is left to anaerobically self-digest for a certain time. The process can then be restarted.

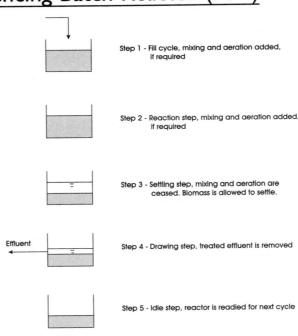

Step 1 - Fill cycle, mixing and aeration added, if required

Step 2 - Reaction step, mixing and aeration added, if required

Step 3 - Settling step, mixing and aeration are ceased. Biomass is allowed to settle.

Effluent

Step 4 - Drawing step, treated effluent is removed

Step 5 - Idle step, reactor is readied for next cycle

Figure 4.7 Sequencing Batch Reactor (SBR)

Box 4-3 Overview of Sequencing Batch Reactors

Applications	Suitable for treatment of organic chemical, textile, municipal sewage, petrochemical, steel making, and pulp and paper industries for dilute compounds (less than 1%) of low toxicity for flow rates less than 19,000 m³/day.
Cost	Operating: $0.21 to $0.53/m³, Capital: $410 to $1,300/m³ per day
Advantages	Equalization, primary and secondary clarification, treatment in one reactor
	Small footprint
	Operation is flexible and easily controlled
Disadvantages	High level of sophistication and maintenance
	Potential for plugging of aerators
	Sludge can be discharged during decanting step
Organic loading	0.5 to 2.0 kg BOD/m³-day
Hydraulic retention time	6 to 14 h (municipal wastewater), variable for industrial wastewater
Other considerations:	Highly chlorinated organics, aliphatics, amines, and aromatic compounds are not efficiently treated by this process. Alcohols and ketones can form filamentous organisms and can, therefore, be difficult to settle.

Operating Conditions and Design

The cycle can last from between 4 and 48 h and the SRTs from 15 to 80 days (AWWA 1992). The ratio of F/M can be from 0.15 to 0.4/day for municipal wastewater and 0.15 to 0.5/day for industrial wastewater, depending on the cycle length. Pharmaceutical, pulp and paper, corn, metal, and potato processing wastewaters are some the wastewaters that have been treated using SBR technology. Suspended solids concentrations vary from 2,000 to 2,500 mg/L for municipal wastewater and 2,000 to 4,000 mg/L for industrial wastewaters (Mikkelson 1995). BOD removal rates are usually 85 to 95%.

Costs

Operating costs are dependent on the site but usually vary from $210 to $530/1,000 m³ capacity. This is similar to the activated sludge cost. Installed equipment costs vary according to the size of the installation (Mikkelson 1995):

- $510 to $1,320/m³ for 1,900 to 3,800 m³/day
- $480 to $710/m³ for 4,200 to 5,700 m³/day
- $430 to $870/m³ for 5,700 to 7,600 m³/day

Case Study

One of the largest SBR type wastewater treatment plants is located in Kunming, People's Republic of China, and has been in operation since November 1997. The ABJ™ Intermittent Cycle Extended Aeration System(ICEAS®) treats 300,000 m³/day of domestic and light industrial wastewater. Within the same basin, aeration, anoxic/anaerobic mixing, mixing, settling, and decantation take place in sequence (Anonymous 1998).

Aerobic Biological Treatment: Aerated Lagoons and Stabilization Ponds

Box 4-4 Overview of Lagoons and Ponds

Applications	Used for low concentration municipal and industrial applications such as petrochemical, textile, pulp and paper, refinery wastes, and some inorganics.
Cost	Operating: $0.01 to $0.04/m³, Capital: $50 to $790/m³ per day
Advantages	Low maintenance
	Low operating and capital costs
	Relatively little waste sludge
Disadvantages	Susceptible to shock loadings
	High land requirements
	No operational control
Organic loading	14 to 50 kg BOD/ha-day
Hydraulic retention time	Seven to 20 days
Other considerations:	Most suitable for warm climates

Description

Stabilization ponds are simple, large, shallow ponds used to promote the growth of algae and bacteria. They are used for municipal and industrial applications such as petrochemical, textile, pulp and paper, and refinery wastes. The algae, which grow well in the excellent light and oxygen conditions at all depths, are useful for the bacteria for oxygen production via photosynthesis.

Methods

Facultative Ponds

Facultative ponds are the most common type of pond and have depths of 1.2 to 2.5 m (Figure 4.8). They are not as deep as anaerobic ponds and are able to handle intermediate levels of organic matter. The upper layers of the pond are of large surface area and contain significant oxygen levels. The lower layers, where the conversion of the solids to soluble components and methane occurs, are anaerobic. The soluble components can then be degraded in the upper levels.

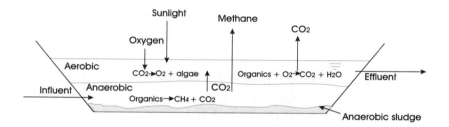

Figure 4.8 Schematic of a Facultative Lagoon. Various Biological Processes are Depicted within the Lagoon

Aerobic Ponds

Aerobic ponds are very shallow with depths of 30 to 45 cm. They often receive the effluent from facultative ponds. They are fed low organic loadings and are completely aerobic throughout the pond. These ponds are useful for polishing by enabling suspended solids to settle and removing low concentrations of organics.

Partial-mix aerated ponds involve addition of sufficient oxygen for bacterial growth but not enough for mixing as in complete-mix systems. A complete-mix model with first order reaction kinetics is used to calculate BOD removal (Reed et al. 1995). Oxygen requirements for partial-mix systems are based on the influent BOD concentrations. These types of systems require ten times less power than complete-mix ones.

Aerated ponds (Figure 4.9) are stabilization ponds with aeration or mixing to ensure aerobic conditions since they are deeper (2 to 6 m). Odor problems are reduced and smaller areas are required since they are more efficient than the non-aerated ponds. Mechanically-agitated ponds are often referred to as lagoons.

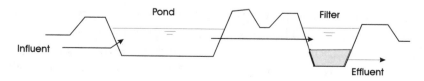

Figure 4.9 An Aerated Pond with Filtration Pond

The growth of algae is not encouraged since algae do not settle, but they can supply oxygen to the pond. This sludge, however, must be removed occasionally to avoid filling up the lagoon with the sludge.

Operating Conditions

Oxygen can also be supplied by the wind and surface mixing. The mixed liquor volatile suspended solids (MLVSS) are usually in the range of 50 to 150 mg/L. The bacteria then can degrade the organic material in the wastewater. The facultative pond has both aerobic and anaerobic reactions taking place within a depth of 2 m as shown in Figure 4.8. Organic solids settle on the bottom and are degraded anaerobically while the soluble components are decomposed by aerobic and facultative bacteria. Sludge is allowed to settle in a designated area of the pond. The sludge forms a blanket which creates anaerobic conditions for the promotion of self-digestion.

Facultative ponds

Loading rates for facultative ponds depend on the temperature. In the northern United States and Canada, loading rates in the first cell are less than 40 kg BOD/ha-day and HRT in the order of 120 to 180 days at temperatures of 0°C. The WEF (1992) suggests loading rates of 14 to 50 kg BOD/ha-day with retention times between 80 and 180 days. Hydraulic loading rates of 3 to 14 m³/ha-day are also suggested. Organic loadings can be greater than 100 kg BOD/ha-day at temperatures above 15°C. In the Northwest Territories, retention times can be up to one year. A minimum of two stages is recommended (Heinke et al. 1991). The first stage is of a short retention time (three days) to allow settling of suspended solids. Depths are in the range of 3 m and sludge is removed every five years. The effluent is then discharged to the long detention ponds for continuous or intermittent discharge. Continuous discharge ponds function in the winter, whereas intermittent discharge ponds discharge during warmer temperatures and dilute conditions, approximately 40 days after ice breakup.

Aerobic ponds

Sludge removal is difficult from aerobic ponds since it is usually not preferable to drain the pond. Other problems can include biomass that does not settle but is carried into the effluent and the accumulation of metals and toxins in the sludge, leading to a decrease in microbial activity and sludge disposal problems. Sludge retention times are in the order of months to years. Compounds that are not easily degradable can be treated this way, which avoids the use of large reactors. In continents other than North America, fish are grown in the ponds, including species such as *Talipia* and the carp family (Wrigley et al. 1988).

Design and Maintenance

Retention time of the wastewater can be a few days in warm climates to several months or more in cold climates. The long detention time enables flow and quality equalizations. Since sludge accumulation is slow, desludging for ponds of small communities, such as those for eastern Ontario, is required only every 15 to 20 years (Droste 1997). Maintenance is required on a more regular basis to remove aquatic plants such as cattails that can take up a large amount of the pond's volume.

Facultative Ponds

Designs of the facultative ponds are based on the average temperature for the coldest month. Temperatures of the pond can be calculated by heat transfer at the surface of the pond via evaporation, convection, and radiation. Wind velocity, heat from the sun, and water vapor pressure all affect the heat balance. Heat transfer rates are usually higher for aerated lagoons than stabilization ponds. High heat transfer coefficients result in pond temperatures near ambient.

Aerobic Ponds

For aerobic ponds, predictions from models can vary by factors of up to ten. BOD is the basis for many of the models. The U.S. EPA (1983) compared several design approaches but didn't recommend a specific model. Correlations are based on plug flow regimes up to completely mixed systems. Mixing assists bacterial and organic contact. Power input of 3 to 10.8 kW per 1,000 m^3 for lagoons of 2.4 to 5.5 m in depth is adequate to keep solids suspended (U.S. EPA 1983). To ensure good distribution of oxygen and mixing, the aerators are placed so that their zones of influence overlap. Completely mixed models for the design can be used when mixing is sufficient. The lowest operating temperature is chosen to design the pond. Rate constants from pilot or other similar applications should be used for design purposes.

Applications

The ponds are often used for small, remote communities or industries located where land is inexpensive. They can also be used in warm climates and in developing countries for sewage treatment where low maintenance is desired. Northern communities also favor their use due to low maintenance requirements and their simple design and operation (Smith and Knoll 1986).

There are several problems, however, with the use of these ponds.
*   They require large areas of land due to their shallow nature;
*   Since there is no control, hydrogen sulfide and other odors can be produced if there is insufficient oxygen;
*   Flies and mosquitoes can be attracted to the ponds;
*   Efficiency can decrease significantly in the winter;
*   Rainstorms can cause overflows, biomass washout, and seepage into the soil of undegraded toxic compounds; and
*   Algae cause problems if they are found in the effluent since they can increase the TSS, nitrogen, and phosphorus levels significantly. Algae removal can increase the operating cost of the pond system.

Costs

The capital costs for a facultative pond are in the order of $80,000 to $160,000/ha which is approximately $500 to $1,000 per cubic meter per day (WPCF 1990). Operating and maintenance costs are in the order of $0.07 to $0.13 /m³ of water treated. The treated water from a facultative pond usually contains 100 mg/L of TSS and, thus, must be tertiary treated.

Aerobic Biological Treatment: Fixed-Film or Trickling Filter Reactors

Box 4-5 Overview of Trickling Filter Processes

Applications	Used in areas where large areas of land are not available, for high concentrations of organics, and for small to medium-sized populations
Cost	Operating: $0.01 to $0.45/m³, Capital $170 to $400/m³/day for 190 to 38,000 m³/day
Advantages	Simple, reliable
	Moderate level of skill needed to operate the system
	Low power requirements
Disadvantages	Highly chlorinated organics, aliphatics, amines, and aromatic compounds are difficult to treat while heavy metals may inhibit, kill, or accumulate in the biofilm.
	Slow to adapt to shock loadings and environmental changes
	Clogging can occur regularly
Organic load	100 to 300 g BOD/m³-day for low-rate and up to 1 kg BOD/m³-day for super-rate
Hydraulic load	1 to 4 m³/m²-day for low-rate to 40 to 200 m³/m²-day for super-rate
Other considerations:	Hydrogen sulfide may also be produced.

Description

Fixed-film processes involve the formation of biofilm on a fixed surface. This surface material can be rocks, wood, sand, or plastic beads of approximately 25 to 100 mm in size (Belhateche 1997). The surface area to volume ratio can be increased by using plastic. High void volumes are also desirable for increasing treatment efficiencies. Wastewater can be sprayed either on top of the medium so it can trickle to the bottom or it can pass through a packed bed reactor which is submerged in water (Figure 4.10). At the bottom, the treated wastewater is collected through the perforated floor. Openings in the walls near the floor allow air to enter.

Methods

Biofilters

Biofilters are a new type of trickle filter that use an absorbent natural or synthetic medium for microbial attachment. The University of Waterloo (Guelph, Ontario) has developed a system (Waterloo Biofilter™) that can be used for the secondary and tertiary

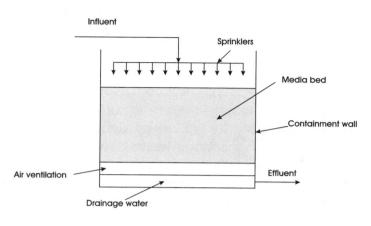

Figure 4.10 Trickling Filter

treatment of municipal or communal wastewater, individual domestic sewage, landfill leachate (Jowett 1997; Jowett 1999), food-processing wastewater, and some farm animal wastewater applications. The plastic foam medium (about the size of a fist) is chosen to obtain high biomass retention times, optimal microbial attachment, high porosity, and separate paths for air and water. Typical removal rates are 95% TSS, 90 to 95% BOD, 20 to 50% total nitrogen, and 90 to 99% coliform bacteria.

Biotowers

Biotowers are mainly used for high-strength wastewaters such as in the dairy industry. They are vertical reactors of fiber reinforced plastic (FRP), concrete, or steel and contain plastic rings or other media to grow the bacteria. They can produce from 20 to 50% less sludge than suspended growth systems. Eimco-K.C.P. (Chennai, India) is also a supplier of biotowers.

The Aerothane Process developed by Biothane is a granular aerobic sludge system (GASS) for partial nitrification, odor treatment, and treatment of effluents from anaerobic reactors to remove sulfide before effluent discharge. The reactors have bubble aeration and internal baffle plates to create gaslift loops for maximum mixing efficiency. A settler within the reactor retains the sludge so that hydraulic retention times of 1 to 4 h are sufficient.

Submerged Filters

Submerged filters differ from trickling filters in that the surface of the filter medium is below water. The media can be either stationary or movable. In the case of stationary medium, oxygen is supplied at the bottom of the reactor to allow the bubbles to rise through the filter. The filter can be operated in the upflow or downflow mode.

Movable medium filters were developed in the 1970s, later than trickling filters. They can be operated as an expanded bed where 30 to 40% of the bed expands, as a fluidized bed where 100% expansion occurs, or as a suspended-growth, completely mixed reactor with impellers. In the first type of reactor, the particles lift slightly but constantly grate against each other. This may not be ideal for biofilm formation, however. In the fluidized bed system, there is a balance between gravitational and hydrodynamic forces with the particles moving separately from each other in the turbulent flow.

Rotating Biological Contactors (RBC)

A rotating biological contactor involves a rotating cylinder of styrofoam, polystyrene, polyvinyl chloride, high density plastic, or another lightweight material in a semicircular holding tank (Figure 4.11). The plastic cylinder builds up a biomass layer as it rotates (1 to 2 rev/ min) between the wastewater and air with 40% submersion. This biomass film is then able to degrade the organic matter in the wastewater.

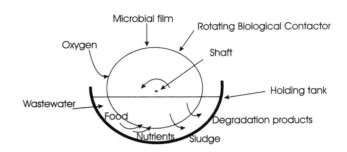

Figure 4.11 Diagram of a Rotating Biological Contactor

Operating Conditions

The main elements influencing the design and operation of the trickling filter system are wastewater composition and biodegradability, type and depth of the media, recirculation ratio and setup, hydraulic and organic loading, and temperature and the distribution of the wastewater. In the trickle bed reactor, the medium is usually angular like limestone and is loosely packed to allow the water to slowly follow diverse pathways. The tank containing the medium can be cylindrical or rectangular and is generally 5 to 50 m in diameter and 1 to 2.5 m in depth (Kiely 1997). Oxygen is supplied via natural convection due to the temperature difference between the top and bottom of the column. In the summer, however, this temperature difference can be small, causing problems with oxygen supply.

Substrates and oxygen diffuse through the layer. This layer builds up as the microorganisms grow until the rate of diffusion decreases with the thickness. The organisms on the surface may die due to lack of oxygen and food, and washing off of the biomass or sloughing off may occur. A downstream settling unit is used to remove these solids from the water. Unlike the activated sludge process, recycling is not used since the material cannot be reattached to the material. However, the effluent is recycled to increase treatment efficiency and dilute the incoming influent.

Since diffusion is a slow process, the hydraulic loading rates are very large—between 10,000 and 40,000 m^3/m^2 of filter area per day for low-rate conventional systems up to between 600,000 and 1,800,000 m^3/m^2 for roughing filters. Organic loading rates can vary from between 100 and 300 g BOD/m^3 per day for low rate systems up to almost 1 kg BOD/ m^3 per day for super rate systems (Harremoes and Henze 1996). Low rate filters remove 80 to 90% BOD_5, intermediate rate filters remove 50 to 70% BOD_5, and high-rate filters remove 65 to 85% BOD_5 (Environmental Engineers Handbook 1997). Sometimes they are used as a pretreatment step to remove some of the organics for activated sludge processes or biological nitrification processes for the removal of ammonia. In these cases, the settling step can be eliminated and the roughing filter can be smaller than if they were used as the only unit.

Recirculation ratios of 0.4 to 4.0 are generally used. BOD_5 removal rates are lower (40 to 65%). The recirculated stream may be taken directly from the reactor effluent or the clarifier and reenter the reactor by meeting the influent stream before or at the filter. At high recirculation ratios, it may be possible to treat higher BOD wastewater. Industrial wastewaters such as milk, aldehyde, acetic acid, acetone, formaldehyde, and phenol have been treated by trickling filters. The media can vary from stone at 1 m depth for municipal and other dilute wastewaters to a very light plastic with 6 m in depth for more highly concentrated industrial wastewaters.

Temperature decreases, particularly in the winter, can decrease the filter efficiency significantly and can be shown as:

$$E_t = E_{20}a^{t-20}$$

where E_t is the efficiency of the filter at temperature t, E_{20} is the efficiency at 20°C, and a is the constant of 1.035.

Another factor is the distribution of the wastewater in the filter. In the past, distributing arms at a rate of 1 rev/min and spraying every 30 seconds were used. However, it has recently been shown that slower distributor rates (0.5 to 2 rpm) and increased time between spraying led to better filter performance since this reduces problems with flies and the occurrence of sloughing off (Metcalf and Eddy 1991). The arms are placed at a minimum of 15 cm above the filter bed. If there is a chance of ice, the arms should be higher. The diameter of the arms is 4.5 to 70 m with ports of 95 mm for water discharge.

Biofilters

A sprinkler system distributes the liquid over the foam. The bacteria remain in the foam bubble spaces, and the treated water passes through the filter and is discharged. A portion of the effluent can be recycled to increase nitrogen removal. Natural convection or a small fan is used for aeration. However, forced ventilation can improve BOD and solids removal by 5% and can significantly improve nitrification. A timed pump can be used to balance the load over 24 h. Loading rates are in the range of 50 to 80 cm/day which can be compared to sand filters which typically handle around 4.7 cm/day.

Submerged Filters

Biomass is formed separately and can reach up to 40 kg VSS/m^3. The highest loading rate that has been achieved is 10 kg BOD/m^3-day. There is no long-term experience in operating such reactors which are not simple. The suspended biofilm reactors do not use upflow for stirring but instead use mechanical mixing. This is similar to the activated sludge systems with the exception that inert particles are added. Although it can assist sedimentation, it also adds more sludge to be handled and more energy may be required for mixing in the aeration basin. This system is still experimental. Oxitron (Anonymous 1979) has tested a fluidized bed reactor in Iowa for the treatment of effluents from corn wet milling. For an effluent of 3,000 mg/L, 95% BOD removal was achieved (Cheremisinoff 1996).

Rotating Biological Contactors (RBC)

Energy requirements are very low (2 to 4 kW per shaft) which is approximately one-third to one-half the power of activated sludge systems. The speed of the rotating disc controls the biofilm thickness (not less than 0.3 m/s or 1 to 2 rpm), and the spacing between discs is usually about 1.5 to 3.5 cm. Loading rates vary from 5 to 26 g BOD/m²-d (Henze 1997). Results are consistent, and expansion and retrofit costs are low. The reactors must be covered in the cold climates since the biomass would be cooled in the air.

Design

Although fixed-film processes are easy to operate, care must be taken to avoid shock loadings and upsets. Since the slime layer can take months to develop, a long period of time would then be required to accumulate a new slime layer and stable system performance. Modeling of filter systems is not well advanced due to the irregular shape of the media in the filter. The United States National Research Council Model (1946) and the British Manual of Practices Model (1988) have been used for design purposes. German design criteria are shown in Table 4.4 and are based on organic and hydraulic surface loading rates of municipal wastewater treatment, which is relatively homogeneous. Surface area, filter medium, influent concentration, and the presence of other chemicals can all influence the treatment efficiency. According to the German guide (Abwassertechnishe Vereinigung 1985), the expected efficiency can be calculated as:

$$E = 93 - 0.017 \times \text{volumetric load rate for rates below } \frac{1{,}000 \text{ g BOD}}{\text{m}^3\text{-day}}$$

where E is the treatment efficiency in percent.

Biotower

A design equation was developed by Eckenfelder and Barnhart (1963) as follows:

$$\frac{S_e}{S_o} = \exp\left(\frac{kD}{Q^n}\right)$$

where S_e is the effluent BOD concentration (mg/L), S_o is the influent BOD concentration (mg/L), D is the media depth, Q is the hydraulic loading rate (m³/m²-s), k is the treatability constant, and n is the medium characteristic constant. BOD removal rates of 80% have been achieved (Techknow Database 1998).

Rotating Biological Contactors (RBC)

For modeling RBC performance, an equation similar to the Schultz-Germain formula for trickling filters is used (Spengel and Dzombak 1992):

$$S_e = S_i^{k\left(\frac{V}{Q}\right)^{0.5}}$$

where S_i is the influent substrate concentration, S_e is the effluent substrate concentration, V is the volume in m^3, and Q is the water flow rate in m^3/s. The constant k is in the range of 90 to 112 for temperatures of 13°C and is based on actual performance.

Scale-up can be problematic for this type of reactor (Spengel and Dzombak 1992). If the same tip speed is used on the small and full-scale systems, the shear levels and turbulence will not be the same. Film thicknesses will also be different since the biomass is not exposed to the air and water at the same frequency in the two systems.

Maintenance and Monitoring

Psychoda flies are the most common operational problem since they can breed in the filter bed and can be a problem for the operators close to the filter. Flooding of the media for a full day during each summer month is the most effective way to kill the fly larvae.

Odors can also be a problem for low-rate filters where there is no recycle. This does not occur for high-rate filters since recycle is continuously used throughout the day and the water does not become stale.

Submerged filters

Since clogging can be a problem, the best way to deal with it is to use filter media with enough space and good hydraulic conditions. This reduces the available surface area and leads to large treatment systems. Another solution is backwashing which is accomplished by increasing the water flow. This removes small particles, such as sand, and the biofilm.

Applications

Trickling filters are mainly used in sewage treatment plants where they are effective for the treatment of low concentrations (below 1%) of acetaldehyde, acetic acid, benzene, chlorinated hydrocarbons, cyanides, formaldehyde, ketones, and resins. Highly chlorinated organics, aliphatics, amines, and aromatic compounds are difficult to treat while heavy metals may inhibit, kill, or accumulate in the biofilm. Hydrogen sulfide gas is a possible by-product.

Biofilters

Numerous field trials have been performed since 1991 at a wide range of temperatures (-40 to 30°C) for municipal wastewater in Ontario and Saskatchewan and for domestic wastewater under a number of conditions and landfill leachates at a field-site in Owen Sound, Ontario. Biofilters have been approved for use in Ontario and Massachusetts and can operate with flows less than 1,000 L/day or up to more than 100,000 L per day of wastewater.

Rotating Biological Contactors (RBC)

Although this process was very popular in the 1960s and 1970s, other methods are now preferred in the U.S. due to problems with biomass buildup on the shafts and medium. In Germany and Switzerland, these systems are popular in small treatment plants. Excess biomass that breaks off can be removed downstream in a sedimentation tank. In general, it is beneficial for the treatment of waste water containing alcohols, phenols, phthalates, cyanides, and ammonia at concentrations less than 1%. It can be operated in a series, first for organic removal and then for nitrogen removal. These systems are not suitable for aliphatics, amines, chlorinated organics, and aromatic compounds. Heavy metals and organics of low biodegradability can kill or inhibit the microorganisms or accumulate in the sludge.

Table 4.4 Suggested Loading Rates for Trickling Filters with Rocks/Stones as Filtering Media for Municipal Wastewater Treatment

Parameter	Low rate	Intermediate rate	High rate	Super rate (roughing)
Organic loading rate (kg BOD/m^3-day)	0.08–0.32	0.20–0.48	0.32–1.0	0.8–6.0
Hydraulic surface loading rate (m/day)	1–4	4–10	10–40	40–200
Estimated efficiency (%)	88 ± 8	60 ± 12	75 ± 10	55 ± 10
Effluent concentration (g BOD/m^3)	<20	<25	20–40	30–80
Depth (m)	1.5–3.0	1.25–2.5	1.0–2.0	1–4 m
Power requirements (kW per 1,000m^3)	2–4	2–8	6–10	10–20
Effluent	Fully nitrified	Partially nitrified	Nitrified at low loads	Nitrified at low loads

Adapted from Triebel (1975) and Metcalf and Eddy (1991)

Costs

Although costs vary from site to site, the following information has been determined by Martin and Martin (1991). Operating and maintenance costs are the highest ($0.04/1,000m³) for lower capacity filters (3,800 m³/day) and lowest ($0.01/1,000 m³) as the capacity increases by a factor of ten. Construction costs vary between $0.2/1,000 m³-day for low capacity filters to $0.13/1,000 m³-day for higher capacity (190,000 m³/day).

Costs for a small biofilter system for the home including septic tank, pump, and controls, and a small disposal system are approximately $5,200 to $6,500. The required space for a 1,500 L/day system would be 3.3 m² and 1.2 to 1.5 m in depth.

The costs for RBC were compared to activated sludge systems by Berktay and Ellis (1997). At low wastewater flows of 190 m³/day, capital costs ($617/m³ per day) were comparable to the activated sludge ($526/m³). However at higher flows (1,140 m³/day), the capital costs of the RBC were significantly higher ($425/m³ per day) than the costs for the activated sludge ($213/m³ per day). Operating costs were less for the RBC at low and high flows ($0.26/m³) compared to $0.63/m³ at low and $0.4/m³ at higher flows for the activated sludge.

Case Studies

BioTrol (Eden Prairie, MN) performed a nine-month field test of a new fixed-film reactor (the BioTrol aqueous treatment system, BATS) at a wood preserving facility (EPA 1991). The patented process consists of a mix tank for pH and/or temperature adjustment and nutrient addition. The water is pumped to the multi-cell bioreactor where bacteria are immobilized on a highly porous packing material. Bubble membrane diffusers are located at the bottom of each cell to supply oxygen. The effluent may then be discharged to a municipal wastewater facility. The system has been applied for the treatment of groundwater, lagoon, and process water that contained pentachlorophenol (PCP), creosote, gasoline, fuel oil, chlorinated hydrocarbons, phenolics, and other solvents. At the site containing PCP, concentrations were decreased to below 1 ppm from 45 ppm. Operating costs were $0.64 per 1,000 liters of water treated for a 6,800 L/h system. Over 20 full-scale units have been installed.

Another system called Ecoflo® has been developed by Premier Tech Ltd. (Rivière-du-Loup, Quebec) for residential and commercial septic tanks (Premier Tech 1999). It consists of a fiberglass shell containing a peat biofilter and an infiltration zone or an outflow pipe. The system has been used to treat effluents from the residential sector but can also be used for communities and businesses such as hotels and restaurants if more units are added. Typically, 89% removal of the TSS, 94% removal of BOD, and 99% removal of coliform bacteria is obtained so that effluents contain on average 11 mg/L BOD, 7 mg/L TSS, and 12,000 tCFU (fecal coliform bacteria)/100 mL. More than a thousand units operate in Canada, Europe, and the United States, mainly as residential systems.

BIOFOR® Biological filtration (Degremont, France) is used to remove BOD as a secondary or tertiary treatment. It is a fixed-film system that uses a biological contactor and a filter. The media for biomass support and solids removal is proprietary. Compared to conventional activated sludge processes, space requirements are reduced, secondary clarifiers are no longer needed, and treatment efficiencies for cold and dilute wastewaters are improved. Start-up is rapid and operation is automated. It is an upflow biofilter (Figure 4.12a, b) with both air and wastewater flowing upward. This allows water with the highest BOD load to be in contact with oxygen of the highest concentration. Bubbles do not coalesce and flow is homogeneous across the biofilter. Depths of the filters are usually about 300 cm and backwashing is performed every 24 to 48 h. Installations of 380 to 420,000 m³ per day are in operation. BIOFOR C has been developed for carbon removal only while BIOFOR N is for nitrification and BIOFOR CN is for both.

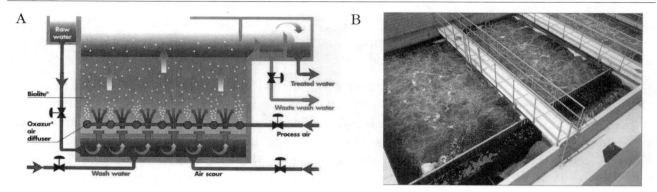

Figure 4.12 Schematic (a) and Photograph (b) of a BIOFOR Biofilter (courtesy of Infilco Degremont)

A wastewater system was built in for a town in Ontario with a flow rate of 9,658 m³/day with a BOD concentration of 64 mg/L, TSS of 54 mg/L, TKN of 15 mg/L, and 1.1 mg/L total phosphorus (TP). The system was designed with four BIOFOR C cells of 40 m², 2.5 m of Biolite N® of 2.7 mm, and four BIOFOR N cells with the same characteristics. Effluent criteria had to meet levels less than 10 mg/L BOD and TSS, less than 1 mg/L TP, and TKN levels less than 5 mg/L in the summer and 10 mg/L in the winter. The cells contained blowers and backwash pumps with fully automatic operation (Anonymous 1998). After adaptation, the system operated well even in winter temperatures of 10°C.

Aerobic Biological Treatment: Biological Nutrient Removal

Box 4-6 Overview of Nutrient Removal Processes

Applications	Nitrogen and/or phosphorus removal from industrial and municipal wastewaters
Cost	Operating: $0.07 to $0.67/m³, Capital: $83 to $480/m³/day
Advantages	Low sludge production
	Floc settleability is improved
	Energy consumption is reduced
	Alkalinity can be recovered in nitrifying systems
Disadvantages	Limited design criteria available
	High levels of phosphorus removal difficult to achieve
	Retention times can be long
Other considerations:	Significant advances in last few years

Description

Biological nutrient removal (BNR), once thought of as an emerging technology, has made significant advances in the past few years. This process is now an economical alternative since it has been studied and is now better understood. Over 300 systems treating more than 2,300 m³/day of wastewater are now in operation in North America (R.V. Anderson and Associates, Ltd. 1998). Not only do they remove nutrients, but also less sludge is produced which reduces clarifier solids loadings, improves floc settleability, reduces energy consumption, and, in nitrifying systems, recovers alkalinity.

Methods

Nitrogen Removal

Nitrogen removal is performed as a two-step process: the aerobic nitrification step followed by the anoxic denitrification step. Nitrifying bacteria (*Nitrosomonas* and *Nitrobacter*) are required for nitrification or ammonia oxidation via the equation:

$$NH_4^+ + 1.682O_2 + CO_2 + 0.0455HCO_3^- \rightarrow 0.0455C_5H_7NO_2 + 0.955NO_3^- + 0.909H_2O + 1.909H^+$$

Denitrification follows nitrification and takes place in the absence of dissolved oxygen using denitrifying bacteria such as *Pseudomonas*. The equation for denitrification is as follows:

$$NO_3^- + 1.08CH_3OH + H^+ \rightarrow 0.065C_5H_7NO_2 + 0.47N_2 + 0.76CO_2 + 2.44H_2O$$

Phosphorus Removal

Since phosphorus removal is not normally performed at high efficiencies (less than 20%), bacteria have to be forced to remove more phosphorus. This is accomplished by using anaerobic followed by aerobic conditions as in the A/O (licensed by I. Kruger AF, Denmark) or Phoredox processes (Figure 4.13) The A/O Process combines oxidation and phosphorous removal and is operated as an activated sludge system. An anoxic step is proceeded by an aerobic one causing stress on the microorganisms and enhanced uptake of phosphorus. Phosphorus is released from the biomass under anoxic conditions in the first tank. Under aerobic conditions in the second tank, the phosphorus is taken up by the biomass. If the ratio of BOD/P in the water is greater than 20 to 25:1, effluent phosphorus levels of less than 1 mg/L can be obtained. Wastage of the sludge prevents accumulation of phosphorus in the sludge.

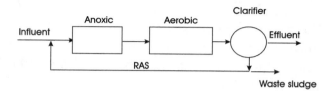

Figure 4.13 Biological Phosphorus Removal

Combined Nitrogen and Phosphorus Removal

Phosphorus removal can also be combined with a nitrification-denitrification process and the addition of an easily degradable carbon source such as acetate or an internal carbon source. The carbon source must be supplied so that the ratio of phosphorus to easily degradable COD is 1:10. The ratio of COD to nitrate will be 4 to 6:1 (Henze 1996c).

Operating Conditions and Design

Nitrogen removal

The main factors influencing this process include temperature, ammonia concentration, organic matter type and concentration, pH, cell retention time, and dissolved oxygen concentration. According to the AWWA (1992), 4.18 kg of oxygen, 14.1 kg of alkalinity in the form of $CaCO_3$, 0.15 kg of new cells, and 0.09 kg of inorganic carbon are required for the oxidation of 1 kg of ammonia to nitrate. The AWWA (1992) has developed various equations for process design. Usually, however, the design of a system is not based only on nitrification requirements but on many other factors; thus, there are no uniform criteria for design. A minimum sludge age of seven days at 10°C is required for nitrification (U.S. EPA 1993).

In the denitrification step, anoxic activated sludge or fixed-film systems can be used. Carbon must be present, and if it is not, methanol or another substrate is added. The pH is maintained between 7.0 and 7.5. In this step, nitrate is reduced to nitrogen. Approximately 2.9 kg of oxygen and 3.0 kg of carbonate are recovered for each kg of nitrate reduced (AWWA 1992). Biomass is also produced during this process (0.4 kg of VSS/kg of COD removed).

Due to the recovery of oxygen during denitrification, approximately 25% oxygen can be saved if denitrification follows nitrification. In addition, since denitrification requires a carbon source, cycling of the organics from the nitrification process can also provide savings. This type of integration is demonstrated in Figure 4.14. Fluidized sand beds based on the Dyna-Sand™ process (Koopman et al. 1990) or bacteria on activated carbon can be used. In this combined process, the wastewater passes into the anoxic denitrification reactor. The influent to this tank is high in BOD, ammonia, and nitrate levels and low in oxygen,

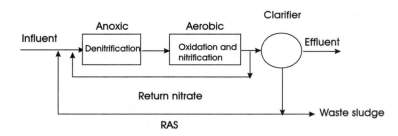

Figure 4.14 Combined Denitrification/Nitrification System for Nitrogen Removal. RAS Denotes Return Activated Sludge

conditions which promote denitrification. The effluent from the denitrification reactor is fed to the aerobic reactor, which is maintained at oxygen levels higher than 2 mg/L. A sufficient retention time in this reactor will enable carbon oxidation to take place. A further increase in the retention time will allow nitrification of ammonia to nitrate. The effluent is then fed to the clarifier for separation of the biomass and the nitrified water. Nitrification and denitrification can occur at 2°C (Oleskiewicz and Berquist 1988) but the rate is approximately one-fifth and one-quarter, respectively, for each process at 15°C.

The oxidation ditch (Figure 4.15) can also be used for denitrification-nitrification. In the aerobic zone, the oxygen level is maintained at levels greater than 0.5 mg/L with greater than 2 mg/L preferred. Solids retention is ten to 50 days with an HRT of 24 h. The depth of the ditch is 120 to 180 cm. As it reaches the anoxic zone, the concentrations are reduced to almost zero. The wastewater is added at this point, and the effluent is removed at the aerobic area. The Bardenpho process is a modification where a second anoxic zone has been added. Design equations are described by Metcalf and Eddy (1991).

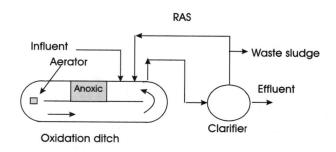

Figure 4.15 Oxidation Ditch for Denitrification/Nitrification

At a wastewater treatment facility in Lake Geneva, Wisconsin, oxidation ditches are used to convert ammonia to nitrates and then to remove the nitrates. Six aerators are used to mix oxygen into the wastewater. After aeration, the wastewater is sent to clarifiers for settling of the activated sludge. Sludge is then either returned to the oxidation ditch or treated in an aerobic digester before land application. A bituminous-lined lagoon is used for sludge storage if necessary. The secondary effluent is pumped to eight seepage cells which are a couple of miles away. The water then passes into the aquifer. The treatment facility treats 1,007 kg BOD/day, 1,181 kg/day TSS, 202 kg Kjeldahl nitrogen /day, and 118 kg ammonia nitrogen/ day for a population of 17,800 which generates 6,500 m³/day of wastewater on average. An effluent quality of 50 mg BOD /L and 10 mg nitrogen /L must be achieved.

Alternating anoxic/aerobic zones with step feed is often called ABNR (advanced biological nutrient removal) or the Tallman Island Process since this was primarily where the process was developed. The process is shown in Figure 4.16. External or recycled carbon sources are not required. In addition, energy requirements can be reduced by up to 30%. Anoxic conditions are maintained by adding only 50 to 70% of the oxygen required. The incorporation of the sequencing batch reactor can further minimize tank requirements.

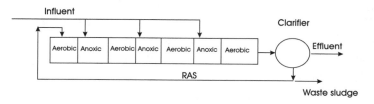

Figure 4.16 Tallman Island ABNR Process

Phosphorus Removal

In the A/O process, the anaerobic stage follows the aerobic stage in a multiple-compartment reactor. The two stages are of equal size. Hydraulic retention times of 1 to 3 h are used. The phosphorus in the wastewater is solubilized in the anaerobic zone, and the phosphorus is taken up by the bacteria in the aerobic zone. The biomass of low aerobic sludge age can then accumulate up to 4 to 12% phosphorus on a total dry basis compared to the usual 1 to 2%. *Acinetobacter* sp. are responsible for most of the phosphorus removal under anaerobic conditions coupled with low nitrate levels. Phosphorus levels can be reduced from 10 to 2 or 3 mg/L or 70 to 80%. Chemical requirements are significantly reduced since only 3 to 6 mg/L of iron would be required to decrease the phosphorus below 1 m/L, compared to 25 mg/L if biological removal was not used (Kiely 1997). Sludge production is also reduced, which decreases the tank size requirements. Further phosphorus reduction can be biologically achieved to less than 1 mg/L if the BOD/P ratio is greater than ten (Metcalf and Eddy 1991).

Another phosphorus removal method is called the PhoStrip Process (Figure 4.17). Instead of removing the phosphorus by adsorption into the sludge, a supernatant containing high levels of phosphorus is produced, which is then chemically treated. Unlike most activated sludge processes, a portion of the sludge is sent to an anaerobic phosphorus stripper where it is retained for 8 to 10 h. The phosphorus is released from the sludge into the supernatant which is then treated by lime precipitation before returning to the aeration basin. These processes are summarized in Table 4.5.

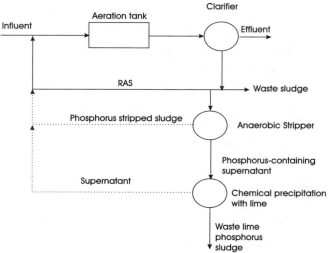

Figure 4.17 Pho Strip Process for Phosphorus Removal

Combined Nitrogen and Phosphorus Removal

The Bardenpho process (Eimco, Salt Lake City, Utah), which consists of five stages, is one example of a combined nitrogen and phosphorus removal system (Figure 4.18). The four-stage process involves ammonia, organic, nitrite, and nitrate removal with no phosphorus removal. Volatile acids are produced anaerobically in the first step. In the second step, some denitrification and BOD removal take place. In the third reactor, phosphorus (orthophosphate, polyphosphate, and organic phosphorus) removal and nitrification occur. Next, further denitrification occurs as the biomass decays. Finally, the nitrogen gas is stripped and oxygen is supplied to the bacteria to reduce phosphorus loss in the clarifier. There is a recycle from the end to the beginning as in other activated sludge processes. Randall et al.

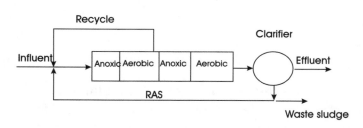

Figure 4.18 Four Stage Bardenpho Phosphorus Removal Process

(1992) has described the design equations. Another system, the Virginia Initiative Process (VIP), was developed by the Hampton Roads Sanitation District and CH2M Hill. The process has two recycle streams to enable phosphorus and nitrogen removal (Figure 4.19).

Since there is limited experience in these types of treatment plants, design of the reactors is difficult. Many proprietary nutrient removal processes have been developed over the past two decades in South Africa, Europe, and the United States (AWWA 1992; Metcalf and Eddy 1991; Randall et al. 1992).

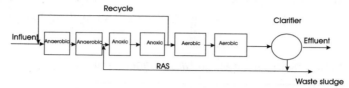

Figure 4.19 Virginia Initiative Process (VIP)

Table 4.5 Operating Conditions for Biological Phosphorus Removal

Parameter	Process			
	PhoStrip	A/O	A/O with nitrification	Bardenpho
F/M (kg BOD/kg VSS/day)		0.2–0.7	0.15–0.25	0.1–0.2
SRT (days)		2–6	4–8	10–30
MLSS (mg/L)	600–1,000	2,000–4,000	3,000–5,000	2,000–4,000
HRT (h)	1–10			
Anaerobic		0.5–1.5	0.5–1.5	1–2
Anoxic 1			0.5–1.0	2–4
Aerobic 1		1–3	3.5–6.0	4–12
Anoxic 2				2–4
Aerobic 2				0.5–1.0
Sludge recycle (% of influent)		25–40	20–50	100
HRT in stripper	5–20			
Stripper flow (% of influent)	20–30			
Lime dosage in clarifier (mg/L)	100–300			

Adapted from U.S. EPA (1987b)

Denitrification

Table 4.6 gives the optimal pH, hydraulic retention time, and mean cell residence time for a three-stage process. The EPA (1975) described the design kinetic parameters in a process design manual as follows. Approximately 3 kg of metal is required per kg of nitrate removed, and 3 mg of alkalinity as $CaCO_3$ is produced per milligram of nitrate reduced.

Table 4.6 Operational Parameters for Three-stage Nitrification/Denitrification (Reynolds and Richards 1996)

Stage number	pH	Mean cell retention time, $_c$ (days)	Hydraulic retention time (h)
1st	6.5 to 8.0	Less than 5	Less than 3
2nd	7.8 to 9.0	10 to 25	Less than 3
3rd	6.5 to 7.5	Less than 5	Less than 2

The minimum mean cell retention time (MCRT) can be described as:

$$\frac{1}{\theta_c^{min}} = Y\bar{q} - k_e$$

where θ_c^{min} = minimum cell retention time for denitrification, days; Y = gross yield, mg VSS/mg nitrate removed; $\bar{q}$ = peak nitrate removal rate, mg nitrate removed/g VSS/day; and k_e = decay coefficient, day^{-1}.

The design retention time is then determined by the desired final effluent nitrate concentration:

$$\frac{1}{\theta_c} = \frac{Y\bar{q}(D_o - D_1)}{\left((D_o - D_1) + K_D \ln\left(\frac{D_o}{D_1}\right)\right)} - k_e$$

where θ_c = design mean cell retention time, days; K_D = is the half-saturation constant, mg nitrate/ L; D_o = influent nitrate concentration, mg/L; and D_1 = effluent nitrate concentration, mg/L.

Through rearrangement, the design nitrate removal rate (mg nitrate removed per mg VSS/day), q, is given by:

$$q = \frac{1}{Y}\left(\frac{1}{\theta_c} + k_e\right)$$

The hydraulic retention time, θ, is given by:

$$\theta = \frac{D_o - D_1}{X_1 q}$$

where θ = hydraulic retention time, days; and X_1 = volatile suspended solids in the denitrification reactor, mg/L.

The mass of sludge leaving per day is determined by:

$$\text{mass MLVSS leaving per day} = \frac{X_1}{\theta_c}$$

The mass of sludge wasted per day, X_w is given by:

X_w = mass MLVSS leaving per day - mass of sludge leaving in the effluent per day.

In general, the half saturation constant varies from 0.08 mg nitrate per liter without cell recycle to 0.16 mg nitrate per liter with cell recycle (U.S. EPA 1975). The denitrification cell yield can vary from 0.6 to 1.20 kg VSS/kg nitrate removed. The decay coefficient is 0.04 day^{-1}. The peak denitrification rates (kg nitrate removed per kg per day) depend on the temperature as follows:
- 0.05 to 0.10 at 10°C;
- 0.05 to 0.17 at 15°C;
- 0.15 to 0.43 at 20°C; and
- 0.20 to 0.48 at 25°C.

Costs

The costs of two VIP plants (150,000 and 114,000 m³/day) and an anaerobic reactor prior to an oxidation ditch (11,350 m³/day) were compared to a chemical plant with ferric chloride addition (68,000 m³/day) in the Chesapeake Bay Watershed (Randall and Cokgor 2000). Capital costs of the three biological plants were $475, $433, and $30 per m³ per day, respectively compared to $528 per m³ per day for the chemical plant. Operating costs per 1,000 m³ were $79, $82, and $665 for the biological and $118 for the chemical plant. The larger biological plants compare favorably to the chemical plant in terms of capital and operating costs. In contrast, the capital costs for the smaller biological plant were the lowest and its operating costs were the highest.

Case Studies

The A²O process of Kruger Inc. (Denmark) is an example of a high-rate biological nutrient removal process using an oxidation ditch. It is an anaerobic/anoxic/oxic process. The process installed at Titusville, Florida, is designed to treat up to 15,140 m³/day of waste-water at a cost of $12 million U.S. (Kruger 2001). The effluent limits of 5-5-3-1 (CBOD-TSS-TN-TP) were easily met. There was some trouble with phosphate levels on start-up, but once the flows to the plant were more stable, levels of 0.2 to 0.3 mg/L could be obtained. Effluent quality is high while energy requirements are minimized.

Aerobic Biological Treatment: Other Aerobic Processes

Activated Biofilter (ABF)

Description

Activated biofilters are similar to activated sludge and trickling filter systems. Microorganisms are grown on redwood slats or synthetic media as is done in trickling filters. However, unlike trickling filters, the settled sludge is recycled to the ABF unit. This enables both suspended and fixed cell growth to occur. The depth of the ABF is generally around 360 cm. The wastewater distributors on the top of the ABF and the trickling of the wastewater over the support supplies oxygen to the biomass. Sometimes a completely mixed aeration basin is used after the ABF to improve organic oxidation and settling of the sludge. A second clarifier is then used before sludge recycling to the ABF.

PACT Process

Description

For the treatment of recalcitrant or volatile organics, US Filter, Zimpro Products, and Du Pont (Du Pont 1982) developed a system in the 1970s called PACT®, a proprietary activated sludge process enhanced by the addition of powdered activated carbon of less than 200 mesh diameter. The addition of the activated carbon improves stability to shock loads; improves sludge settling and dewatering; improves oil and grease, BOD, COD, and refractory organic removal; lowers effluent toxicity to fish; and decreases the tendency of the sludge to foam. Traditional

Figure 4.20 PACT Process. (Courtesy of US Filter)

pretreatment processes including sulfide pretreatment, heavy metal precipitation, oil/water separation, and steam stripping are used. The wastewater (1,000 to 20,000 mg BOD/L) is then neutralized, if required, to pH 7 and the solids are removed by flocculation and settling. The solids of 3 to 5% solids are sent back to the aeration tank. Dewatering of the excess solids is accomplished by filter presses. The PACT process consists of two steps, carbon adsorption and biological treatment together (Figure 4.20). Dupont operates a 151,000 m³ per day unit at Chamber Works in Deepwater, NJ (Du Pont 1982). The bacterial consortium is able to degrade the organics adsorbed onto the PAC, thus liberating sites on the carbon for further adsorption. The activated carbon and biomass that settle to the bottom of the tanks are recovered for subsequent recycle to the aerators.

Applications

These systems have been applied to wastewaters including municipal, joint municipal-industrial, industrial, surface runoff with COD of greater than 50 mg/L, groundwater, and landfill leachates. Most installations are in the refinery, petrochemical, and organic chemical industries, in addition to those for leachate and groundwater treatment. More than 100 systems have been installed throughout the world. Units built in the factory are able to treat 7.5 to 380 m³/day of wastewater while units erected on the field treat from 3.8 to 190 thousand m³/day of water. Besides removal of organic matter, color and odor removal are also possible, in addition to improved settleability/dewatering and reduction in toxicity.

Costs

Capital costs are usually less than $200,000, and operating costs are $0.13/m³ to $1.30/m³, according to U.S. Filter/Zimpro (2000).

Deep Shaft Reactor

Description

The deep shaft reactor developed by Imperial Chemical Industries (London, England) was designed in 1975 to allow high transfer rates of oxygen with minimal land area requirements due to its vertical development (Anonymous 1978). The reactor is 50 to 150 m in depth with a 0.5 to 10 m diameter. The influent enters the top of the downflow where it is mixed with air. It then rises in an adjacent section (Figure 4.21). Excess sludge is removed from the treated effluent and returned to the reactor. Consumption of electricity per cubic meter of effluent treated is very low.

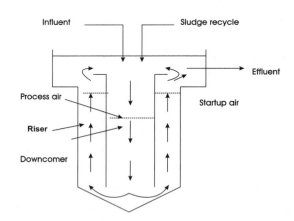

Figure 4.21 Deep Shaft Process

Air-Lift Reactor

Description

The CIRCOX® reactor, developed by PAQUES, Gist-Brocades, TNO, and the Technical University of Delft, is an air lift reactor with an outer wall and an internal cylindrical riser. The influent is fed into the bottom of the reactor and is brought into the riser by the air. The bacteria are attached as a biofilm to a carrier like sand, preventing biomass washout.

Operating Conditions and Design

Biological loads of 4 to 10 kg COD/m^3-day can be treated. Hydraulic retention times in the order of 0.5 to 4 h are used. Biomass settling velocities are 50 m/h and the biomass concentrations are 15 to 30 g VSS/L. Due to the high sludge age, compounds which are difficult to degrade, such as nitrogen compounds, can be treated.

Applications

Pilot studies show that nitrogen levels can be reduced to less than 10 mg/L in brewery wastewater. The reactor can be converted to a denitrifying reactor by the addition of an anoxic compartment. The sludge then converts the ammonia into nitrate and then to nitrogen gas.

Membrane Bioreactor

Description

Another variation of an activated sludge process is the incorporation of ultrafiltration membranes with a biological reactor to increase sludge retention while decreasing hydraulic retention times. Zenon Environmental (Burlington, Ontario) has developed a system called the ZenoGem® (Figure 4.22a, b). In the process, wastewater that enters the reactor for biological treatment is passed to the ultrafiltration step where the biomass and high molecular weight soluble components are separated from the treated water. The retained components are then recycled to the bioreactor. The process can nitrify or denitrify organics when an anoxic reactor is added. It reduces construction costs, land area, operator labor, sludge volumes, odor, and chemical costs.

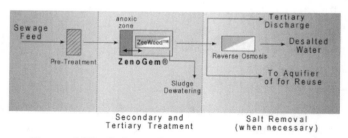

Figure 4.22a Schematic of the ZenoGem Process
(courtesy of Zenon Environmental)

Figure 4.22b ZenoGem Membrane Unit
(courtesy of Zenon Environmental)

Applications

In a typical sewage treatment plant, the sludge would be removed by the membranes, thus eliminating the need for clarifiers and sludge digestion. It also can be used for other types of wastewater ranging from oil wastewater, metal finishing wastes, landfill leachates, alcohol-based cleaning solutions, detergents, aqueous paint-stripping wastes, deicing fluids, soil washing effluents, and contaminated groundwater to high-strength and variable feed wastewater. A three-month site demonstration took place in 1992. Results showed 99.9% removal of methyl methacrylate from Plexiglass production (Zenon 1995). Sludge volumes were reduced by 60% and solids contents increased from 1.6 to 3.6%. The system could be left unattended at night and on the weekends. Recently a unit treating 2,275 m^3 of wastewater per day was designed and installed on Baffin Island, in the Canadian Arctic.

Degremont of France also supplies a membrane reactor called the BRM. It combines an activated sludge reactor with ultrafiltration as the polishing process.

Case Study

Triqua of the Netherlands has developed a membrane reactor for the treatment of bilge water (with its attendant high salt content) for the Navy (Bakx et al. 2000). The reactor has been placed near the harbor in a building of 11.5 x 6 m and has a volume of 22 m^3 and a capacity of 1.5 m^3/h. A flocculent pretreatment is used. The membrane has a surface area of 32 m^2.

Anaerobic Processes

Box 4-7 Overview of Anaerobic Processes

Applications	Suitable for treatment of low to high concentration organic chemical, textile, petrochemical, food, pulp and paper industry wastewater, chlorinated organics and inorganics
Cost	Operating: $0.04 to $0.08 / m^3, Capital: $320 to $540 per m^3/day (based on 2 kg COD/m^3)
Advantages	Low production of sludge
	Production of methane useful for energy purposes
	Removal of chlorinated wastes
Disadvantages	Complicated to start-up and operate
	Susceptible to fluctuations in influent, composition, and toxins
	Optimal operation at 30 to 40°C
Organic loading rate	2–6 kg COD/m^3-day for contact processes, 3–10 kg COD/m^3-day for high rate
Hydraulic retention time	4 to 48 h
Other considerations:	Proper reaction design and operation highly important in overcoming potential disadvantages

Description

Anaerobic processes have become popular since 1980 when the Dutch UASB system was introduced. Anaerobic processes result in lower electricity costs than that of aerobic processes. Anaerobic processes also produce biogas, which can be used as a fuel, and have enzymes to remove chlorine from chlorinated compounds. Complete degradation of chlorinated compounds can be accomplished by anaerobic processes followed by aerobic processes. Anaerobic processes do not require oxygen and they produce methane, carbon dioxide, and low molecular-weight end products. Methane production, once it is handled with care, can be useful for heating purposes. Sludge production in anaerobic processes is greatly decreased compared with aerobic processes, reducing disposal problems and costs. Degradation times may be longer, however, since anaerobic metabolism is a slower process. Anaerobic systems remove mainly organic matter, not nutrients. The food and beverage industry produce many effluents high in organic content at temperatures higher than ambient, which is ideal for anaerobic treatment.

Anaerobic degradation is performed by a consortium of bacteria from three different categories (Figure 4.23). The first step in the degradation process is hydrolysis to break down large molecules and dissolve suspended solids which enables easier transport into the bacterial cell for metabolism. Extracellular enzymes from primary fermentative bacteria are capable of performing this task. Next the same bacteria form the hydrolysis products into organic acids (mainly acetic), hydrogen, and carbon dioxide. Acetogenic bacteria produce acetic acid and hydrogen by acetogenesis. Propionic and butyric acids are also formed by other bacteria.

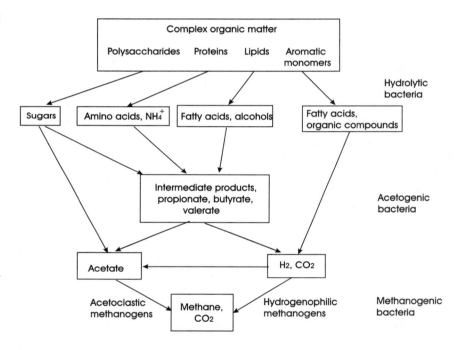

Figure 4.23 Conversion of Organic Matter by the Different Types of Anaerobic Bacteria

Their pH optimum is in the range of 5 to 6 but they tolerate pH of 7. Methanogenesis is the formation of methane from acetic acid (methane and carbon dioxide are formed) or acetic acid and hydrogen (formation of methane and water). The pH optimum of methane forming bacteria is 7 with a substantial decrease in activity below 6 or above 8.

There are two main types of anaerobic processes: low rate and high rate. For the low rate, fresh sludge is added two or three times a day. Three layers of sludge form including the upper layer of scum, the middle layer of supernatant, and the lower sludge layer. The sludge has an active top layer and a stabilized bottom layer. The high-rate systems have one or two stages (stabilization followed by settling and thickening). Gas production is usually represented as volume of gas produced per VSS produced.

Anaerobic metabolism can use sulfate, carbon dioxide, nitrate, and some organics instead of oxygen. Oxygen can be toxic to anaerobic microorganisms and must be eliminated in the wastewater for growth to occur. Temperature should be maintained in the range of 30 to 40°C for mesophilic organisms or 50 to 75°C for thermophilic. Temperature shifts of more than 2°C per day can upset the microbial activity (Belhateche 1997). The ideal pH for methane producing organisms (methanogens) is 7. Since organic acid production decreases the pH, maintenance of the buffering capacity (alkalinity) of the wastewater is very important.

Anaerobic organisms can be disturbed by changes in the composition of the wastewater, the presence of toxins, or buildup of degradation products. The percentage of methane in the biogas is dependent on the substrate type. Systems include sludge blanket reactors and fluidized filters. Pretreatment can be performed by screening, filtering, flotation, or filtration.

The contact process resembles the activated sludge process. The first one was built in Denmark in 1929 (Henze 1996a) and was used for the treatment of wastewater from a yeast factory. Today it is also used for the sugar and spirits industries. The sludge is mixed mechanically or by the production of the gas. The tank for the biological degradation process is usually about 5 to 10 m deep. The sludge is removed by lamella sedimentation. Sometimes the gas must be removed to facilitate settling. This can be accomplished by vacuum or by mechanical or hydraulic means. The sludge could also be cooled before settling to slow the gas production process.

Methods

Conventional Anaerobic Treatment

Conventional treatment processes are well-mixed reactors without recycle (Figure 4.24). The SRT is equal to the HRT. In general, HRT of ten to 30 days are used for temperatures of 35°C. Mixing can be achieved by recirculation of biogas or by the use of mechanical mixers (U.S. EPA 1987a). The mixers must be resistant to corrosion in the reactor.

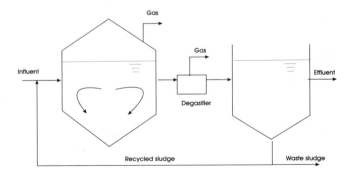

Figure 4.24 Conventional Anaerobic Process without Recycle

Loading rates are usually based on VSS. Rates of 0.5 to 6.0 kg VSS/m³-day are typically used with the lower part of the range for lower temperatures. In a reactor without recycle, decreases in the influent COD can affect the bacterial balance. As the loading rate reaches normal values, the acid-forming bacteria will predominate over the methane-forming bacteria since the acid-forming bacteria recover faster. Conventional aerobic systems are more susceptible to upsets than others. Sometimes a second reactor is added for sedimentation or additional digestion; it can then serve as a reserve for the first reactor. The second reactor may or may not be heated and/or mixed. It can help to reduce the reactor volume in the event of dilute sludges.

Contact Processes

Contact processes have well-mixed reactors with solids recycle to increase the SRT. Because the SRT can be controlled, the HRT can be significantly reduced. Since it is similar to activated sludge processes, those models can be used. Contact processes are applicable for influent COD values of 2,000 to 100,000 mg/L with suspended solid contents of 10,000 to 20,000 mg/L.

Depending on the characteristics of the influent stream, HRT of one to seven days are usually sufficient. This can significantly reduce the volume required compared to the conventional system. Loading ranges are in the range of 2 to 10 kg COD/m³-day. High loads can be applied with decreased efficiencies.

Gas formation can cause settling problems because of the gas bubbles attached to the flocs. Using flotation separators or thickeners can solve this problem. Adding oxygen or decreasing the sludge temperature to 5 to 10°C can temporarily reduce gas formation to allow the use of gravity sedimentation without harming the bacteria. Although lamella or tube settlers are efficient devices, the sludge can cause clogging.

Continually stirred tank reactors called Biobulk CSTR supplied by Biothane Corporation (The Netherlands) have a medium capacity load of 2 to 5 kg COD/m³-day. This is achieved by using a special internal mixing design capable of TSS and oil and grease (FOG) removal. They are cost effective when flows are below 570 m³/day, according to Biothane (Biothane 2001), since longer retention times can be allowed. These types of systems have been used for ice cream and other food processing plant effluents with high levels of oil and grease. Typical results are 90% removal of COD, FOG, and BOD with 75% removal of VSS. Another type of contact process is the SBR which was described earlier as an aerobic system. Sludge is internally recycled. Mixing may be necessary during the fill and reaction phases.

Degremont (France) is a supplier of the Analift reactor, an anaerobic contact reactor that mixes and de-aerates and has a separate settler for industrial wastewater treatment. They have also developed a suspended growth blanket reactor with a pulsed feed and settling tank called the Anapulse system.

Low-Rate Anaerobic Systems

ADI Inc. (New Brunswick, Canada) developed a low-rate anaerobic reactor called ADI-BVF® to treat high concentrations of fats, oils, and grease, such as from the potato or meat processing and dairy industries. It is also suitable for winery, distillery, soft drink, and food processing wastewaters with BOD of greater than 2,000 mg/L and flow rates of more than 500 m³/day. There is little sludge production. They can be designed for aboveground use with steel or concrete tanks or for in-ground basins of earth or concrete. In this case, geomembrane covers are used to contain the biogas and to control the temperature and odor. Payback periods for these systems range from three to five years.

Upflow Anaerobic Sludge Blanket (UASB) Reactor

A variation of the anaerobic contact process is the sludge blanket process (UASB) which is a biological tank with upflow and settling tanks developed in the Netherlands (Lettinga et al. 1980). Granules consisting of high concentrations of biomass are produced during the degradation of the easily degradable organic matter (Figure 4.25). These granules are permanently formed and remain in the reactor. The wastewater enters the bottom of the reactor and passes through the granules. The organic matter is converted to methane and carbon dioxide which leads to the formation of gas bubbles that can provide adequate mixing and wastewater/biomass contact. The granules rise in the reactor because of the bubbles; however, the granules will settle in the tank since their settling velocities are greater than the upflow velocity (typically 1 m/h). An adequate settling zone is provided (van Haandel and Lettinga 1994). Since the concentrations of sludge can be up to 5 to 15 kg VSS/m^3, generally twice that of contact processes, recycling is not required. UASBs are the most common type of high-rate process in the world today because they can perform at higher efficiencies than anaerobic fixed-film and continuous flow aerobic reactors (Latkar and Chakrabarti 1994). UASBs are used in the majority of all anaerobic treatment processes.

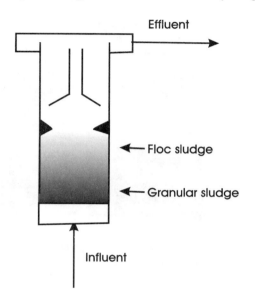

Figure 4.25 Upflow Anaerobic Sludge Blanket (UASB) Reactor

Separation of gas and solids is very important in these types of systems because significant solid losses can significantly decrease the performance of the reactor. The top of the reactor has a clarifier to accomplish this task. Different suppliers such as Biothane and Paques use different designs. Newer designs include flexirings to prevent solids losses and to promote biofilm formation. Thus, the reactor becomes a sludge blanket fixed-film reactor.

Expanded Granular Sludge Blanket (EGSB) Reactors

For treating sewage at lower temperatures (4 to 20°C), it was determined in pilot tests that the UASB reactors did not provide adequate influent distribution. The expanded granular sludge bed (EGSB) reactor was developed to incorporate higher superficial velocities (greater than 4 m/h). This is accomplished by the higher height to diameter ratios and the use of effluent recirculation. The higher upflow velocities expand the bed, eliminate dead zones, and improve wastewater/biomass contact (van der Last and Lettinga 1992). Soluble pollutants are efficiently treated by these reactors, but suspended solids or colloidal matter are not.

The Multiplate Reactor

The multiplate reactor technology was developed by SNC-Lavalin (Montreal, Canada). The first full-scale unit (450 m^3) was constructed in 1991 at a dairy plant in the province of

Quebec (Canada) (Mulligan et al. 1996). The reactor is composed of a shell, plates, parallel feed entrances, and lateral gas exits (Figure 4.26). The reactor works in the following manner:

- The liquid to be treated enters the reactor by parallel entrances;
- The gas formed during the treatment rises through the biomass and hits the plate;
- A portion of the gas continues its ascent to another compartment while the rest of the gas leaves the reactor by a gas exit situated under the same plate;
- The gas that passes through the plate continuously cleans passage points and displaces all points of the biomass bed.

This movement prevents the formation of dead spots and short-circuiting in the reactor. Typical total COD removal rates of greater than 93% and soluble COD removal rates of 98% are achieved. Other effluents have also been pilot tested with this reactor as shown in Table 4.7.

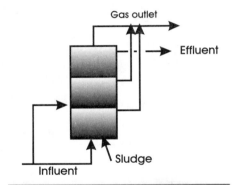

Figure 4.26 Schematic and Photograph of a Full-scale 450 m³ SNC Multiplate Reactor at a Dairy Plant

Table 4.7 Performance of the SNC-Lavalin Anaerobic Reactor for Various Types of Wastewater (Mulligan et al. 1997)

Parameter	Brewery	De-icing agent	Potato processing	VOCs	Whey permeate (full scale)
Organic load (kg COD/m³-day)	17.0	18.6	16.2	13.8	14.5
Inlet total COD (mg/L)	10,000	—	8,080	—	32,900
Inlet soluble COD (mg/L)	8,800	8,256	2,300	4,854	30,271
Inlet total BOD (mg/L)	5,747	6,500	4,450	2,910	—
Outlet total COD (mg/L)	1,142	—	1,160	—	2,208
Outlet soluble COD (mg/L)	542	656	370	395	628
Outlet total BOD (mg/L)	245	383	280	158	—
Total COD removal (%)	88.6	—	85.6	—	93.3
Soluble COD removal (%)	93.8	92.0	83.9	91.8	97.9
Total BOD removal (%)	95.7	94.0	89.3	94.6	—
Gas production (m³/kg COD converted)	0.39	0.44	0.27	0.38	0.41
Methane (%)	80	80	74	76	70

Anaerobic Filter

Since there is only a small amount of solids in the effluent from anaerobic filter processes, settling tanks are often not required. In addition, these filters are used as pretreatment process. The most common type is the fixed filter in which plastic filter media is submerged and operated in the upflow mode. Other less common types include the fluidized bed, expanded bed, and rotating disc. A large recycle is used to keep the fluidized and expanded bed processes moving. However, some hydraulic mixing can occur due to the formation of gas bubbles, which can inhibit separation of water and granules at the top of the filter or cause sloughing off of the biofilm from the carrier. Retention times are in the order of one to two days for wastewater with COD up to 20,000 mg/L. Up to 85% removal is achieved.

Fixed-film Reactors

In fixed-film reactors, the bacteria are attached and grown on a surface which allows SRT of up to 100 days to be achieved. Thus, these types of systems are very stable, despite varying flow rates and influent characteristics. They can work on a five-day feeding schedule (i.e., weekend shutdowns) and can be left dormant for periods greater than six months. Activity can then be returned within one or two weeks. They can even perform well at low temperatures for influent of low concentration. Therefore, full-scale domestic sewage treatment is promising. Start-up is slower than for conventional suspended growth systems but faster than UASB reactors.

Upflow Fixed-film Reactors

Upflow fixed-film reactors are also called biofilters even though filtration is not significant. A large surface area and high porosity, both desirable characteristics, are provided by rocks or plastic support media. Wastewater at low flow rates enters the bottom of the reactor and flows upward. Bacteria grow as clumps within the channels and as a biofilm on the surface of the media.

Suppliers of the upflow fixed-film, random-packed design reactor are Hoeschst Celanese Chemical Group (Princeton, New Jersey) and Raytheon Engineers & Constructors (Lexington, Massachusetts). In an upflow fixed-film reactor, the wastewater is pumped from the equalization tank and mixed with recycled reactor effluent. The mixture is heated, and pH buffering and nutrients are added as required. The water is pumped into the bottom of the reactor where the microorganisms are attached to the packing. This ensures a long solid retention time. A biomass control technology is used to prevent biomass over accumulation. The biogas (with a heat value of 6,300–7,100 kcal/m³) is compressed and used in a furnace or boiler or sent to a flare. The treated wastewater is either recycled or discharged. The economics has been estimated by Celanese and Raytheon in 1995 for an 18,000 kg COD/day system as $143/kg COD per day. Caustic use per kg/day was estimated at 0.05 kg, electricity at 0.13 kwh, and steam at 0.3 kg.

Downflow Fixed-film Reactors

The downflow stationary fixed-film reactor (DSFF) was developed by the National Research Council of Canada in the 1980s by Dr. Bert Van den Berg (Hade 1998). The influent is fed into the top over bacteria which grow on vertical surfaces, and the effluent is removed at the bottom. The vertical surfaces are usually made of clay or fibrous polyester. Operating conditions are similar to the upflow reactor. However, there are fewer clogging problems due to the presence of large channels of 1 to 2.5 cm with specific surface areas of 100 to 150 m^2/m^3. Void volumes are 60 to 90% (Kennedy and Droste 1991). Gas production can provide mixing to the system and prevent solids from settling on the bottom. At solids concentrations higher than 3%, recycling may be necessary, however, to prevent settling. The maximum amount of biomass that can accumulate is in the order of 15 kg VSS per cubic meter of reactor volume.

Fluidized Bed Reactors

Expanded or fluidized bed reactors are the most recently developed anaerobic technologies. Particles such as sand, high density plastic beads, styrene and polyvinylbenzene beads, crushed rock, or granular activated carbon (1.35 g/cm^3) of diameters 0.1 to 0.7 mm are suspended by the liquid velocity (Iza 1991; Fox et al. 1990). As the velocity of the liquid increases, the greater the extent the expansion of the particles. Fluidized beds are expanded to a greater degree and have more particle movement than expanded bed reactors. Due to the intense movement and friction between the particles, biofilm formation is kept to a minimum. This increases the mass transfer of substrates, nutrients, and microbial by-products across the biofilm. Expansion also reduces the possibility of plugging and provides a large surface for bacterial growth. Disadvantages for this type of reactor include the extended period for biofilm formation, the sophistication of the process, the increased energy costs of the pumping system, and the perfectly uniform fluidization of liquid distribution necessary.

Anaerobic Sequencing Batch Reactor

The Anaerobic Sequencing Batch Reactor (ASBR) is an anaerobic version of the conventional SBR technology (Lemna Technology, St. Paul, Minnesota) that is applicable for high-strength wastewaters. ASBR can remove 75 to 94% COD with hydraulic retention times of 8 to 24 h. The age of the biomass is 60 to 70 days. The four cycles of fill, react, settle, and decant operate on 3 to 12 h cycles. Operation is based on timing. Due to the batch-fed operation, short circuiting does not occur. The biomass is highly granulated and contains many bacterial species and fungi with mineral deposits. These granules settle rapidly at a rate of a meter per minute.

Two-Phase Digestion

Since the pH optima of acid-forming and methane-forming bacteria are different, separate reactors can be designed for each step for optimal performance. The formation of the volatile acids would occur in the first reactor (pH between 5 and 6) and the methane production in the second (pH around 7). This appears to increase overall treatment efficiency (Ghosh 1990). Since there is an increase in the capital cost due to an additional reactor, there are only a few installations employing the two-phase approach.

Anaerobic Ponds

Anaerobic ponds contain insignificant levels of oxygen. Like the anaerobic reactors, acid formation occurs followed by methane production. They are used when the wastewater concentrations are high (up to 800 g BOD/m³-day) and as a first step to remove up to 50% of the organics—more in warm climates. Temperatures below 10°C significantly reduce metabolism. Detention times up to 50 days may be necessary to achieve 50% BOD removal at loads of 40 g BOD/m³-day. Facultative ponds can become anaerobic if the loading rate increases.

Operating Conditions

Upflow Fixed-film Reactor

Pall rings, rock or plastic balls, activated carbon, sintered glass, Raschig rings, and flexirings are examples of various types of media. Crossflow or tubular media can also be used. Void volumes of rock or plastic balls (20 mm diameter) are about 40%, but this can be increased to 90% by using the more expensive flexirings. Velocities of up to 20 m/d have been achieved with sintered glass media (Anderson et al. 1994). Biomass washout was not significant. In general, diameters for the media are usually in the 20 to 170 mm range (Table 4.8).

Table 4.8 Operating Conditions of Full-scale Upflow Anaerobic Reactors in North America

Parameter (unit)	Range
Reactor height (m)	3 to 12.2
Temperature (°C)	32 to 37
Influent COD (mg/L)	2,500 to 24,000
Loading (kg COD/m³-day)	1.5 to 15
Recycled flow: influent flow	0 to 10.1
HRT (h)	20 to 96
COD removal efficiency (%)	61 to 90

Anaerobic ponds

Typical operating conditions for anaerobic ponds are shown in Table 4.9. Oswald (1968) determined that loads above 85 kg BOD$_5$/ha-day lead to anaerobic conditions. Anaerobic ponds are usually for industrial wastes, not municipal wastewaters. They are 2.5 to 5 m in depth and black with bubbles of methane and carbon dioxide at the surface. Odors can be a problem, particularly from hydrogen sulfide, which can also be toxic to the bacteria if levels are too high.

Table 4.9 Typical Operating Conditions for Anaerobic Ponds (Henze 1997)

Loading rate (kg/ha-day)	Depth (m)	Retention time (day)	Removal (%)	Application
1,120–3,360	2.4–3.0	2–7	70–80	Various ponds of industrial wastes in India
404	0.9	60	70	Chemical, Texas
439	1.8	15	51	Canning
1,411	2.2	16	80	Meat and poultry
388	1.8	18.4	50	Paper
1,604	1.8	3.5	44	Textile
179	1.8	245	37	Rendering
3.360	1.3	6.2	68	Leather

Anaerobic processes consist of three stages: hydrolysis, acid production, and methane production. A disturbance in any of these steps can influence the whole process. For example, an immediate introduction of a higher concentration of organics can cause problems with methane production since the acid producing stage may not be sufficient. This phenomenon can be seen within one retention time after the disturbance. Methane bacteria can also wash out when their sludge age becomes low. This can occur due to higher contents of suspended solids in the influent, loss of sludge from inadequate sludge retention, or sloughing off of the biomass from the filter material.

Inhibition of the bacteria can result from several means. Higher concentrations of volatile acids can be from increased organic load, temperature variations, pH increase or decrease, or inhibition by hydrogen sulfide, ammonia, or other toxic substances. If the bacteria can adjust quickly to the introduction of a substance such as cyanide, it can tolerate multiple doses of high concentration (e.g., greater than 100 g/m^3) better than single doses (greater than 5 g/m^3). If the bacteria cannot adapt, as in the case of nickel and other metals, high concentrations of a single dose (greater than 200 g/m^3) can be tolerated more easily than multiple doses at lower concentrations (around 50 g/m^3).

Temperature

Optimal operating temperatures are usually in the range of 30 to 40°C for mesophilic methane-producing bacteria or 50 to 60°C for thermophilic bacteria. Methane production can occur at temperatures lower than 10°C; however, above 20°C is preferred. In the mesophilic range, each increase of 10°C doubles the methane production. Due to high energy requirements, operation is rarely in the thermophilic range. Stable temperatures are a requirement for stable operation; as the temperature decreases, so do the loading rates required to obtain the same level of treatment efficiency.

pH

As previously mentioned, the optimal pH range for methanogenic bacteria is between 6 and 8, while other anaerobic bacteria prefer close to pH 7. Due to the production of organic acids and carbon dioxide, alkalinity of pH control must be sufficient to prevent shifts in pH. Volatile acid concentrations between 100 mg/L and 5,000 mg/L can be tolerated if pH control is adequate. Alkalinity can be provided with lime, sodium bicarbonate, sodium carbonate, or sodium hydroxide. Mixing is important for distribution of buffering agents for pH control.

Ammonia and Sulfide Control

High concentrations of ammonia can inhibit anaerobic treatment processes, as can fluctuating conditions. Ammonia concentrations of 50 to 200 mg/L are beneficial, of 200 to 1,000 mg/L do not affect the process, of 1,500 to 3,000 mg/L are inhibitory at pH above 7.4, and above 3,000 mg/L are toxic (Kugelman and Chin 1971). Free ammonia causes an increase in pH and is more toxic than ammonium ions. Acid addition can offset the detrimental effects. Ammonia is produced during the degradation of proteins. Alkalinity can be added to the reactor influent by recycling the reactor effluent from protein degradation. Sufficient alkalinity can increase the tolerance of methanogens up to ammonia concentrations of 5,000 mg/L. Although proteins do not usually increase the pH significantly, waste containing blood can produce high levels of ammonium bicarbonate; thus, the addition of acid is required. The effect of other elements is shown in Table 4.10.

Since sulfate serves as an electron acceptor in anaerobic processes, sulfide is produced, which can inhibit methane-production. Fixed-film processes are more stable than contact processes at higher hydrogen sulfide concentrations. For example, 60 to 75 mg/L of hydrogen sulfide can affect contact processes with organic loads of 0.25 to 0.50 kg COD/m³-day compared to filter systems, which are only affected at hydrogen sulfide concentrations above 200 mg/L at organic loads of 1 to 5 kg COD/m³-day (Maillacheruvu et al. 1993). Wastes high in sulfate (200 mg/L) can also cause toxicity problems (Parkin and Owen 1986). However, the addition of metals such as iron can precipitate the sulfides to decrease sulfide toxicity (Gupta et al. 1994).

Table 4.10 Elements Toxic to Anaerobic Treatment

Element	Concentration (mg/L)		
	Beneficial	Little effect	Inhibitory
Calcium	100–200	2,500–4,500	8,000
Magnesium	75–150	1,000–1,500	3,000
Potassium	200–400	2,500–4,500	1,2000
Sodium	100–200	3,500–5,500	8,000

Adapted from Samson and Guiot (1990)

Nutrients

Compared to aerobic bacteria, nutrient requirements are lower for anaerobic bacteria. Although the composition of aerobic and anaerobic bacteria are about the same ($C_5H_7NO_2$ with phosphorus one-fifth that of nitrogen on a mass basis), sludge production is significantly less for anaerobic processes (approximately 20%). On a mass basis, the COD:N:P ratio for aerobic activated sludge is 100:5:1 whereas for anaerobic systems, COD:N is around 700:5 or 250:5 for highly-loaded processes (0.2 to 1.2 kg COD/kg VSS-day). Trace elements such as nickel and cobalt can enhance methane production (Murray and van den Berg 1981). Many wastes, however, contain sufficient amounts of these elements.

Design

Process design is usually based on a COD volumetric load or COD sludge load. To calculate the volume of the anaerobic tank required, V in m³, the following equation can be used:

$$V = \frac{Q \times C_{COD}}{B_v}$$

where Q is the flow rate of the wastewater (m³/day), C_{COD} is the concentration of the wastewater in COD (kg/day), and B_v is the volumetric load (kg COD/m³-day). Typical loading rates for the various types of reactors are shown in Table 4.11.

Table 4.11 Typical Volumetric Loads of Various Types of Anaerobic Reactors

Reactor type	Volumetric load (kg/m³-day)		
	15–25°C	30–35°C	50–60°C
Contact	0.5–2	2–6	3–9
Sludge blanket	1–3	3–10	5–15
Fixed filter	1–3	3–10	5–15
Rotating disc	1–3	3–10	5–15
Fluidized filter	1–3	4–12	6–18
Expanded filter	1–4	4–12	6–18

Adapted from Henze 1996b

Designing by sludge load can be performed by using the following equations:

$$Bx_{COD} = Q \times \frac{C_{COD}}{Mx}$$

where Bx $_{COD}$ is the sludge load based on COD (kg COD/kg VSS-day), Q is the flow rate of the wastewater (m³/day), C$_{COD}$ is the concentration of the wastewater (kg COD/m³), and Mx is the sludge mass (kg VSS). The mass of the sludge can be determined by:

$$Mx = VxX$$

where X is the sludge concentration in the reactor (kg VSS/m³). Therefore, the reactor volume can be calculated from:

$$V = Q \times \frac{C_{COD}}{X \times Bx_{COD}}$$

The sludge load does not vary according to the reactor type but is dependent on the type of wastewater to be treated. For example, to achieve 80 to 90% COD removal efficiency at 30 to 35°C, sludge loading rates of between 5 and 10 kg COD/kg VSS-day are used for acetic acid, 0.7 to 1.5 kg COD/kg VSS-day for dissolved organic matter, and between 0.1 to 0.3 kg COD/kg VSS-day for wastewater containing suspended matter (Henze 1996b).

The concept of sludge age or solids retention time (SRT) is also very important since this determines how long the sludge is in the reactor. Increasing the SRT can be accomplished by using recycle. This was not done until the 1950s with the implementation of the anaerobic contact (suspended growth recycle process). Without recycle, the HRT and the SRT are the same and very large reactors are required. However, with the new processes, the SRT can be controlled by promoting high SRTs which enables the sludge concentration to increase in the reactor. The microorganisms must be allowed to grow without a washout. As in the aerobic systems, SRT = Mass of sludge in the reactor/ mass removal rate of sludge from the reactor. In a suspended growth reactor, the parameters for anaerobic sludge formation are as follows:

$$\frac{1}{SRT} = \frac{Q_w X_w}{V X_v}$$

where Q$_w$ is the volumetric flow rate of sludge from the reactor, X$_w$ is the concentration of sludge in Q$_w$, V is the volume of the reactor, and X$_v$ is the VSS in the reactor. Q$_w$ is equal to Q, the flow rate into the reactor, unless solids recycle is used. The minimum SRT that can be used for methane-producing bacteria is three to five days at 35°C. In practice, a safety factor of three to 20 times the SRT is used for stable operation (Lawrence and McCarty 1969).

Increasing the solids in an anaerobic reactor can be accomplished by separating the solids from the reactor effluent and recycling to the reactor, fixing the bacteria on solid supports within the reactor, developing a dense sludge blanket which is retained by gravity in the reactor, or operating the reactor at long retention times. The first three methods have been used successfully in recent years. The first two methods are used in aerobic activated sludge and trickling filter processes, respectively. The third method is used only in anaerobic processes. Proper operating conditions must be used to obtain very dense granules. One disadvantage of the third method is that inert solids may also accumulate in the reactor. The last one is typical of conventional sludge digestion units and is not practical for low-strength wastewaters since large reactor volumes would be required.

Gas Production

During anaerobic processes, methane makes up 60 to 80% and carbon dioxide the remaining 20 to 40% of the total gases produced by the bacteria. Hydrogen, hydrogen sulfide, and nitrogen are also produced but in much smaller quantities. From a COD balance, approximately 0.35 m³ methane is produced from 1 kg of COD. In general, approximately 90 to 95% of the COD is converted to methane while the remaining is used for sludge production. Due to the high concentration of methane in the biogas, it can be used for generation of electricity or heat. The production of methane can be determined by the equation:

$$Q_m = Q(COD_{in} - COD_{out})M$$

where Q_m is the amount of methane produced per time, Q is the influent flow rate, COD_{in} is the total amount of COD in the influent, COD_{out} is the total amount of COD in the effluent, and M is the volume of methane produced per unit of COD removed. Values for M can vary from 0.1 to 0.35 m³/kg COD.

Overall treatment efficiencies can vary from 60 to 90% depending on the type of COD to be treated. For example, if the COD consists of seeds and skins, these cannot be removed by anaerobic treatment processes. Laboratory treatability studies are always a good practice to determine the biochemical methane production potential (BMP) test and toxicity assay (Owen et al. 1979).

A minimum influent COD of 1,000 mg/L is generally accepted in determining an appropriate wastewater for anaerobic treatment. Below this level solid losses can be critical due to the low substrate concentrations. Increasing the substrate concentration will enhance the treatment efficiency and allow more methane to be produced per unit volume per unit time.

Start-up Procedures

Start-up of anaerobic processes takes between 30 and 60 days, which corresponds to two to four times the sludge age. The reactor is first seeded with the sludge from another reactor and water is recycled, usually for about a week. Then an organic load is fed to the reactor at 10% of the maximum load. The load is increased by 50 to 100% as the volatile fatty acid content of the effluent is around 100 to 400 mg/L (acetic acid basis).

Applications

Organic loads of 5 to 30 kg COD/m³-day and media with specific surface areas of 100 m²/m³ are usually employed. HRT values are usually about one day. The reactors work best on influents with COD concentrations between 1,000 and 30,000 mg/L, with soluble to insoluble COD ratios greater than one and with suspended solids levels lower than 500 mg/L.

Other information on North American installations is shown in Table 4.12. Applications for the treatment of toxic components are shown in Table 4.13.

Low-rate Anaerobic Reactor

The ADI-BVF® reactor has been applied for the treatment of wastewater from a candy bar manufacturing plant in Canada (Cocci et al. 2001). The average wastewater flow is 1,135 m³/day with BOD of 1,600 mg/L, TSS of 410 mg/L, and oil and grease of 200 mg/L. With the system, the effluent limits of 300 mg/L BOD and 350 mg/L TSS are achieved.

Another system comprising a screen, raw waste pumping station, equalization tank, calamity tank, ADI-BVF reactor, and ADI-SBR reactor was installed at a dairy in Virginia, USA (Cocci et al. 2001). The pH of the influent wastewater ranged from 2 to 12, and the temperature from 15 to 31°C. The average TSS was 1,000 mg/L, and oil and grease was approximately 1,200 mg/L. Despite these conditions, the effluent conditions were 30 mg/L BOD, 30 mg/L TSS, and 20 mg/L oil and grease. This enabled the effluent to be discharged to a publicly owned treatment works (POTW) for wastewater treatment.

Table 4.12 Several Anaerobic Treatment Processes and the Companies That Developed Them

Type of process	Name of process	Company Name
Contact	ANAMET	A.C.
Contact	—	Biotechnics
Contact	—	ADI
Low-rate	ADI-BVF®	ADI
Contact	Bioenergy	Biomechanics
Contact	MARS	Dorr-Oliver
Fixed-film	Bacardi	Bacardi
Filter	Celrobic	Badger
UASB	Biopak	PAQUES
UASB	Biothane	Biothane
EGSB	BIOBED	Biothane
Sequencing batch	ASRB	Lemna
Fluidized bed	Anaflux	Degremont
Fluidized bed	ANITRON	Dorr-Oliver
Fluidized bed	Hy-Flo	Ecolotrol
Fluidized bed	Gist-Brocades	Gist-Brocodes
Multiplate reactor	Multiplate Reactor	SNC-Lavalin

Table 4.13 Applications of Anaerobic Systems Treating Toxic Components

Component	Reference
Aliphatic hydrocarbons and chlorinated alcohols	Blum and Speece (1991)
Pentachlorophenol	Wu et al. (1991)
Nitroaromatic compounds	Donlon et al. (1996)
N-substituted aromatics	Razo-Flores et al. (1996)
Alkylphenols	Razo-Flores et al. (1997a)
Azo dyes	Donlon et al. (1997)
	Razo-Flores et al. (1997b)

UASB

The use of UASB systems has been well established for a wide variety of applications including the treatment of brewery, fruit and soft drink, yeast, pulp and paper, starch, potato processing, sugar, wine and alcohol, pharmaceutical, and other food-processing wastewaters. Disadvantages for UASB processes include long start-up, production of odorous compounds, and bacterial sensitivity to pH, temperature, and toxic compounds. Although chemical addition may be necessary for industrial effluent treatment, this is not usually the case for domestic wastewater and sewage (van Haandel and Lettinga 1994). The bacteria adapt well to low temperatures and can tolerate some toxicants such as aliphatic hydrocarbons and chlorinated alcohols even better than aerobic bacteria (Blum and Speece 1991). UASB reactors have been used to degrade pentachlorophenol (PCP) with up to 99% efficiency (Hendrikson et al. 1992). They have also been used for nitroaromatic compounds (Donlon 1996). Start-up times can be reduced by using adequate inoculum such as digested sludge or biomass from operating anaerobic reactors, particularly if lower operating temperatures are used (Singh et al. 1997).

UASB reactors are suitable for organic loads of 0.5 to 20 kg/m^3-day which is higher than aerobic processes (Kato et al. 1994). This reduces reactor volume and space requirements. UASB reactors can be used for high-strength wastewaters with VSS:COD ratios less than one and with COD concentrations between 500 and 20,000 mg/L. The HRT can be less than 24 h. Cases in which wastewaters are lower strength with less than 1,000 mg/L have seldom been reported. They have been used for domestic wastewater treatment with 4 h retention times at 18 to 28°C (Barbosa and Sant'Anna 1989). TSS concentrations of 1,000 to 2,000 mg/L are ideal for granule formation and, thus, for reactor functioning. Ammonia nitrogen loads in excess of 2.4 to 3.4 kg NH_4^+-N/m^3-day and low divalent concentrations can be unsuitable for the reactor performance. The UASB reactor can replace the primary settler, anaerobic sludge digester, the aerobic step, and secondary settler in a conventional aerobic treatment plant. The UASB reactor effluent, though, usually requires polishing to remove some organic matter, nutrients, and pathogens. This can be done in conventional aerobic systems such as stabilization ponds.

In Kanpur, India, a full-scale UASB reactor has been treating 5,000 m³ of raw sewage per day since 1989 (Draaijer et al.1992). Another unit was subsequently built in the same town to treat 36,000 m³/day (Haskoning 1996a). Average loadings were 2.5 kg COD/m³-day with COD, BOD, and TSS removals of 50 to 70%, 50 to 65%, and 45 to 60%, respectively. Based on these results, a pond with one-day retention was added for installation in another Indian town (Mirzapur) (Haskoning 1996b). With loading rates of 0.95 kg COD/m³-day on the UASB reactor and 0.13 kg COD/m³-day for the polishing ponds, the final effluent conditions of 30 mg/L and TSS of 10 mg/L could be achieved. Temperatures were between 18 and 32°C. Overall COD, BOD, TSS removal rates were 81, 86 and 89%, respectively. Effluents with COD values ranging from 1,000 to 120,000 mg/L have been treated with 55 to 90% efficiencies and with loading rates between 5 and 35 kg/m³-day.

Expanded Granular Sludge Bed (EGSB)

A 205 m³ unit has been operating in the Netherlands since 1989 at temperatures between 16 and 19°C with HRT of between 1.5 and 5.8 h COD and BOD removal rates of 30 and 40%, respectively, and with no TSS removal (van der Last and Lettinga 1992). Low-strength wastewaters (Kato et al. 1994) as well as high-strength wastewaters, which are diluted due to the recirculation stream, are treated efficiently in these reactors. For the low-strength wastewaters such as sewage, recirculation is not necessary. UASB reactors behave like static beds whereas the EGSB reactors are similar to mixed tanks (Rinzema 1988). This phenomenon increases the organic loading that can be handled by the latter type of reactor.

The trademarked Biobed system of Biothane is an EGSB reactor. Unlike other EGSB reactors, no carrier material is used to retain the biomass within the reactor. Therefore, it is considered as a high-rate UASB reactor or a fluidized bed reactor. Since it is higher rate than the UASB reactors, its space requirements are less. It can be used for brewery, chemical, fermentation products, and pharmaceutical industries. It is simple, containing only the settlers on the top of the reactor and the feed distributors at the bottom. Loadings are between 15 and 30 kg COD/m³-day. Due to the high circulation rates, inhibitory components are diluted and more easily treated. There are approximately ten installations in the United States. The highest capacity reactor (26 kg COD/m³-day) was installed in 1996 at the Redhood Brewery while the largest reactor in the U.S. is at an Anheuser-Busch Brewery (7,800 m³) which treats 137,000 kg COD/day at a capacity of 17.6 kg COD/m³-day.

The Internal Circulation (IC®) reactor technology of Paques B.V. (The Netherlands) is a type of EGSB reactor particularly applicable for the pulp and paper industry and the brewery industry. It consists of two UASB reactors, one on top of the other (Figure 4.27). One reactor is for high loads, and the second is for low loads. The four basic processes in the reactor are the mixing section, expanded bed section, the polishing section, and the recirculation system. The gas from the first stage drives a gas lift and internal circulation. The biogas lifts the sludge and water to the upper section toward the gas/liquid separator at the top where the gas separates from the water/sludge and leaves the reactor. The water/sludge then proceeds downward. The reactor can be up to 25 m in height with a small surface area. The upper velocity in the first section is 10 to 20 m/h and 2 to 10 m/h in the next section. The biogas is collected in the top of the reactor.

At a kraft mill in Alabama, the reactor treats foul condensates from pulp digesters and evaporators (Cocci et al. 2001). The reactor has a 6.5 m diameter and a water depth of 20 m. The wastewater contains mainly methanol, sulfur compounds, and terpenes. Methanol removal was 98% during a retention time of 6 to 7 h. The influent COD is from 4,500 to 5,000 mg/L and is reduced by 75 to 80%. The effluent is sent to aerated lagoons at the mill.

Anaerobic filter

The Hyuandai (Korea) anaerobic filter reactor (HAF) is a high-rate reactor to treat high-strength wastewater from petrochemical, canneries, dairies, breweries, and so on. The water is collected in an equalization tank for nutrient addition and/or pH adjustment (Figure 4.28). The water passes through a heat exchanger for temperature control before entering the reactor at the bottom. The reactor is of cylindrical or tetrahedral shape with packed media. Methane bacteria adhere to the media providing a long solids retention time (100 to 300 days). The treated water is then passed into a buffer and sedimentation tank before discharge. The biogas can be used as a boiler fuel gas after compression due to its high heat value (4,800 to 5,100 kcal/m³). Typical BOD removal rates are 92 to 98% for 5,000 to 50,000 mg BOD/L effluents.

Downflow Fixed-film Reactor

There are installations of 45 and 410 m³ in Canada and 15,300 m³ in Puerto Rico

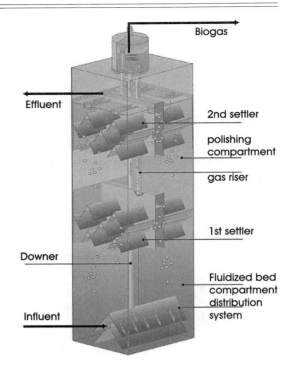

Figure 4.27 Expanded Granular Sludge Bed (EGSB) IC Reactor Developed by Paques (Courtesy of Paques)

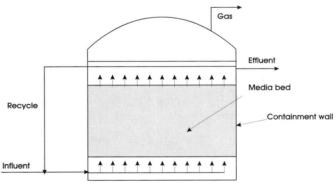

Figure 4.28 Anaerobic Filter Reactor

(Bacardi Corp.) (Hall et al. 1986). The 410 m³ installation in Canada used 18,500 blocks of cut earth of 250 mm x 250 mm x 290 mm as bacterial supports (Hade 1998). The canal openings were 42 mm x 42 mm to give 10% porosity. Bacteria were inoculated from a municipal anaerobic digester. A retention time of 12 h was employed to give an organic load of 7 kg COD/m³-day of dairy wastewater fed into the reactor. At the beginning, the COD removal rate was 70% but this decreased as the useful volume decreased to 30% due to difficulties in arranging the blocks. It was then decided to convert the reactor to a UASB.

Fluidized Bed

Fluidized beds have been tested at laboratory scale for the treatment of domestic sewage at ambient temperatures. They are applicable for the treatment of influents with COD concentrations in the range of 1,000 to 30,000 mg/L with soluble to insoluble COD ratios greater than one. The reactors operate best at loads between 1 to 30 kg COD/m^3-day with HRT of 9 to 24 h. Full-scale fluidized bed reactors have been installed in Denmark (Enso-Fenox by Enso-Gutzeit, Finland), a 360 m^3 ANITRON (Dorr Oliver) in the United States, a HY-FLO of 120 m^3 in the United States, two 250 m^3 and 100 m^3 in the Netherlands by Gist-Brocades (The Netherlands), and a 210 m^3 Anaflux (by Dégremont, France) in France.

Costs

Paques B.V. (The Netherlands), one of the major suppliers of UASB reactors, has over 250 full-scale units worldwide. In 1996, they estimated that capital costs for their systems were in the order of $160 to $260 per kg COD/day. Annual operating costs are $7 to $10 per kg COD/day. The biogas produced is 80% methane with a calorific content of 6,100 kcal/m^3. Richards (1996) also performed an analysis of the costs of a UASB system for a brewery (18,900 m^3/day with a BOD of 1800 mg/L). He estimated that the operating costs would be in the order of $0.08/$m^3$, and the capital costs would be $449 per m^3/day. Revenues for gas production would exceed the operating costs ($0.11 /$m^3$). The process has been used for chemical, pulp and paper, and many other industries.

The cost of the anaerobic filter is approximately $230 to $280 per kg BOD per day. Twelve full-scale units have been constructed for the treatment of ethylene glycol, polyester, and teraphthalic acid wastewaters.

Case Study

A full-scale multiplate reactor (450 m^3) was designed, constructed, and started-up in June 1992 at the Nutrinor plant (Mulligan et al. 1996). This was the first full-scale unit of the reactor that was first developed at the Ecole Polytechnique of the University of Montreal (Montreal, Quebec). In 1986, SNC Research Corporation obtained the commercialization rights from the university and subsequently built and tested the pilot unit.

The process includes pumping the domestic wastewater and whey permeate to the treatment unit from the plant. The wastewater is pumped via the pumphouse to the buffer tank after passing through a rotating 0.5 mm screen and settling tank. The permeate produced from the ultrafiltration of cheese whey can be stored adjacent to the rectangular buffer tank (constructed of concrete lined with epoxy).

The permeate is transferred to the buffer tank at a specific ratio of wastewater to permeate. The retention time is 24 h and constitutes the phase where the acidification takes place. The first stage of the biological conversion includes the hydrolysis of the polysaccharides and fatty material in the whey permeate to volatile fatty acids including acetic, butyric, and propionic acids. The pH is controlled via a pH analyzer/transmitter and by soda ash addition. The temperature is maintained at 37°C with steam injection as required.

The effluent is then pumped to the anaerobic multiplate reactor with a working volume of 400 m³. The reactor is cylindrical with a diameter of 8 m and a height of 9 m and is constructed of stainless steel coated with epoxy. Three horizontal plates separate the reactor into four sections.

Construction of the demonstration unit at Nutrinor began August 1991 and was completed by February 1992 when the process of start-up was initiated. By March of 1992, several modifications and improvements were completed. The first stages of experimentation took place in April, and by July continuous operation was under way. By then 219 m³ (12,000 kg VSS) of granular anaerobic sludge was loaded into the reactor.

Batch operation was started by filling the buffer tank (200 m³) with wastewater and permeate leaving the mixture for 24 h, and then feeding the mixture into the reactor where it remained two or three days. This process was repeated as the cycle finished.

Continuous operation was initiated by filling the buffer tank with wastewater (100 m³) and 10 m³ of permeate for a period of three to five days per week due to a limitation in the quantity of permeate. Once this problem was resolved, seven-day operation began. Organic loading subsequently increased from 2.5 to 15 kg COD/m³-day by increasing the ratio of permeate to wastewater from 0.2 to 1.3. The objective of 6,000 kg COD/day was reached at this feed rate.

The total and soluble COD were monitored in the buffer tank and at the effluent of the reactor for a period of 24 days from day 99 to day 123 of continuous operation. The results (shown in Table 4.7) are comparable to the results during the transition phase. Reducing the pH of the buffer tank from 6.5 to 4.5 decreased the quantity of soda ash from 1.5 tonnes per day to less than a tonne. However, methane content decreased from 67.5% to 57.5% and the flow of gas increased from 98.0 m³/h to 126.7 m³/h. A couple of years later, the pH was increased in the buffer tank to 6.5 and a recycle from the reactor effluent to the reactor base was introduced to conserve soda ash consumption.

Hydrological testing to determine the presence of dead spots and short-circuiting in the reactor was performed. The results of the tests indicated that the reactor is mixed similarly to a series of perfectly mixed-reactors in a cascade due to the parallel feeding. In addition, there were no dead zones in the reactor at start-up and throughout the functioning of the reactor. The following are several conclusions:

- The design of the reactor is simple, without dead spots or short-circuiting;
- Start-up of the reactor proceeded without significant problems;
- COD conversion of the plant wastewater and whey permeate was above 90% for an organic loading of up to 15 kg COD/m³-day. Since the COD removal values are similar to those of the mobile unit, it is expected that a much higher organic loading could be handled by the full-scale reactor;
- Operation of the reactor was very stable, resistant to organic shock, and performed in a superior manner to other previously reported on reactors;
- The Multiplate Reactor can be successfully scaled-up from 1.2 to 450 m³; and

- Due to the large amount of methane produced (1.2 million cubic meters per year), which can be recycled to heat the plant, the operation of the reactor provides a clear economic incentive.

The system has operated successfully with no problems for the past nine years.

Comparison to Other Processes

The advantages and disadvantages of the various anaerobic systems are summarized in Table 4.14. Full-scale systems have been quite successful.

There are many misconceptions about anaerobic treatment that limit its use. These problems have arisen due to poor designs before the 1950s and a lack of understanding of anaerobic processes. Many times anaerobic sludge digestion is compared to aerobic wastewater treatment. Anaerobic digestion is a slow process since biological solids are treated whereas in the aerobic process—also a slow process—easily degraded soluble organics are removed. Treatment of soluble organic matter by anaerobic processes can be as fast as or faster than aerobic ones.

On an energy basis, even though aerobic bacteria can extract 14 times more energy from a sugar substrate than anaerobic bacteria, the aerobes use the energy to produce sludge which often becomes a disposal problem. The chemical energy is transformed by anaerobic bacteria into methane, a good fuel. In addition, anaerobes do not require oxygen, thus reducing operating costs.

In terms of treatment efficiency, there is little difference in the biodegradability of many wastewaters by aerobic or anaerobic means. Although the effluent quality is usually poorer for anaerobic processes, the overall removal is about the same, particularly for high-strength wastes.

Anaerobic processes have also been thought to be unstable. This was the case in the older reactors where mixing was inadequate. Poor mixing, dead zones, or short-circuiting can reduce treatment efficiencies and stabilities significantly and may even lead to complete failure of the system. Instability can also be caused by insufficient detention times and low microbial concentrations. The newer designs promote stability with large inventories of microorganisms, which also eliminates additional mixing requirements.

Another misconception about anaerobic processes is that they are more sensitive to toxic substances than aerobic ones. This concept also originated from anaerobic sludge digestion processes for municipal wastewater systems. However, when a wide variety of aerobic heterotrophs and methanogens were compared for their sensitivities to various chemicals, there was no significant difference between them for most chemicals. Methanogens were only more sensitive to chlorinated compounds and alcohols. Methanogens can tolerate some toxicants such as heavy metals better than aerobic bacteria since they form sulfides that remove the metals by precipitation or complexation. In addition, anaerobic systems are usually used for food and beverage industries where toxicity is not a problem.

Table 4.14 Comparison of Anaerobic Treatment Processes

Type of process	Advantages	Disadvantages
Anaerobic contact	Suitable for concentrated wastewaters; easy to mix; high effluent quality achievable	Ability of biomass to settle is critical; system is mechanically complex; suitable for low to moderate levels of TSS
UASB	High biomass concentrations; small bioreactor volumes; high-quality effluent; mechanically simple; well mixed	Special bioreactor configurations required; performance depends on settleable biomass; little process control; little dilution of inhibitors
Anaerobic filter	High biomass concentrations; small reactor volumes; high-quality effluent; mechanically simple; well mixed; not dependent on settleable biomass	Suspended solids can accumulate and plug reactor; not suitable for high levels of TSS; little process control; high cost of media and support; little dilution of inhibitors
Downflow stationary fixed-film (DSFF)	High biomass concentrations; small reactor volumes; high-quality effluent; mechanically simple; compact system; performance not affected by high levels of TSS and not dependent on settleable biomass; well mixed	Biodegradable solids not degraded; high cost of media and support; little process control; little dilution of inhibitors
Fluidized bed/expanded bed	High biomass concentrations; small reactor volumes; excellent mass transfer; often better quality effluent than other processes; most compact of all processes; performance does not depend on settleable solids; well mixed; good process control	Long start-up period; high power requirements; not suitable for high levels of TSS; mechanically complex; cost of media is high; little dilution of inhibitors

Adapted from Grady et al. (1999)

Because of the requirement for a balance between acid-forming and methane-forming, anaerobic processes can be more complex than aerobic ones. Adequate buffering, sludge retention times, and pH control will enable stable operation to be achieved and maintained. Thus, although more monitoring may be required for anaerobic processes (pH, COD, VSS, TSS, alkalinity, volatile acids concentrations, and gas composition), this is not substantial since most monitoring can be done by readily available equipment. The ratio of volatile acids to alkalinity is important and should not be above 0.3 to 0.4 (Hartwig 1981).

It is also believed that anaerobic treatment processes require higher temperatures since the optimal range for methanogens is either in the 30 to 45°C range or between 65 and 70°C. Although aerobic systems are designed to work at ambient temperatures, this is for economic purposes and is not the optimal operating temperature. High temperatures for anaerobic processes require large amounts of energy which may not always be necessary. Full-scale operation in the thermophilic range for anaerobic reactors has not been frequent. Methane production does not vary with temperature, and excellent treatment efficiencies can occur at 20°C (Massé and Droste 1993).

Start-up can indeed take longer periods of time for anaerobic processes due to their lower growth rates. However, start-up time can be minimized by good start-up procedures (Kennedy 1985). Once started however, anaerobic processes can recover quicker from shutdown than aerobic processes. Aerobic bacteria can take substantial periods of time to recover.

A major advantage for anaerobic processes over aerobic is the decreased rate of sludge production. Sludge production can be between three and 20 times less than for aerobic processes (Rittmann and Baskin 1985). The costs of disposal of large amounts of sludge can be substantial. In addition, the processing of the sludge before disposal is energy intensive unless gravity or flotation thickening is feasible. Chemical conditioning and energy-intensive processes such as drying beds, vacuum or pressure filters, and centrifuges may also be required and are expensive. Overall, proper reactor design and operation can overcome any disadvantages of anaerobic treatment. Table 4.15 summarizes aerobic processes and compares them with anaerobic ones.

Natural Wastewater Treatment Systems

Alternatives for wastewater treatment when land is available at low cost are natural or artificial wetlands. In this section, however, only constructed wetlands will be discussed. Municipal wastewaters can be treated in this manner, as can other wastewaters with low concentrations of hazardous materials. These include agricultural runoff, settled sewage, leachate, acid mine wastes, and stormwater runoff. Natural wastewater systems can be useful for polishing wastewaters treated by other processes. Significant BOD and TSS removal are obtained.

Table 4.15 Comparison of Biological Wastewater Treatment Processes

Process	Applications	Advantages	Disadvantages	Cost rating	Time requirement
Activated sludge	Low-concentration organics, refinery, petrochemical wastewaters, some inorganics	Removal of dissolved components; Low maintenance; Destruction process; Relatively safe Low capital costs; Relatively easy to operate	Volatile emissions; Waste sludge disposal; Relatively high energy costs; Susceptible to shock loadings and toxins; Susceptible to climatic changes	+++	*
Aerated lagoons Stabilization ponds	Low-concentration biodegradable organics, some inorganics	Removal of dissolved components; Low maintenance; Destruction process; Relatively safe; Low capital and energy costs; Infrequent waste sludge	Volatile emissions; Susceptible to shock loadings and toxins; Susceptible to climatic changes; High land requirement; No operational control	+	**
Trickling filters, Fixed-film Reactors	Low-concentration biodegradable organics, acetaldehyde, benzene, chlorinated hydrocarbons, nylon, rocket fuel, some inorganics	Removal of dissolved components; Low maintenance; Destruction process; Relatively safe Relatively little waste sludge	Volatile emissions; Waste sludge disposal; Relatively high capital costs; Susceptible to shock loadings and toxins; Susceptible to climatic changes; Susceptible to fouling	+	*
Anaerobic reactors	High-concentration biodegradable organics	Production of methane for energy; Proven technology; Low sludge production	Sensitive to temperature variations; High capital costs	++	**

Cost rating: +denotes operating cost of less than $0.1 /m^3, ++ denotes operating cost of $0.1 to $0.2/m^3, and +++ denotes operating cost of $0.2 to $0.8/m^3. Time requirement rating: * denotes < 1 day, ** denotes 2-10 days, *** denotes >10 days.

Source: Adapted from Belhateche (1997)

Natural Wastewater Treatment Systems: Wetland Systems
Box 4-8 Overview of Wetland Processes

Applications	Suitable for agricultural runoff, sewage, leachate, acid mine wastes, and stormwater runoff and for polishing of secondary effluents of less than 227,100 L/day
Cost	Operating: $0.03 to $0.90 /m^3, Capital: $75,000 to $170,000/ha
Advantages	Low-cost process when adequate land available
	Natural process including plants, peat, microorganisms, and vegetation
	Easy to operate and maintain since passive process
Disadvantages	Reduced treatment rates in cold climates
	Can require large areas of land
	Accumulation of phosphorus, metals, and some organics over time
Organic loading	66 to 154 kg/ha-day
Hydraulic loading	230 to 3,000 m^3/ha-day
Other considerations:	More experience in Europe than North America

Description

Natural wetlands are areas of land with the water surface near that of the land, thus maintaining saturated soil conditions, peat, wildlife, microbial cultures, and vegetation including cattails (*Typha* spp.), reeds (*Phragmites* spp.), sedges (*Carex* spp.), bulrushes (*Scirpus* spp.), rushes (*Juncus*, spp.), water hyacinth (*Eichhornia crassipes*), duckweeds (*Lemna* spp.), grasses, and others detailed by Mitsch and Gosselink (1992). Constructed wetlands have been specifically designed to include these species for the removal of BOD, suspended solids (SS), nutrients, and heavy metals for optimal performance. Denitrification also occurs due to the anaerobic conditions in the water. It was reported by Reed et al. (1995) that 1,000 managed wetlands are in operation throughout the world. In Canada, organizations such as Ducks Unlimited have designed and constructed 1,100 wetlands in Ontario, Canada.

Design

The wetland systems can be designed either as surface flow with a free water surface or as subsurface flow (Figure 4.29) in which the water must enter after passing through a permeable medium. The latter systems are usually more efficient and less prone to mosquito problems.

Water balance is calculated as:

$$P - A + Q_i + G_i - Q_o - G_o - ET = \frac{dV}{dt}$$

where P is precipitation, A is the abstraction from precipitation, Q_i is the surface inflow, G_i is the groundwater inflow, Q_o is the surface outflow, G_o is the groundwater outflow, ET is evapotranspiration, V is the water stored in the wetland, and t is time. All of these factors should be considered over a long period of time when designing a system, particularly if sewage is to be treated. Compacted soil or an impermeable liner is used on the bottom of the wetland to prevent contamination of groundwater (U.S. EPA 1988). Organic soil can significantly reduce infiltration rates. Stormwater runoff can be treated

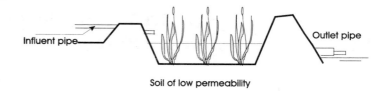

Surface flow

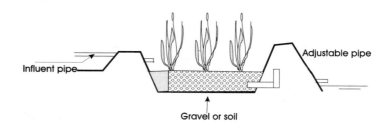

Subsurface flow

Figure 4.29 Main Types of Wetland Systems: Surface Flow and Subsurface Flow Treatment Systems. (adapted from Knight et al. 1999)

via this method. Recently, grassed swales and vegetated filters were used to remove pollutants including up to 85% of total suspended solids from highway and urban area runoff (Barrett et al. 1998).

Hydraulic loading rates are frequently used for wetland design. Average hydraulic loadings are in the range of 500 m³/ha-day. Greater than 60% BOD and TSS removal can be achieved at these rates (Knight 1993). In general, subsurface systems can be loaded at higher rates than surface ones. Loading rates also can be adjusted depending on the type of treatment (Watson et al. 1989). For example, for surface flow systems, hydraulic loading rates of 120 to 470 m³/ha-day and 190 to 940 m³/ha-day are suggested for secondary and polishing treatments. For subsurface flow, 230 to 620 m³/ha-day, 470 to 1,870 m³/ha-day, and 470 to 1,870 m³/ha-day can be used for basic, secondary, and polishing treatments, respectively. To remove nitrogen and phosphorus at rates of greater than 50%, the hydraulic loading rate should be lowered to between 10 and 40 m³/ha-day (Richardson and Nichols 1985).

A wetland consists of inlets and outlets, a water surface exposed to the atmosphere, aquatic vegetation, and soil for the roots. According to the WPCF (1990), surface flow wetlands (also known as free-water surface wetlands) usually have depths in the order of 0.3 to 0.5 m but can vary from a few centimeters to 0.8 m. At this depth, if a hydraulic loading rate of 500 m³/ha-day is used, the detention time in the wetlands would be ten days. Usually five days is considered sufficient for high nitrogen removal. Increasing the depth to between 0.7 and 1.0 m results in the replacement of emergent vegetation with floating plants. Water flow rates range from 4 to 75,000 m³/day.

The following expression is frequently used to determine surface flow wetland performance:

$$C = C_o e^{\frac{-kAD}{Q}}$$

where C_o and C are the influent and effluent BOD_5 concentrations, respectively, D is the depth of the bed, k is the rate constant, A is the surface area, and Q is the influent flow rate. This expression is plug flow first order which will be approximated by wetlands constructed as a series of cells. Although other models have been postulated, they have not been extensively tested. The rate constant is dependent on temperature and is determined (usually through pilot tests) for temperatures above freezing. At very low temperatures, the treatment efficiency will decrease significantly, even though the water will still move under the ice (Kadlec 1989). These types of wetlands are an inexpensive way to treat acid mine drainage and ash pile drainage in coal-mining areas.

For subsurface flow systems, a basin is excavated and filled with gravel or other porous media and then with water to reach below the top of the gravel. The depth is generally between 0.3 and 0.6 m. A liner may be necessary to protect the groundwater. Systems can be used to treat single swellings or large municipalities (up to 13,000 m³/day). The species found in these systems are similar to the surface flow wetlands. However, since biological reactions play a large role in both systems, the subsurface flow ones have higher reaction rates due to the presence of more surface area on the gravel. Also, since the water level is low, mosquitoes are not a problem, and the mosquitoes are less susceptible to cold weather. These systems can then be placed next to schools, parks, commercial, and public buildings (U.S. EPA 1993). The delivery and placement of the gravel increase the cost of the system. Surface flow systems of greater than 4,000 m³ per day are generally less expensive. There are approximately 500 of the subsurface flow systems in the United States (Reed et al. 1995).

The water flow will be plug flow as it is in the surface flow wetlands. The following equation can be used for the rate constant (Reed et al. 1995):

$$k_{20} = k_o (37.31e^{4.172})$$

where k_{20} is the rate constant at 20°C in day⁻¹, k_o is the optimal rate constant (1.839 day⁻¹ for municipal wastewater and 0.198 day⁻¹ for industrial wastewater with high COD), and e is the total porosity (0.18 to 0.35 for coarse to fine gravel).

The cross-sectional area of the bed can be calculated using Darcy's law:

$$Q = KA\frac{\Delta h}{\Delta L}$$

where Q is the flow rate in m³/s, K is the hydraulic conductivity of the bed in m/s, A is the cross-sectional bed area in m², and $\Delta h/\Delta L$ is the bed slope. For a flat, shallow bed with an overflow weir, the slope should be 0.001 or less according to Reed et al. (1995), but up to 4 to 5% can be employed. In the United Kingdom, most beds have been designed with conductivities between 10^{-4} and 3×10^{-4} m/s even though the beds can be fairly wide (Cooper and Hobson 1989). The use of gravel in the bed can increase the conductivity to 10^{-3} m/s. Some guidelines for the engineering of wetlands are shown in Figure 4.30 (API 1998).

Maintenance

Two aspects related to wetlands that must be controlled are mosquitos and weeds. Maintenance of aerobic conditions by limiting BOD loading rates below 110 kg/ha-day allows dragonflies and water beetles, the predators of mosquitos, to flourish. In addition, uniform distribution of the incoming wastewater can assist in mosquito control. Weeds can impede the growth of beneficial plant species. Weed control in subsurface systems, which are preferably flat, can be obtained by flooding the bed (Cooper and Hobson 1989).

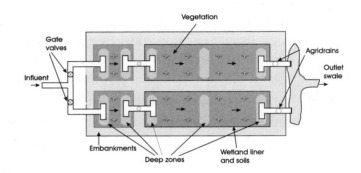

Figure 4.30 Design of a Constructed Wetland (adapted from API 1998)

The following are parameters that must be considered for wetland design:

- pretreatment used;
- siting of the wetland;
- wetland area, aspect ratio, slope, porosity, and substrate;
- type and flow of wastewater and hydraulic conductivity;
- hydraulic loading rate;
- nominal retention time;
- inlet and outlet BOD, DO, TSS, nitrogen, phosphorus and fecal coliform, and streptococcus;
- bacteria;
- design of inlets and outlets;
- type of plants used;
- operation and maintenance requirements;
- monitoring requirements;
- performance of the system (percent removal of contaminants);
- temperature, rainfall, and other environmental influences;
- capital, operating, and maintenance costs; and
- problems encountered during operation (ie., mosquitos, odor, etc.) and their solutions.

Operating and maintenance of the wetlands is mainly related to mowing the grass on the dikes and inspecting for damage caused by rodents. Removal of vegetation residues may be required if they interfere with the flow of surface water systems. Mosquito control can also be a problem for surface water wetlands, particularly as the organic content of the water increases. Insecticides or, during warm weather, *Gambusia* fish can be used to control mosquito populations.

Some future research needs have recently been identified related to the long-term performance of wetlands: the optimization of design procedures and the need for input from multidisciplinary teams (Steffler 2000).

Applications

Subsurface flow wetlands are generally used for thermal protection in northern climates and for on-site domestic wastewater treatment because personal contact with the wastewater is avoided and mosquitos are not a problem for this process. If nitrogen removal is required, reeds and bulrushes should be used in a bed of 0.6m. Cattails in a 0.3 m bed can be used in warm weather. Otherwise, ornamental plants or shrubs can be employed. For larger units, two parallel cells are recommended.

For municipal wastewater treatment, the required capacity determines whether subsurface or surface flow systems are chosen. The subsurface systems are more efficient due to higher BOD removal rates, but, depending on the location, transport and placement of the gravel can be expensive. Free water systems are certainly economically favored for water flows above 4,000 m³/day. As for domestic systems, if nitrogen removal is required, reeds and bulrushes or cattails should be used. Submerged plants could also be used to enhance water quality if deeper zones are added to increase the hydraulic retention and oxygen transfer from the atmosphere. Duckweed mats must be avoided, however, since they can block sunlight transmission into the water. Temperature analyses must be performed to determine the effect of winter temperatures. In some surface flow systems in northwestern Canada, lagoons are used for winter storage, and wetlands are employed in the summer.

Both types of wetland systems can be applied to commercial and industrial wastewaters. Characterization of the wastewater is necessary to determine if the waters are high-strength, high or low pH, and low in nutrients and if the waters have toxic or inhibitory components for the wetlands. Nutrient addition might be necessary or BOD and ammonia removal rates could be an order of magnitude lower. High-strength wastewaters are usually treated anaerobically first. Both types of wetlands have been used for wastewater from food processing, pulp and paper processing, chemical production, and oil refineries. Pilot tests may be preferable if inadequate treatment data exists.

Wetlands have also been used for the removal of sediments from urban stormwater from landscapes, streets, and parking lots. They also have some benefit for BOD, nitrate, phosphate, and trace metal removal. The basis of these systems is a combination of shallow marshes and deep ponds to which wet meadows and shrub areas can be added. Due to the high solids content and highly-variable flows, the submerged flow systems are not used due to the potential for clogging. There is usually in an inlet ditch or basin for sedimentation, a spreader swale or weir for flow distribution if there is a wet meadow or marsh, then a deep pond, and finally an outlet device to allow overflows during storms or to slow discharge to maintain a shallow depth in the marsh (the "datum"). Usually 13 mm is maintained between the overflow and the datum. Cattails, bulrushes, and reeds can temporarily survive in water up to 1 m. The surface area of the surface flow system is based on the flow during the 5-year storm event. From the frequency and intensity of storm events, the hydraulic retention time can be calculated between events, and from this value, the removal efficiency can be determined.

The U.S. Soil Conservation Service (1993) developed a process for the treatment of agricultural runoff that consists of a wet meadow followed by a marsh and pond and is capable of removing sediments and nutrients such as phosphorus. A vegetated polishing area is optional. The wet meadow with a slope of 0.5 to 5% consists of permeable soils with cool-season grasses. Drains are placed perpendicular to the flow for discharge into the marsh below the water surface. The depth of the marsh varies from zero at the surface of the meadow to 0.46 m at the deep pond. Cattails are recommended. The deep pond performs as a biological filter for the removal of nutrients and sediments. Fish such as common or golden shiners should be included in the pond to feed on the plankton. Average sediment and phosphorus removal in a system for potato growing in northern Maine over two seasons were determined to be 96 and 87%, respectively (Higgens et al. 1993).

Surface flow wetlands have been used after the anaerobic treatment of high-strength, high-solid, and high-ammonia wastewaters from feed lots, dairy barns, swine barns, and poultry operations. In a two-cell surface water system for a 500-animal swine operation, the 90 kg BOD/day was reduced to 36 kg BOD/day over a 3,600 m^2 wetland area (Hammer et al. 1993).

Landfill leachate has been treated by both surface and subsurface types of wetlands. Wastewater characterization is required to determine if the nutrient levels such as phosphorus, potassium, or others are sufficient to maintain plant growth. It is also necessary to know the levels of BOD, COD, ammonia, iron, and toxic compounds in the leachate. These systems may have an equalization pond as a pretreatment before the wetland, particularly if BOD levels are greater than 500 mg/L. Metal and priority substance removal are the main requirements. In general, surface systems are preferred since they are less expensive, unless the climate is a problem, in which case the subsurface system will be used.

For the treatment of mine drainage, iron and manganese removal are the key objectives. Preference is usually given to surface systems since they can be aerated more efficiently and there is an increased risk of clogging in subsurface systems as a result of iron and manganese precipitation. Brodie et al. (1993) described several systems for mine drainage. Since the pH decreases from 6 to 3, the Tennessee Valley Authority has developed an anoxic limestone drain (ALD), a high-calcium limestone aggregate (20 to 40 mm) placed in a trench of 3 to 5 m wide and 0.6 m to 1.5 m in depth. The anoxic conditions in the trench are ensured by backfilling with clay. A plastic geotextile is placed between the clay and limestone. The inlet of the trench is placed at the source of the acid mine drainage. The ALD is beneficial under the following conditions:

• Alkalinity of greater than 80 mg/L and an iron content of greater than 20 mg/L;

- Alkalinity less than 80 mg/L and iron content less than and greater than 20 mg/L;
- Negligible alkalinity and iron content approaching 20 mg/L.

At conditions where the alkalinity is greater than 80 mg/L and less than 20 mg/L, ALD is not necessary. However, when the oxygen in the drainage is greater than 2 mg/L or the pH is greater than 6 and the redox potential is greater than 100 mV, use of the ALD is detrimental due to the formation of oxide coatings. The use of a sedimentation pond before treatment in a wetland with or without ALD is preferred since it is easier to remove iron precipitation from the pond than the wetlands.

Hydraulic loadings of between 15×10^{-3} and 42×10^{-3} m³/m²-day have been recommended by the TVA, whereas Witthar (1993) suggested that up to 0.14 m³/m²-day could be applied to wetlands. The number of wetland cells should conform to each 50 mg/L level of iron.

In a partial list of the wetlands in the North American Wetland Treatment System Database, 154 were for the treatment of municipal wastewater, nine for industrial, six for agricultural wastewater, and seven were for stormwater (Kadlec and Knight 1996). These systems treat more than 190 m³/day. Of these, 120 are surface flow systems and 48 are subsurface systems while eight are both. In 1988, 142 North American wetland systems were used for acid-mine drainage (Wieder 1989). The performance of these systems is summarized in Table 4.16.

Table 4.16 Summary of the Performance of North American Wetland Treatment Systems

Parameter	Concentration (mg/L)			Lowest achievable effluent concentration (mg/L)
	In	Out	Efficiency (%)	
BOD	29.8	8.1	73	1–15
TSS	46.0	13.0	72	2–20
NH_4-N	4.97	2.41	52	<0.1
Total N	9.67	4.53	53	1–3
Total P	3.8	1.68	56	<0.1

Adapted from Kadlec and Knight (1996) and API (1998)

In Europe, however, where constructed wetlands were developed, small-scale subsurface flow systems are the norm. These systems are primarily for domestic use. Denmark for example, has over 150 systems and Poland has approximately 100 (Cole 1998).

Wetland treatment systems are becoming more and more accepted for municipal, agricultural, and industrial wastewaters, as well as for stormwater management. In general, average removal efficiencies are 50 to 80% (API 1998). Phenols can be reduced by 70% in the petroleum industry and VOCs up to 95%. Even 50% metal removal efficiencies have been achieved for aluminum, cadmium, copper, iron, lead, mercury, nickel, silver, and zinc. This is due to growing confidence in the performance of wetland systems and the shortage of affordable technologies (Kadlec and Knight 1996). In the United States, wetlands do not come under the Clean Water Act because they are integrated with other treatment systems. If rare or endangered wildlife species are attracted to the wetlands, the Endangered Species Act and the Migratory Bird Treaty will come into effect.

Costs

For surface wetland systems, land, a liner, and dikes or berms are required. Costs for these can range from $75,000 to $170,000 per hectare (Reed et al. 1995). The per unit cost of subsurface systems is approximately 50% higher. However, the increased efficiency and lower land requirements must be taken into account. For example, for a 1,900 m^3/day, a surface water system with a HRT of 14 days and four hectares of land would be required at costs of $311,400 with a clay liner or $839,400 with a plastic liner (Gearhart 1993). For the same water flow, only 2.6 hectares would be required for a subsurface system, thus reducing the cost of the land. The requirement for gravel must also be taken into account. If 15,300 m^3 of gravel at $18/$m^3$ or less for a 60 cm of depth is required, then the subsurface wetland system would be less expensive for a plastic liner system. For a clay liner system, the cost of the gravel would have to be less than $6/$m^3$ for the subsurface system to be less expensive. Operating and maintenance costs are low at $0.03 to $0.09/$m^3$ (WPCF 1990).

A recent study by the EPA (2000a, b) compared the capital and operating costs for free water and subsurface systems to SBR systems treating 378,000 L/day. The capital costs of the free water and subsurface systems were $259,000 and $466,700, respectively, compared to $1,104,500 for the SBR system. Operating costs per cubic meter were $0.11 and $0.19 for free water and subsurface systems, respectively, compared to $0.81 for the SBR system.

Case Studies

An installation at Tompkins County, New York is an example of a subsurface system (Peverly et al. 1994). Two cells were used, the first with a retention time of 15 days and the second of seven days for a total of 22 days. The leachate flow rate was 1 m^3/day. Although an effluent BOD of less than 5 mg/L and an ammonia level 72 mg/L were predicted, only 124 mg BOD/L and 136 mg ammonia/L were achieved (33% and 46% removal rates, respectively). It was believed that the levels were much lower than expected due to low phosphorus concentrations (only 0.15 mg/L). Since normal biological treatment requires approximately 0.006 kg P/kg BOD removed, then the phosphorus level should have been about ten times higher.

Natural Wastewater Treatment Systems: Aquatic Treatment Systems

Box 4-9 Overview of Aquatic Treatment Systems

Applications	Suitable for secondary or tertiary treatment of dilute compounds (less than 1%) of low toxicity.
Cost	Operating: $0.01 to $0.14/m^3, Capital: $100 to $1,000/m^3/day
Advantages	Simple to start-up and operate
	Low capital and operating costs
	Low maintenance requirements
Disadvantages	Lack of design criteria
	Large land requirements necessary
Organic loading	100 to 300 kg BOD/ha-day
Hydraulic retention time	30 to 50 days
Other considerations:	Still under development

Description

Aquatic treatment systems involve the use of plants or animals for wastewater treatment. Floating plants such as water hyacinth (*Eichhornia crassipes*), duckweeds, and pennywort have been tested in either pilot or full-scale treatment systems. Some tests in the laboratory and greenhouse have involved submerged plants such as pondweed (*Potamogeton* sp.), water milfoil (*Myriophyllum heterophyllum*), water weed (*Elodea* sp.), coontail (*Ceratophyllum demersum*), and fanwort (*Cabomba caroliniana*). Results have been mixed and problems with algae have occurred since the conditions for plant growth are the same as for algae. Treatment occurs in aquatic systems by microbial metabolism on the plant roots, sedimentation of dead plants, and microbe and nutrient incorporation into the plants. There may be some applications as a final polishing after wetland units. Aquatic animals such as the crustaceans *Daphnia*, brine shrimp, clams, oysters, lobsters, and fish have been evaluated for suspended solid, nitrogen, or algae removal. Cost effectiveness has not been proven at this point.

Methods

Hyacinth Ponds

The water hyacinth is a perennial freshwater aquatic macrophyte with round, shiny green leaves and lavendar flowers. They can measure 50 to 120 cm from the flower top to the root tips. The roots can be 10 cm in length if the nutrients in the water are sufficient. These plants are sensitive to freezing temperatures, which will limit their applicability in northern climates unless greenhouses are used, but this may be too costly (Leslie 1983).

On the surface, hyacinths prevent algae growth by shading the surface and helping to maintain neutral pH. Because they also minimize turbulence by reducing wind effects, the water temperature does not fluctuate significantly, and surface oxygen is low. Oxygen is not necessary, however, for the plants since oxygen travels from the leaves to the roots. Bacteria, fungi, predators, filter feeders, and detritovores (Reed et al. 1995) grow well on the roots.

Duckweed Ponds

Three genera of duckweed, *Lemna* sp., *Spirodela* sp. and *Wolffia* sp., have either been tested or used in wastewater treatment systems (Reed et al. 1995). They are small, green freshwater plants with small fronds (a few millimeters wide) and short roots (less than a centimeter long). These plants have a higher tolerance for cold water than hyacinths and can survive to around 7°C (Leslie 1983).

Operating Conditions

Hyacinth Ponds

If 100% of the water surface is covered with hyacinths, then the loading is approximately 222 kg BOD/ha-day for water depth of 1 to 2 m. Suspended solids are entrapped within the plants' roots. Since the water is less turbulent, sedimentation is more easily achieved than in open water. Nitrogen removal occurs via plant uptake, nitrification/denitrification, and volatilization of ammonia. Weber and Tchobanoglous (1985) observed that 19 kg nitrogen/ha-day could be removed at loading rates between 9 and 42 kg/ha-day. Phosphorus removal occurs only by plant uptake and will not usually exceed 25%. Metal removal, however, is quite efficient via uptake, chemical precipitation, and surface adsorption. Mature plants with the accumulated metals can become part of the sediment sludge as root matter is sloughed off (Reed et al. 1995). Organic removal may occur through bacterial degradation and plant uptake.

Duckweed Ponds

Duckweed ponds can operate on a seasonal basis. Plants should be harvested before freezing to avoid high BOD levels in the effluent. They do not have to be reseeded in the spring.

Design

Hyacinth Ponds

Design of hyacinth ponds is usually based on organic loading. The same considerations for facultative ponds can be used if secondary treatment for BOD and TSS removal and prevention of algae growth are the major objectives. Multiple cells are usually required. For advanced secondary systems, aeration can be added to enhance treatment efficiencies. Tertiary systems are primarily used for nutrient removal. Typical engineering criteria for secondary, advanced secondary, and tertiary treatment are shown in Table 4.17.

Table 4.17 Design Consideration for Secondary, Advanced Secondary,
and Tertiary Treatment by Hyacinth Ponds

Factor	Secondary treatment	Advanced secondary treatment	Tertiary
Effluent requirements	BOD <30 mg/L, TSS <30 mg/L	BOD <10 mg/L TSS <10 mg/L	BOD <10mg/L, TSS <10 mg/L, N&P <5 mg/L
Wastewater input	Untreated	Primary treated	Secondary treated
Organic loading (kg BOD/ha-day)			
First cell	50	100	<50
Entire system	100	300	<150
Aeration	Not required	Similar to partial mix aerated pond	
Water depth (m)	<1.5	<0.9	<0.9
Maximum area (ha)	0.4	<0.4	<0.4
Total detention time (days)	>40	>6	< 6
Hydraulic loading (m³/ha-day)	>200	<800	<800
Water temperature (°C)	>10	>20	>20
Basin shape (L:W)	Rectangular (>3:1)	Rectangular (>3:1)	Rectangular (>3:1)
Influent flow diffusers	Recommended	Required	Required
Mosquito control	Necessary	Necessary	Necessary
Harvest schedule	Seasonal or annual	>Monthly	Every few weeks
Multiple cells	Required, two sets of three basins	Req., two parallel sets of three basins each	Req., two parallel sets of three basins each
Effluent collection manifold	Not required	Required	Required

Duckweed Ponds

The design of the duckweed ponds can follow the same principles of facultative ponds. Duckweeds do not remove pollutants like hyacinths but function mainly as surface cover for the pond. Inlet diffusers and mosquito control are not required. An effluent manifold is necessary for complete utilization of the basin width for treatment. A screen or baffling system at the outlet prevents the plants from leaving with the effluent. Harvesting is not as frequent as for hyacinths since they serve as a water cover. The plants can be disposed of by composting, landfilling, or by using as animal feed after drying.

Maintenance

The major operational parameters are related to mosquito control, plant harvest, and sludge management or disposal. Other requirements are similar to pond or lagoon systems. The design of the basin should facilitate plant harvesting. Different flow distribution is available and must be able to provide good wastewater distribution, maintenance of aerobic conditions, and avoidance of frost during the cold weather (Tchobanoglous and Angelakis 1996). Mosquito control can be achieved by surface-feeding such as *Gambusia*. Anoxic conditions must be avoided since the fish require oxygen to survive. Once the plants are harvested by aquatic plant harvesters, front-end loaders, backhoes, rakes, boats or other means, the plants must be dried since they contain up to 95% moisture. Sun drying is a common drying method. The plants can then be landfilled, if permitted, composted for production of soil conditioner/fertilizer, or digested anaerobically for methane production.

Applications

Hyacinths can remove BOD, suspended solids, metals, nitrogen, and trace organics. Thus, they are useful for upgrading existing systems, secondary, advanced secondary, or tertiary treatment. BOD loadings of 6.7×10^{-4} kg per kg of wet plant mass per day have been recommended for treating effluents from facultative ponds (Wolverton 1979).

By 1992, approximately 15 duckweed systems for BOD and TSS removal were operating in the United States. On a system in Mississippi, organic loads were 22 kg BOD/ha-day for a 22-day detention time basin with an anaerobic effluent of 15 mg/L BOD. This can be compared to a lower range conventional facultative pond (Wolverton and McDonald 1979).

Costs

The capital costs for floating plants systems are $270,000 per hectare or $500 to $1,000 per m^3/day of water treated (Kadlec and Knight 1996). Operating costs are about $0.12 to 0.14/m^3.

Case Study

A variation of the wetland and pond systems called the Living Machine™ has been developed for the biological treatment of industrial wastewater and sewage (Living Machines, Burlington, VT)(Figure 4.31a, b). It uses a reactor system in similar fashion to ponds and marshes to treat water with bacteria, plants, snails, and fish. Depending on the climate, the system can be employed in the open air or in a greenhouse. These biological systems have been installed at several locations. The performances of a few are indicated in Table 4.18.

At the Ethel M. Chocolates Factory (Las Vegas, NV), a unit was installed for the treatment of wastewater which varied from 1,300 to 2,000 mg BOD/L (Living Technologies, Inc. 2000). The target was 300 mg BOD/L. Fats and oils were 300 to 350 mg/L; the standard was 250 mg/L. Suspended solids were 600 mg/L; the standard was 300 mg/L. The installed system included a grease trap and three closed aerobic reactors. A biofilter was used to treat the odors. The wastewater was pumped into three open cells with vegetation. After exiting these reactors, the effluent passed through a clarifier and ecological fluidized beds to remove organics and suspended solids. The water was then stored in a wetland for use in irrigation and then treatment by ultraviolet light. The full-scale unit was able to treat 121 m^3/day. The effluent had a BOD of below 10 mg/L, an oil and grease content of less than 1 mg/L, and a TSS level of about 10mg/L. The cost was about half of an activated sludge process. Approximately $8,000 per year of bacteria are added to the systems, and electrical charges are about $30/day. The plant has zero discharge since the sludge is composted and the water is used for irrigation.

Figure 4.31a Photograph of a Living Machine at Ethel M. Chocolates (Courtesy of Living Machines)

Figure 4.31b Photograph of a Canopy at Ethel M. Chocolates (Courtesy of Living Machines)

The process in Scotland can be described as follows. Wastewater is fed into buried anaerobic reactors that begin the digestion process. Next, an aerobic reactor is used, followed by open tanks with a wide range of plant and animal life in a greenhouse completing the process. Clarifiers are used to settled the solids for return to the anaerobic reactors. The effluent is polished in fixed-film reactors called "Ecological Fluidized Beds," patented by Living Machines. Re-use of the water is then possible after ultraviolet disinfection. In the second case, the process consists of two parallel trains of nine tanks in a double-glazed greenhouse. The first six tanks are aerated to promote the growth of bacteria on the rafted plants for BOD degradation. Clarification occurs in the eighth tank before final polishing in the Ecological Fluidized Beds.

Table 4.18 A Sampling of the Performances of the Living Machine™

	Location			
	Finhorn, Scotland	Littlehampton, England	South Burlington, Vermont	Wyong, Australia
Type of wastewater	Sewage	Cosmetics	Sewage	Food Processing
Capacity (m³/day)	65	25–50	30	200
Influent COD (mg/L)	500	3,701	454	
Effluent COD (mg/L)	37	346	31.3	
Influent BOD (mg/L)	300	1,762	219	2,334
Effluent BOD (mg/L)	2.7	28	5.9	7
Influent TSS (mg/L)	250		174	1,100
Effluent TSS (mg/L)	10		4.8	26
Influent NH₄ (mg/L)	25		14	
Effluent NH₄ (mg/L)	3.4		0.25	

The process for the third case was similar to the second one in which two trains were used with five aerobic reactors, a clarifier, and three Ecological Fluidized Beds. The aerobic tanks contain floating plant racks. For the last case, to achieve the performance in Table 4.18, dissolved air flotation was first, followed by a three-stage UASB reactor and two aerobic trains with five planted reactors, a clarifier, and an Ecological Fluidized Bed. The Living Machine was located outdoors where frost is a possibility.

Metal Treatment Processes

Metal Treatment Processes: Biosorption

Description

Biosorption involves the removal of metals from wastewater via adsorption on living or dead biomass. The biomass can include bacteria (*Bacillus subtilis*, *Bacillus licheniformis*), yeast (*Candida tropicalis*), fungus (*Aspergillus niger*, *Penicillium chryosogenum*, *Rhizopus arrhizus*), algae (*Sargassum natans*, *Ascophyllum rodosum*, *Fucus vesiculosus*), and plant material (peat moss, wood chips, and pine cones).

Box 4-10 Overview of Biosorption Processes

Applications	Suitable for dilute concentrations of metals such as silver, gold, cadmium, chromium, copper, iron, mercury, nickel, lead, zinc, radioactive metals, and sulfur
Cost	Capital cost for sorbent: $1 to $4/kg
Advantages	Recovery of metals and sulfur for resale
	Easy to operate and maintain
	Sorbents are low cost
Disadvantages	Recycling of sorbents can be problematic and not proven industrially
	Lack of cost and design data
Other considerations:	Biosorption is still under development

Design

Algal biomass (*Sargassum natans*) can uptake metals up to almost 40% of its dry weight. Many metals can be adsorbed, such as silver, gold, cadmium, chromium, copper, iron, mercury, nickel, lead, zinc, and radioactive metals. Various types of materials have been used for immobilization including alginate, polyacrylamide, polysulfone, silica gel, cellulose, and glutaraldehyde. Biosorption processes which include granulated, pelletized, or immobilized biomass are similar to ion exhange ones. Therefore, for commercial systems, biosorption processes can take place in batch or continuous-stirred tanks, fixed-packed beds, and fluidized beds (Volesky 1990).

Biosorbent use depends on biosorption capacity, availability of the biosorbent, cost, ease of regeneration, and use in various reactor configurations. Eluents such as dilute acids or carbonate can be used to desorb the adsorbed contaminants. *Aspergillus, Penicillium,* and *Saccharomyces* can withstand ten cycles of regeneration without decreased adsorption capacity.

Applications

Biosorption can be useful for radionuclides from dilute streams such as mine leachates. *Aspergillus niger* can adsorb between 31 to 214 mg/g of uranium, *Rhizopus arrhizus* can adsorb about 200 mg/g, and *Saccharomyces cerevisiae* can adsorb 150 mg/g. This can be compared to traditional adsorbers such as ion exchange resin IRA-400 (79 mg/g) and Activated Carbon F-400 (145 mg/g).

B.V. Sorbex (Montreal, Canada) has also developed numerous biosorbents that are capable of working over a wide pH and metal range but are not affected by calcium, magnesium, or organics. Flows for fluidized bed or pulsed bed systems of up to 100 m³/day are possible (Volesky 1990). Potential applications of these biosorbents include industrial effluent polishing and metal removal from dilute effluents.

Costs

The costs of macroalgal biosorbents were estimated to be $1 to $2/kg and microalgal biosorbents at $3 to $4/kg (Volesky 1990). This is compared to ion exchange resins which cost about $3 to $7/kg. In the case of *Chlorella* biomass, the cost ($0.55/kg) is known but not the production cost for this biosorbent (Kuyucak 1990).

Case Studies

A few biosorbents have been commercialized. BIO-FIX® is immobilized biomass in beads of high density polysulfone. It has been produced by U.S. Bureau of Mines (Golden, CO), tested for the treatment of acid mine waste, and found to adsorb heavy metals over alkaline earth metals (Eccles and Holroyd 1993). The preparation method consists of dissolving polysulphone pellets in dimethylformide and mixing in dried biomass. The slurry is sprayed into water through an atomizer to form beads of 0.5 to 2.5 mm. Over 50 types of wastewater have been tested, such as mine waters and groundwaters containing metal concentrations of 0.5 to 100 mg/L. Up to 170 regeneration cycles were possible with peat, algae, and other types of biomass. Gupta et al. (2000) demonstrated that the BIO-FIX beads worked from pH 3 to 8 with increasing metal affinities in this order: Mg<Ca<Mn<Fe<Zn<Cu<Cd<Al. Compared to lime precipitation for a wastewater stream of 6,400 L/min (pH 6.9) containing 60 mg/L of Zn, Mn, and Cd, costs were similar per liter of wastewater treated by a three-column packed bed of BIO-FIX beads (Eccles and Holroyd 1993). This three-column system uses a lead column, a scavenger column, and an elutant column (Figure 4.32). The elutant column is regenerated while the other two columns are loaded. At the end of a loading cycle, the scavenger column becomes the lead, the elutant one becomes the scavenger, and the lead one is regenerated.

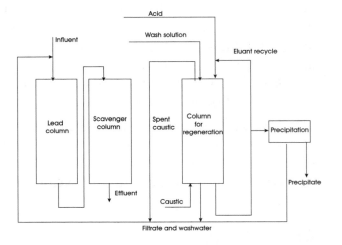

Figure 4.32 Flowsheet of BIOFIX Biosorption Process for Metal Removal

AMT-BIOCLAIM™ (MRA) (Brierley et al. 1986) is a sorbent composed of immobilized *Bacillus subtilis* bacteria in beads of porous polyethyleneimine and glutaraldehyde that was commercialized by Visa Tech Ltd. Fixed bed or fluidized bed reactors can be used with this type of sorbent. In tests, gold, cadmium, and zinc were adsorbed from cyanide solutions. This sorbent could also accumulate lead (601 mg/L), copper (152 mg/L), zinc (137 mg/L), cadmium (214 mg/L), and silver (86 mg/L) from dilute solutions, thereby achieving 99% metal removal (Kuyucak 1990). Platinum and palladium could also be removed from solution by this sorbent (Brierley and Vance 1988). A determination of the capital cost for a 190 m³/

day system indicated a 50% savings over alkaline precipitation and 28% over ion exchange systems (Eccles and Holroyd 1993).

Another sorbent, AlgaSORB© (Resources Management and Recovery, Las Cruces, NM), is an immobilized alga (*Chlorella vulgaris*) in a silica gel matrix. A portable unit can treat waste of 5.5 m³/day. Larger units up to 5,500 m³/day have been manufactured. The immobilization protects the algal cells from destruction by bacteria and allows the beads to withstand the pressures from the water flow. Acids, bases, and other reagents are used to regenerate the algal cells. AlgaSORB can reduce metal concentrations from 100 mg/L to 1 mg/L. More than 100 biosorption-desorption cycles are possible (Kuyucak 1990). It has been demonstrated at a U.S. EPA Superfund Innovative Technology evaluation that mercury could be absorbed from contaminated groundwater in pilot tests. (Barkley 1991). Other applications include metals recovery from wastewaters from electroplating, mining processes, and printed circuit board processing.

Comparison to Other Processes

The advantages of using biosorbents include versatility and flexibility, robustness, cost-effectiveness, selectivity of heavy metals over alkaline earth metals, and ability to reduce metal concentrations to drinking water standards (Garnham et al. 1992). Current constraints are the competition with ion exchange resins, the low capacity of metal fixation in terms of milligrams of metal adsorbed per gram of sorbent, the selectivity, and the ability to regenerate these materials. Engineers, in general, prefer more established processes such as ion exchange, and they have a lack of knowledge concerning biosorbents. A greater understanding of these biological-based systems may help their commercialization potential in the future.

Metal Treatment Processes: Metal and Sulfur Removal

Description

Metal removal

Sulfate reducing bacteria include *Desulfovibrio*, *Desulfotamachulum*, *Desulfobacter*, *Desulfococcus*, *Desulfonema*, and *Desulfosarcina*. These bacteria can remove metals by hydrogen sulfide production, which subsequently precipitates metals. Sulfur oxidizing bacteria include *Thiobacillus thioxidans*, *T. thioplaus*, and *T. denitrificans*. Companies such as Paques have developed reactors (THIOPAQ®) to take advantage of these processes (Figure 4.33). For example, for heavy metal and sulfate removal, metal precipitates are formed by the following reaction:

$$Me^{2+} + SO_4^{2-} + 2CH_2O \rightarrow MeS + 2CO_2 + 2H_2O$$

In a second reactor, the sulfide is converted to sulfur:

$$H_2SO_4 + 2CH_2O + \frac{1}{2}O_2 \rightarrow S^0 + 2CO_2 + 3H_2O$$

Metal reuse is possible if only one metal is used such as zinc. In addition, the sulfur can be recovered with up to a 95% purity for use in sulfur acid production facilities. Metals can be selectively removed.

Sulfur Removal Processes

In the past, sulfur effluents have been treated with lime, which forms gypsum that has to be landfilled. Although lime addition is simple, the sludge is not easy to dewater and frequently sulfate concentrations below 1,500 mg/L cannot be achieved. Biological sulfate removal consists of two steps: sulfate reduction to sulfide and then oxidation of sulfide to elemental sulfur. The latter can be removed from the wastewater by sedimentation. For sulfate reduction, an appropriate electron donor is required such as hydrogen, molasses, ethanol, sewage digest, potato skins, or methanol. Under oxygen limited conditions (less than 0.1 mg/L), sulfur formation occurs. Reactors of low shear are used under autotrophic conditions.

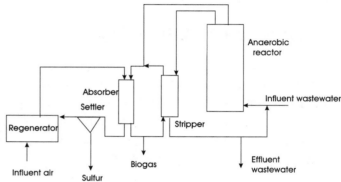

Figure 4.33 Flowsheet of Sulfur Removal Process (Thiopaq). Sulfide is Stripped from Wastewater by Recirculating Biogas and Reacted with the Catalyst Bed to Form Sulfur (Biothane Company Literature)

Operating Conditions

Metal Removal

Metal removal has been accomplished by many types of anaerobic reactors including UASB, completely mixed, gas-lift, and immobilized SRB reactors (Hao 2000). Operation must be done at high sulfate to COD ratios and low retention times. Influent pH is usually neutralized. Iron salts may also be added to reduce H_2S toxicity. Operating conditions are indicated in Table 4.19 for various types of wastewater. Wastewater containing metals, organics and sulfate, or metals and sulfate with an organic supplementation can be treated by SRB bacteria. The sulfide generated in the first reactor is the fed for subsequent precipitation of the metals. Operating problems such as sulfide toxicity, odor generation, high metal contents, lack of organic substrates, and inefficient removal of the metal sulfide precipitation limit full-scale applications.

Sulfate removal

Most types of anaerobic reactors can be used for the removal of sulfate from wastewater. Measures should be made to prevent high concentrations of H_2S in the liquid or gas phases, the precipitation of metal sulfides, and low biogas generation (Lens et al. 2000). Ideally, sulfate reduction should be repressed; however, this has not been accomplished in full-scale reactors. Ratios of COD:sulfate greater than ten have been successful in avoiding excess H_2S generation. Recently, ratios of COD:sulfate of 8:1, 5:1, and 3:1 have also been successful in the seafood processing industry.

Applications

Metal Removal

Applications have been limited for metal removal in industrial wastewater, groundwater, and landfill leachate. The results are summarized in Table 4.19.

Sulfate Removal

Wastewaters from numerous industries include sulfate and organic matter. Such industries include food and fermentation industries, seafood processing, tanneries, and pulping, as shown in Table 4.20. Others, such as acid mine drainage and landfill leachates, contain low organic contents.

These biological precipitation processes have been operated in the treatment of electronic component and electro-plating wastewaters (Philips Semiconductors B.V.) since 1997. Another full-scale application has operated since 1992 for the removal of metals and sulfur from groundwater (300 m³/h) at a zinc smelter (Budelco) where ethanol is used for the second reaction.

Case Studies

Omil et al. (1995) showed that COD:sulfate ratios of 50 and lower than five could be used successfully. Sodium (5 to 12 mg/L) and ammonia (1 to 3 mg/L N) were high. SRB bacteria accounted for over 11% of the COD removal. Total COD removal was 70 to 90%. Sulfate levels of 0.6 to 2.7 g/L were completely removed. The strategy in this case was the optimization of the pH for minimizing the toxic effects of H_2S and ammonia.

A UASB system was used for the anaerobic sulfate reduction reaction followed by the submerged fixed-film reactor for the aerobic production of sulfide to sulfur (Boonstra et al. 1999). Sulfate was reduced from 1,000 mg/L to less than 200 mg/L and zinc from 100 mg/L to less than 0.05 mg/L. Zinc sulfide and sulfur slurries are produced for the further production of zinc and sulfuric acid.

An example of a full-scale sulfate removal installation is at a synthetic fiber production plant (Emmen, The Netherlands) where 40 m³/h of wastewater containing 2 g/L sulfate has been treated since 1995. Approximately 75% of the sulfate is converted to sulfur.

Table 4.19 Metal Removal of Wastewaters Containing Metals (adapted from Hao 2000)

Reactor type	Wastewater	Operating conditions	Influent COD (g/L)	Influent pH	Influent metal conc. (mg/L)	Influent sulfate conc. (mg/L)	Metal removal (%)
Anaerobic filter	Landfill leachate	34 days	11.6	6.5	430 (Fe) 16 (Zn) 5.6 (Cu)	103	97 (Fe) 94 (Zn) 74 (Cu)
Anaerobic filter	Gold mine effluent	10 h	3.4	6.0	6.7 (Ni) 3.2 (Co) 2.9 (Mn)	830	93 (Ni) 72 (Co) 52 (Mn)
Anaerobic filter	Nickel plating	24 h	0.2	7.0	53 (Ni)	60	99 (Ni)
Sludge blanket	Groundwater with ethanol added	11 h at 21°C	33.3	4.9	1,070 (Zn) 18 (Cd) 6.8 (Cu)	1,000	>99 (Zn) >99 (Cd) >99 (Cu)
Suspended system with sludge recycle	Soil leachate with leachate	10 h SRT = 500h	3.2	6.5	7.1 (Cd) 6.3 (Co) 5.3 (Cr) 4.2 (Cu) 83 (Mn) 6.3 (Ni) 1.7 (Zn)	1,600	98 (Cd) 92 (Co) 97 (Cr) 94 (Cu) 90 (Mn) 87 (Ni) 97 (Zn)

The Paques THIOPAQ process converts sulfate to elemental sulfur. Sludge production is six to ten times less than by lime addition. In addition, sulfate concentrations below 500 mg/L can be obtained. The THIOPAQ process involves two steps, the conversion of sulfate to sulfides and then sulfide to sulfur. Since alkalinity is produced during the second step, recirculation of this effluent to the inlet to achieved neutralization is performed. The final sulfur product contains 60% solids with a purity of 95% and can be used for sulfuric acid production or as soil amendments.

Table 4.20 Sulfate Reduction in Anaerobic Reactors (adapted from Lens et al. 2000)

Reactor type	Carbon source	Influent COD (g/L)	Influent sulfate (g/L)	Sulfate reduction (%)
UASB	acetate	1.5 to 2.1	0.7 to 3.4	70
EGSB	volatile fatty acids	0.5 to 2.5	1.2 to 4.6	27 to 68
Upflow sludge bed	volatile fatty acids	0.5 to 6.0	1.0 to 12.0	35
Anaerobic filter	citric acid	25.8	0.84 to 5	93
Hybrid	landfill leachate	19.6 to 42.0	5.9	greater than 90
Central activity digester	sea food	10 to 60	0.6 to 2.7	96.0

References

Abwassertechishe Vereinigung. 1985. *Lehr- und Hanbuch der Aberwassertechnik,* Bd IV (*Textbook for Wastewater Engineering, Vol. IV*) Berlin: Erst & Sons. 900.

American Water Works Association (AWWA). 1992. *Standard Methods for the Examination of Water and Wastewater, 18th ed.*

Anderson, R. V., and Associates Ltd. 1998. The many advantages of biological nutrient removal. *Environmental Science & Engineering.* 28–29.

Anderson, G. K., B. Kasapgil, and O. Ince. 1994. Comparison of porous and non-porous media in upflow anaerobic filters when treating dairy wastewater. *Water Research,* 28:1610–1624.

Anonymous. 1978. Deep shaft process and application. *Effluent and Waste Treatment Journal,* 18(1):33–35.

———. 1979. Space-saving wastewater treatment. *Chemical Engineering,* 86(7):47–48.

———. 1998. Worlds's largest SBR/BNR treatment system begins operation. *Environmental Science & Engineering.* June. 12.

American Petroleum Industry (API). 1998. *Treatment Wetlands for the Petroleum Industry,* Brochure prepared for the API Biomonitoring Task Force by CH2M Hill. Washington, D.C.

Arden, E., and W. T. Lockett. 1914. Experiments on the oxidation of sewage with the aid of filters. *Journal of Society for Chemical Industries,* 33:523, 1122, 1914.

Bakx, T., S. Boom, and H. Ramaekers. 2000. Treatment of oil-contaminated water from naval warships. *Water,* 21(Aug):37–38.

Barbosa, R. A., and G. L. Sant'Anna. 1989. Treatment of raw domestic sewage in an UASB reactor. *Water Research,* 23(12):1483–1490.

Barkley, N. P. 1991. Mercury removal from contaminated groundwater utilizing algae. *Journal of Air and Waste Association,* 41(10):1387–1393.

Barrett, M. E., P. M. Walsh, J. F. Malina, Jr., and R. J. Charbeneau. 1998. Performance of vegetative controls for treating highway runoff. *Journal of Environmental Engineering,* 124(11):1121–1128.

Belhateche, D. H. 1997. Choose appropriate wastewater treatment technologies. *Practical Engineering Perspectives. The Environment: Air Water and Soil.* G. F. Nalven, ed. New York: American Institute of Chemical Engineers.

Berktay, A., and K. V. Ellis. 1997. Comparison of the costs of the pressurized wastewater treatment process with other established treatment processes. *Water Research,* 31(12):2973–297.

Biothane Corp. 2001. Continually stirred tank reactor biobulk. "Contact." Available online: http://www.biothane.com/biobulk.

Blum, D. J. W., and R. E. Speece. 1991. A database of chemical toxicity to environmental bacteria and its use in interspecies comparisons and correlations. *Research Journal Water Pollution Control Federation,* 63(3):198–207.

Boonstra, J., R. Van Lier, G. Janssen, H. Dijkman, and C. J. N. Buisman. 1999. Biological treatment of acid mine drainage. In *Proceedings of the International Biohydrometallurgical Symposium,* Madrid Spain. 559–568.

Brierley, J. A., C. L. Brierley, and G. M. Goyak. 1986. AMT-BIOCLAIM: A new wastewater treatment and metal recovery technology. *Fundamental and Applied Biohydrometallurgy.* R. W. Lawrence, R. M. R. Branion, and H. G. Ebner, eds. Amsterdam: Elsevier. 291.

Brierley, J. A., and D. B. Vance. 1988. Recovery of precious metals by microbial biomass. *Proceedings of International Symposium,* Warwick, England. 1987. P. R. Norris, and D. P. Kelly, eds. Lew: Science and Technology Letters. 477–485.

Brodie, G. A., C. R. Britt, T. M. Toamazewski, and H. N. Taylor. 1993. Anoxic drains to enhance performance of aerobic acid drainage treatment wetlands: Experiences of the Tennessee Valley Authority. *Constructed Wetlands for Water Quality Improvement.* Chelsea, MI: Lewis Publishers. 129–138.

Cheremisinoff, N. P. 1996. *Biotechnology for waste and wastewater treatment.* Noyes Publication, Westwood, N. J.

Cocci, A., C. Smith, and R. Landine, 2001. Anaerobic wastewater treatment reviewed. *Environmental Science & Technology,* 14(2):22–23.

Cole, S. 1998. The emergence of treatment wetlands. *Environmental Science and Technology,* 32(9):218A– 223A.

Cooper, P. F., and J. A. Hobson. 1989. Sewage treatment by reed bed systems: The present situation in the United Kingdom. In *Constructed Wetlands for Wastewater Treatment,* D. A. Hammer, ed. Chelsea, MI: Lewis Publishers. 153–171.

Donlon, B. A., E. Razo-Flores, G. Lettinga, and J. A. Field. 1996. Continuous detoxification, transformation and degradation of nitrophenols in upflow anaerobic sludge blanket reactors. *Biotechnology and Bioengineering,* 51:439–449.

Donlon, B. A., E. Razo-Flores, M. Luitjen, H. Swarts, G. Lettinga, and J. Field. 1997. Detoxification and partial mineralization of the azo dye mordant orange 1 in a continous upflow anaerobic sludge-blanket reactor. *Applied Microbiology and Biotechnology,* 47:83–90.

Draaijer, H., J. A. W. Maas, J. E. Schaaoman, and A. Khan. 1992. Performance of the 5 MLD UASB reactor for sewage treatment at Kanpur, India. *Water Science and Technology*, 25(7):123–133.

Droste, R. L. 1997. *Theory and Practice of Water and Wastewater Treatment*. New York: John Wiley & Sons.

Du Pont. 1982. Protecting the environment. Supplement to *Chemical Week*.

Eccles, H., and C. Holroyd. 1993. Why select a metal biosorption process? *Proceedings of SCI Conference on Biological Removal of Toxic Metals*. Preston. 1–2.

Eckenfelder, W. W., and W. Barnhart. 1963. Performance of a high rate trickling filter using selected media. *Journal of Water Pollution Control Federation*, 35:1535.

Fox, P., M. T. Suidan, and J. T. Bandy. 1990. A comparison of media types in acetate fed expanded-bed anaerobic reactors. *Water Research*, 24(7):827–835.

Garnham, G. W., G. A. Codd, and G. M. Gadd. 1992. Accumulation of cobalt, zinc, and manganese by the estuarine green microalga Chlorella salina immobilized in alginate microbeads. *Environmental Science and Technology*, 26(9):1764–1770.

Gearhart, R. A. 1993. *Construction, Construction Monitoring and Ancillary Benefits*. Arcata, CA: Environmental Engineering Department, Humbolt State University.

Ghosh, S. 1990. Closure discussion on 'Improved Sludge Gasification by Two-Phase Anaerobic Digestion'. *Journal of Environmental Engineering*, ASCE, 116(EE4):786–791.

Grady, C. P. L., G. Daigger, and H. C. Lim. 1999. *Biological Wastewater Treatment*. New York: Marcel Dekker.

Gupta, A., R. V. Flora, M. Gupta, G. D. Sayles, and M. T. Suidan. 1994. Methanogenesis and sulfate reduction in chemostats: I. Kinetic studies and experiments. *Water Research*, 28(4):781–793.

Gupta, R., P. Ahuja, S. Khan, R. K. Saxena, and H. Mohapatra. 2000. Microbial biosorbents: Meeting challenges of heavy metal pollution in aqueous solutions. *Current Science*, 778(8):967–973.

Hade, C. 1998. Traitement anaérobie des eaux usées dans une usine agro-alimentaire. *Vecteur Environnement*, 31(5):31–36.

Hall, E. R., C. F. Prong, P. D. Robson, and A. J. Chmelauskas. 1986. Evaluation of anaerobic treatment of NSSC wastewater. TAPPI Environmental Conference.

Hammer, D. A., B. P. Pullen, T. A. McCaskey, J. Eason, and V. W. E. Payne. 1993. Treating livestock, wastewaters with constructed wetlands. *Constructed Wetlands for Water Quality Improvement*. Chelsea, MI: Lewis Publishers. 343–348.

Hao, O. J. 2000. Metal removal by SRB. *Environmental Technologies to Treat Sulfur Pollution,* P. N. L. Lens and L. W. H. Pol, eds. London: IWA Publishing. 393–414.

Harremoes, P., and M. Henze. 1996. Biofilters. *Environmental Engineering.* U. Forstner, R. J. Murphy, and W. H. Rulkens, eds. Berlin: Springer Verlag. 143–194.

Hartwig, T. L. 1981. Anaerobic sludge digestion of municipal wastewater sludge: Operation and maintenance. *Proceedings: Anaerobic Wastewater and Energy Recovery Seminar,* Duncan Lagnese and Associates, Inc., Pittsburg, PA. July.

Haskoning Consulting Engineers and Architects. 1996a. *36 MLD UASB treatment plant in Kanpur, India, Evaluation Report on Process Performance. Internal Report.*

———. 1996b. *14 MLD UASB treatment plant in Mirzapur, India, Evaluation Report on Process Performance. Internal Report.*

Heinke, G. W., D. W. Smith, and G. R. Finch. 1991. Guidelines for the planning and design of wastewater lagoon systems in cold climates. *Canadian Journal of Civil Engineering,* 18(4):556–567.

Hendrikson, H. V., S. Larsen, and B. K. Ahring. 1992. Anaerobic dechlorination of pentachlorophenol in fixed-film and upflow anaerobic sludge blanket reactors using different inocula. *Biodegradation,* 3:399.

Henze, M. 1996a. Anaerobic wastewater treatment. *Environmental Engineering.* U. Forstener, R. J. Murphy, and W. H. Rulkens, eds. Berlin: Springer Verlag. 273–309.

———. 1996b. Activated sludge treatment plants. *Environmental Engineering.* U. Forstener, R. J. Murphy, and W. H. Rulkens, eds. Berlin: Springer Verlag. 113–142.

———. 1996c. Plants for biological phosphorus removal. *Environmental Engineering.* U. Forstener, R. J. Murphy, and W. H. Rulkens, eds. Berlin: Springer Verlag. 273–284.

———. 1997. Trends in advanced wastewater treatment. *Water Science and Technology,* 35(10):1–3.

Higgens, M. J., C. A. Rock, R. Bouchard, and B. Wnegrezynek. 1993. Controlling agricultural runoff by use of constructed wetlands. *Constructed Wetlands for Water Quality Improvement,* Chelsea, MI: Lewis Publishers. 357–367.

Iza, J. 1991. Fluidized bed reactors for anaerobic wastewater treatment. *Water Science and Technology,* 24(8):109–132.

Jowett, E. C. 1997. Sewage and leachate treatment using the absorbent Waterloo Biofilter™. Site characterization and design of on-site septic systems. *Standard Testing Procedure 1324.* M. S. Bedinger, A. I. Johnson, and J. S. Fleming, eds. West Conshohoken, PA: American Society for Testing Methods. 261–282.

————. 1999. Immediate re-use of treated wastewater for household and irrigation purposes. *Presented at the 10th Northwest Wastewater Conference,* University of Washington, Seattle, Washington. 21 September.

Kadlec, R. H. 1989. Hydrologic factors in wetland water treatment. In *Constructed Wetlands for Wastewater Treatment,* D. A. Hammer, ed. Chelsea, MI: Lewis Publishers. 49–468.

Kadlec, R. H., and R. L. Knight. 1996. *Treatment Wetlands.* Boca Raton: CRC Lewis Publishers.

Kato, M. T., J. A. Field, R. Kleerebezem, and G. Lettinga. 1994. Treatment of low strength soluble wastewater in UASB reactors. *Journal of Fermentation and Bioengineering,* 77(6):679–686.

Kennedy, K. J. 1985. *Start-up and Steady State Kinetics of Anaerobic Downflow Stationary Fixed Film Reactors,* Dept. Of Civil Engineering, University of Ottawa, Ontario.

Kennedy, K. J. and Droste, R. L. 1991. Anaerobic wastewater treatment in downflow stationary fixed film reactors. *Water Science and Technology,* 24(8):157–177.

Kiely, G. 1997. *Environmental Engineering,* London: McGraw-Hill.

Knight, R. L. 1993. Operating experience with constructed wetlands for wastewater treatment. *Tappi Journal,* 76(1):109–112.

Knight, R. L., R. H. Kadlec, and S. Reed. 1999. *Database: North American Wetlands for Water Quality Treatment,* U.S. Environmental Protection Agency, Risk Reduction Environmental Laboratory, Cincinnati, OH. September.

Koopman, B., C. M. Stevens, and C. A. Wonderlick. 1990. Denitrification in a moving bed upflow sand filter. *Research Journal for the Water Pollution Control,* 62(3):239–245.

Kruger. 2001. Anaerobic/anoxic/oxic (A2/O) ditch process. Available online http://www.kruger.dk/doc/Processes/a2o.htm.

Kugelman, I. J., and K. K. Chin. 1971. Toxicity, synergism and antagonism in anaerobic waste treatment process. *Advances in Chemical Series,* 105:55–65.

Kuyucak, N. 1990. Feasibility of biosorbents application. *Biosorption of Heavy Metals.* B. Volesky, ed. Boca Raton: CRC Press. 372–378.

Latkar, M., and T. Chakrabarti. 1994. Performance of upflow anaerobic sludge blanket reactor carrying out biological hydrolysis of urea. *Water Environment Research,* 66(1):12–15.

Lawrence, A. W., and P. L. McCarty. 1969. Unified basis for biological treatment design and operation. *Journal of Sanitary Engineering.* ASCE 96(SA3):757.

Lens, P. N. L., F. Omil, J. M. Lema, and L. W. H. Pol. 2000. Biological treatment of organic sulfate-rich wastewater. *Environmental Technologies to Treat Sulfur Pollution,* P. N. L. Lens and L. W. H. Pol, eds. London: IWA Publishing. 153–173.

Leslie, M. 1983. *Water Hyacinth Wastewater Treatment Systems: Opportunities and Constraints in Cooler Climates*, EPA 600/2-83-075. Washington, D.C.: U.S. EPA.

Lettinga, G. A., A. F. van Velsen, S. W. Hobma, W. de Zeeuw, and A. Klapwijk. 1980. Use of the Upflow Sludge Blanket (USB) reactor concept for biological wastewater treatment, especially for anaerobic treatment. *Biotechnology and Bioengineering*, 22(4):699–734.

Liu, D. H. F., B. G. Lipták, and P. B. Bouis, eds. 1997. *Environmental Engineers' Handbook*. Boca Raton: Lewis Publishers.

Living Technologies, Inc. 2000. *Ethyl M. Chocolates. Henderson, NV*. Downloadable brochure. Available online http://www.livingmachines.com.

Maillacheruvu, K. Y., G. F. Parkin, C. Y. Peng, W. C. Kuo, Z. I. Oonge, and V. Lebduschka. 1993. Sulfide toxicity in anaerobic systems fed sulfate and various organics. *Water Environment Research*, 65(2):100–109.

Massé, D., and Droste, L. 1993. Psychrophilic anaerobic digestion of swine manure slurry in sequencing batch reactors. *Third International Conference on Waste Management in the Chemical and Petrochemical Industries*, IAWQ. Salavador, Brazil. October.

Metcalf and Eddy. 1991. *Wastewater Engineering: Treatment, Disposal and Reuse, 3rd ed*. T. Tchobanaglous and F. Burton, eds. New York: McGraw-Hill.

Mikkelson, K. A. 1995. *AquaSBR Design Manual*. Aqua-Aerobic Systems.

Mitsch, W. J., and J. G. Gosselink. 1992. *Wetlands,* 2nd ed. New York: Van Nostrand Reinhold.

Mulligan, C. N., B. Safi, P. Mercier, and J. Chebib. 1996. Full scale treatment of dairy wastewater using the SNC multiplate reactor. *Environmental Biotechnology*, M. Moo-Young, W. A. Anderson, and A. M. Chakrabarty, eds. Dordrecht: Kluwer Academic Publishers. 544–556.

Mulligan, C. N., J. Chebib, and B. Safi. 1997. Anaerobic treatment using the multiplate reactor. *Canadian Environmental Protection,* 9(7):16–17.

Murray, W. D., and L. van den Berg. 1981. Effects of nickel, cobalt and molybdenum on performance of methanogenic fixed-film reactors. *Applied and Environmental Microbiology*, 42:502–505.

National Research Council. 1946. Trickling filters in sewage treatment at military installations. *Sewage Works Journal*, 18(5).

Omil, F., R. Mendez, and P. J. Reynolds. 1995. Anaerobic treatment of saline wastewaters under high sulfide and ammonia content. *Bioresource Technology,* 54:269–278.

Oswald, W. J. 1968. Advances in anaerobic pond systems design. In *Advances in Water Quality Improvement, Water Resources Symposium*, No. 1. E. F.Gloyna and W. W. Eckenfelder, Jr., eds. Austin, TX: University of Texas. 409–426.

Owen, W. F. 1982. *Energy in Wastewater Treatment*. Toronto: Prentice Hall.

Owen, W. F., D. C. Stuckey, J. B. Healy, L. Y. Young, and P. L. McCarty. 1979. Bioassay for monitoring biochemical methane potential and anaerobic toxicity. *Water Research*, 13(6):495–492.

Peverly, J., W. E. Sanford, T. S. Steenhuis, and J. M. Surface. 1994. *Constructed Wetlands for Municipal Solid Waste Landfill Leachate Treatment*, Report 94-1. Albany, NY: New York State Energy Research and Development Authority.

Premier Tech Ltée. 1999. Assainissement autonome. Système de biofiltration des eaux usées (Ecoflo®) Enviro-Access Technology Fact Sheet, F3-08-96. Available online http://www.enviroaccess.ca.

Randall, C. W., J. L. Barnard, and H. D. Stensel. 1992. *Design and Retrofit of Wastewater Treatment Plants for Biological Nutrient Removal*. London: Technomic Publishers.

Randall, C. W., and E. U. Cokgor. 2000. Performance and economics of BNR plants in the Chesapeake Bay Watershed, USA. *Water Science and Technology*, 41(9):21–29.

Razo-Flores, E., B. A. Donlon, G. A. Lettinga, and J. A. Field. 1996. Biotransformation and biodegradation of N-substituted aromatics in methanogenic granular sludge. *FEMS Microbiology Review*, 20(3):526–538.

Razo-Flores, E., M. Luijten, B. A. Donlon, G. Lettinga, and J. A. Field. 1997a. Complete biodegradation of the azo dye azodiasalicylate under anaerobic conditions. *Environmental Science and Technology*, 31(7):2098–2104.

———. 1997b. Biodegradation of selected azo dyes under methanogenic conditions. *Water Science and Technology*, 36(6):65–75.

Reed, S. C., R. C. Crites, and E. J. Middlebrooks. 1995. *Natural Systems for Waste Management and Treatment*. New York: McGraw-Hill.

Reynolds, T., and P. A. Richards. 1996. *Unit Operations and Processes in Environmental Engineering*. 2nd ed. Boston: PWS Publishing Co.

Richards, E. A. 1996. Bioenergy from anaerobically treated wastewater. Online at: http://www.execpc.com/~drer/anadoc.htm.

Richardson, C. J., and D. S. Nichols. 1985. Ecological analysis of wastewater management criteria in wetland ecosystems. *Ecological Considerations in Wetlands Treatment of Municipal Wastewaters*, P. J. Godfrey, E. R. Kaynor, and S. Pelezarski, eds. New York: Van Nostrand. 351–391.

Rinzema, A. 1988. Anaerobic treatment of wastewater with high concentrations of lipids or sulfate. Doctoral Thesis, Wageningen Agricultural University, Wageningen, The Netherlands.

Rittmann, B. E., and D. E. Baskin. 1985. Theoretical and modeling aspects of anaerobic treatment of sewage. *Proceedings of the Seminar/Workshop on Anaerobic Treatment of Sewage,* Amherst, Massachusetts. M. S. Switzenbaum, ed. 55–94.

Robinson, D. G., J. E. White, and A. J. Callier. 1997. Anaerobic versus anaerobic wastewater treatment. *Chemical Engineering,* 104(4):110–114.

Samson, R., and S. Guiot. 1990. *Les Nouveaux Secteurs à Fort Potentiel de Développement en Digestion Anaérobie,* Study for the centre québecois de valorisation de la biomasse, mai 1990.

Singh, K. S., T. Viraraghavan, S. Karthikeyan, and D. E. Caldwell. 1997. Low temperature start-up of UASB reactors for municipal wastewater treatment. *Proceedings of the 8th International Conference on Anaerobic Digestion,* Sendai, Japan. 3:192–195.

Smith, D. W., and H. Knoll, eds. 1986. *Cold Climate Utilities Manual.* Montreal, Quebec: Canadian Society for Civil Engineering.

Spengel, D. B., and D. A. Dzombak. 1992. Biokinetic modeling and scale-up considerations for biological contractors. *Water Environment Research,* 64(3):223–235.

Steffler, J. R. 2000. Design and operational considerations in the construction of subsurface flow (SSF) constructed wetlands. *6th Environmental Engineering Specialty Conference of the CSCE.* London, Ont. 7–10 June. 530–537.

Tchobanoglous, G. and A. N. Angelakis. 1996. Technologies for wastewater treatment appropriate for reuse: Potential for applications in Greece. *Water Science and Technology,* 33(10):15–24.

Triebel, J. N., ed.1975. *Lehr- und Hanbuch der Abwassertechnik,* Bd II (*Textbook for Wastewater Engineering,* Volume II). Berlin: Verlag von Wilhelm Ernst & Soh.

U. S. Environmental Protection Agency (EPA). 1975. Nitrogen control. *EPA Process Design Manual.* Washington, D.C.: EPA.

————. 1983. *Municipal wastewater stabilization ponds: Design Manual,* No. EPA-625/1-83-015. Cincinnati, OH.: Center for Environmental Research Information.

————. 1986. *Sequencing Batch Reactors,* EPA/625/8-86/011. Cincinnati, Ohio: Center for Environmental Research Information. October.

————. 1987a. Anaerobic digester mixing systems. *Journal of Water Pollution Control Federation,* 59(3):162–170.

————. 1987b. *Phosphorus Removal.* Design Manual No. EPA/625/1-87/001. Cincinnati, OH: Center for Environmental Research Information.

————. 1988. *Constructed Wetlands and Aquatic Plant Systems for Municipal Wastewater Treatment.* Process Design Manual No. EPA-625/1-88-022. Cincinnati, OH: Center for Environmental Research Information.

————. 1991. *Biotrol Biotreatment of Wastewater.* EPA Demonstration Bulletin, EPA/540/A5-91/001.

————. 1993. *Nitrogen Control.* Manual No. EPA/625/1-93/010. Cincinnati, OH: Center for Environmental Research Information.

————. 2000a *Wastewater Technology Fact Sheet, Free Water Surface Wetlands,* EPA 832-F-00-024. September. Washington D.C.: EPA.

————. 2000b. *Wastewater Technology Fact Sheet, Wetlands, Subsurface Flow,* EPA 832-F-00-023. September. Washington, D.C.: EPA.

————. 2000c. Wastewater Technology Fact Sheet, Trickling Filters. EPA 832-F-00-014. Office of Water, Washington, D.C. October.

U.S. Filter/Zimpro. 2000. Company brochure. *Case Study: After 15 years El Paso still reusing effluent.*

U.S. Soil Conservation Service. 1993. *Nutrient and Sediment Control System,* Technical Note N4. Washington, D.C.: U.S. Department of Agriculture. March.

van der Last, A. R. M., and G. Lettinga. 1992. Anaerobic treatment of domestic sewage under moderate climatic (Dutch) conditions using upflow reactors at increased superficial velocities. *Water Science and Technology,* 25(7):167–178.

van Haandel, A.C., and G. Lettinga, G. 1994. *Anaerobic Sewage Treatment. A Practical Guide for Regions with a Hot Climate.* Chichester, United Kingdom: John Wiley and Sons Ltd.

Volesky, B. 1990. Removal and recovery of heavy metals by Biosorption. *Biosorption of Heavy Metals,* B. Volesky, ed. Boca Raton, FL: CRC Press. 8–43.

Water Environment Federation (WEF). 1992. Design of Municipal Water Treatment Plants. *WEF Manual of Practice No. 76,* Volume I: Chapters 1–12, and Volume II: Chapters 13–20. WEF, Alexandria, VA and American Society of Civil Engineers, New York.

Water Pollution Control Federation (WPCF). 1990. *Natural Systems for Wastewater Treatment, Manual of Practice No. FD-16.* Alexandria, VA: Water Environment Federation.

Watson, J. T., S. C. Reed, R. H. Kadlec, R. L. Knight, and A. E. Whitehouse. 1989. Performance expectations and loading rates for constructed wetlands. *Constructed Wetlands for Wastewater Treatment,* D. A. Hammer, ed. Chelsea, MI: Lewis Publishers. 319–351.

Weber, A. S., and G. Tchobanoglous. 1985. Nitrification in water hyacinth treatment systems. *Journal of Environmental Engineering, Div. ASCE,* 11(5):699–713.

Wieder, R. K. 1989. A survey of constructed wetlands for acid coal mine drainage treatment in the Eastern United States. *Wetlands,* 9:299–315.

Witthar, S. R. 1993. Wetland water treatment systems. *Constructed Wetlands for Water Quality Improvement*. Chelsea, MI: Lewis Publishers. 147–156.

Wolverton, B. C., and R. C. McDonald. 1979. Upgrading facultative wastewater lagoons with vacuolar aquatic plants. *Journal of the Water Pollution Control Federation*, 51(2):305–313.

Wrigley, T. J., D. F. Toerien, and I. G. Gaigher. 1988. Fish production in small oxidation ponds. *Water Research*, 22(10):1279–1285.

Wu, W. M., L. Bhatnagar, and G. Zeikus. 1993. Performance of anaerobic granules for degradation of pentachlorophenol. *Applied and Environmental Biotechnology*, 59(2):389–397.

Zenon Environmental, Inc. 1995. ZenoGem Process. EPA Site Demonstration Bulletin, EPA/540/MR-95/503. Washington, D.C.: EPA.

Zinkus, G. A., W. D. Byers, and W. W. Doerr. 1998. Identify appropriate water reclamation technologies. *Chemical Engineering Progress*, 94(5):19–31.

Soil and Groundwater Treatment

Introduction

Contamination of soil originates from numerous sources. In the late 1970s, improved methods of detection enabled contaminants to be identified in the environment at parts per billion levels. Indiscriminate dumping of materials, bankrupt and abandoned manufacturing plants, and insufficient waste storage, treatment, and disposal facilities led to the discovery of many contaminated sites in the 1970s. In the United States, two hazardous waste laws, Resource Conservation Recovery Act (RCRA) and Comprehensive Environmental Response Compensation and Liability Act (CERCLA), were enacted to ensure that waste generators were responsible for long-term impacts due to their waste handling procedures.

Although regulatory agencies classify wastes according to their sources, for remediation purposes, it is more appropriate to classify them in the following manner because it allows similar compounds to be grouped together according to (LaGrega et al. 1994):

- form—liquid or solid;
- composition—organic or inorganic;
- type of chemical—solvent, heavy metal, alkali, acid, etc.; and
- most hazardous form—hexavalent chromium as opposed to trivalent chromium.

Major categories include inorganic aqueous waste (heavy metals, acids/alkalis, cyanide), organic aqueous waste (pesticides), organic liquids (solvents from dry cleaning), oils (lubricating oils, automotive oils, hydraulic oils, fuel oils), inorganic sludges/solids (lime sludge, chromium dust from metal fabrication, sludges from mercury cells used in chlorine production), and organic sludges/solids (painting operations, tars from dyestuffs intermediates). Other sources of dense non-aqueous phase liquids (DNAPL) are shown in Table 5.1. Waste generated from manufacturing usually correlates with production. Most waste originates from cleaning of equipment, accident spills and leaks, residue in used containers, and outdated materials. Smaller generators of waste include landfills that are improperly managed, automobile service establishments, maintenance shops, and photographic film processors. Household wastes such as pesticides, paint products, household cleaners, and automotive products can also contribute significantly as sources (LaGrega et al. 1994). Technologies are required to remove these wastes that have contaminated the soil.

Table 5.1 Sources of DNAPL Contamination (adapted from Brar 1997)

Industrial Manufacturing Sources	Disposal Sources
Wood preservation	Underground tank storage
Coal gas plants	Drum storage
Pesticide and herbicide manufacturing	Solvent loading and unloading
Electronics manufacturing	Landfill disposal
Dry cleaning	Lagoons or ponds
Transformer oil production	Tool and die operations
Steel industry	Metal cleaning/degreasing
Pipeline compressor stations	Paint stripping

Introduction to Soil Treatment Processes

Soil treatment processes are classified as either *in situ* or *ex situ*. The latter type involves excavation of the soil before treatment, which can disrupt or shut down businesses and can cause a negative impact on the public; the former does not. *In situ* bioremediation involves the removal of the floating free-phase nonaqueous phase liquid (NAPL), the removal and treatment of the dissolved contaminants, and the treatment of soils in the aquifer and groundwater. Several variations have been developed such as bioventing, a simple process suitable for volatile and semivolatile contaminants in unsaturated soil. Bioventing is more appropriate than land treatment, slurry bioreactors, incineration, thermal desorption, and landfill because degradable and non-degradable volatile components can be removed. Phytoremediation is a developing process that uses plants—*in situ* or *ex situ*—to remove, contain, or render harmless environmental contaminants. Landfarming is one of the simplest methods for treatment of excavated soil, but time and land requirements can be extensive. The use of bioreactors as an *ex situ* process decreases residence time significantly. Composting, a method for waste treatment, is one of the newest and most promising *ex situ* methods for soil treatment.

Natural Attenuation

According to the U.S. EPA (1996), natural attenuation is the "use of natural processes to contain the spread of the contamination from chemical spills and reduce the concentration and amount of pollutants at contaminated sites." It can also be termed as intrinsic remediation, bioattenuation, and intrinsic bioremediation. In this case, the contaminants are left on the site, and the naturally occurring processes are allowed to clean up the site. The natural processes include biological degradation, dilution of the contaminant, and sorption of the contaminant onto the organic matter and clay minerals in the soil. The last two processes do not degrade the contaminants. Natural attenuation is subject to hydrological changes and can take substantial periods of time. The contaminant plume must not reach humans or wildlife areas. Prior to 1994, natural attenuation was not acceptable; however, by 1996, most U.S. states adopted policies concerning intrinsic bioremediation. These policies specify that the site must be properly assessed with the identification of all potential receptors, fuel must be the main contaminant, and the source should be removed or made harmless. Natural attenuation has become almost as routine as the sole remediation process but has also been used in combination with other processes. Guidelines were recently issued by the NRC (2000) for metal and organic contaminated sites. Natural attenuation is mainly useful, however, for BTEX. During the bioremediation, the contamination must be characterized thoroughly and monitored throughout the process. Modeling of contaminant migration plays a key role in ensuring that there will be no adverse impact. The required sampling and analysis can make this process more expensive than active remediation, the main focus of this chapter.

In Situ Bioremediation

Box 5-1 Overview of In Situ Bioremediation Processes

Applications	Suitable for treatment of low to medium contaminant levels of NAPL, chlorinated compounds, gasoline, solvents, petroleum products below the surface
Cost	Operating: $30 to $100/m^3
Advantages	Excavation not required
	Low capital and operating costs
	Numerous system variations available
Disadvantages	Difficulties with aeration
	Fractured bedrock or poorly defined aquifers can cause incomplete bioremediation
Treatment time	Three to 60 months
Other considerations:	Dependent on hydraulic conductivity and other soil characteristics
	Permeability should be greater than 10^{-4} cm/sec. Soil heterogeneities can also cause problems. Numerous full-scale demonstrations.

Description

In the petroleum industry, *in situ* biodegradation involves the in-place treatment of contaminated soil and groundwater contaminated from spills or leaking facilities. In the subsurface, the contaminants can be found as free product, adsorbed or bound to the soil particles or its constituents, or dissolved in the groundwater. *In situ* bioremediation can be used for all three types of contaminants. Suntech developed the first patented bioremediation process called "Reclamation of Hydrocarbon Contaminated Groundwater" (Raymond 1974).

The first mention of the word "bioremediation" was in 1987 in the *Scientific Citation Index 1974–1996*. During the mid 1980s interest in this technique increased significantly following technical advancements. Thirty applications were noted in 1987 (Schwefer 1988). Naturally-occurring microorganisms are now often used for remediation projects since they readily adapt to environmental change. Laboratory tests are used to determine if sufficient levels of the microorganisms are present and their optimal nutrient requirements.

Percolation of nutrients into the contaminated area is performed by allowing a solution of nutrients to seep into the soil. This method is feasible only for shallow zones since nutrient losses through adsorption would be too great over significant distances. Also, this method cannot be used for oxygen addition because of the limited solubility of oxygen in water. There is also the possibility of contaminant spreading since there is no way to control the vertical flow.

Pump and treat procedures involve pumping the contaminated groundwater to the surface for further treatment. This remediation technique can require periods of 30 years or more. The advantages of pump and treat procedures include low costs, fairly good control of contaminant transport, and minimization of site disruption. However, the flow of the injection can be difficult to control since the aquifer material is not homogeneous and treatment times can be very long, even up to several years. This is due to the difficulty in releasing the contaminants from the sorbed phase. DNAPLs are very difficult to remove by this method. Areas where biodegradation is accelerated (such as close to the nutrient injection wells) can lead to biological fouling. *In situ* bioremediation can be incorporated with pump and treat procedures by the injection of nutrients and oxygen with the pumping water into the contaminated zone. This increases the likelihood that acceptable contaminant levels can be achieved in the aquifer and that remediation rates increase.

In situ bioremediation involves the removal of the floating free-phase NAPL, the removal and treatment of the dissolved contaminants, and the treatment of soils in the aquifer and groundwater. Groundwater can be treated by conventional methods before reinjection or may be discharged elsewhere, depending on regulations and the quality of the groundwater. Injected water is supplemented with water from other sources, if necessary. The free-phase NAPL is removed first since it is highly concentrated and can spread to other uncontaminated areas. The pumping rate is a major design consideration and is selected depending on the hydraulic conductivity, the size of the contaminated zone, and the concentration of the contaminants. Pumping can be used for several purposes: creation of stagnant zones at various locations; creation of gradient barriers for migration of pollutants; control of the movement of the contaminated zone; and interception of the trajectory of the contaminated plume (Schafer 1984). There are four main types of patterns used for well placement: a pair of injection and production wells; several down-gradient pumping wells in a line; injection and production wells placed around the contaminated zone; and the "double-cell" system which consists of an inner cell, an outside recirculation cell, and four cells placed along a line in the direction of the flow that bisects the plume. Trenches can be used instead of wells for shallow groundwater. Samples should be taken during well installation to determine the extent of contamination. Soil samples should also be analyzed for particle size-distribution, porosity, and load capacity.

The volume of water between the injection and recovery wells is defined by the number of pore volumes that is determined by

$$PV = \frac{\text{Volume of flow through soil}}{\text{Volume of void space in soil}}$$

If the contaminant levels are in the range of several hundred mg/kg and the soil is a sand or silty sand, the treatment system will require handling three to 20 pore volumes. From the groundwater flow velocity plus the length of the soil plume divided by the water velocity, an estimation of the amount of time required to process one pore volume can be determined. This is usually between one and three months. To determine the remediation time, the time required for one pore volume is multiplied by the amount of pore volumes. Groundwater flow models can be used to simulate injection and recovery methods such as trenches, wells, leach fields, and others.

Methods

Soil Vapor Extraction

If treatment of VOCs in the vadose zone by air sparging is desirable via soil vapor extraction, suction wells are installed to apply negative pressure (75 to 150 mm Hg) and the contaminated gas is treated by activated carbon sorption, catalytic oxidation, or biofiltration. Extraction rates are usually 1 to 6 m³/min. The distance between the extraction wells depends on the flow rate and the soil type, but typically the radii of the wells are 5 to 10 m. Overlap of the radii of the various wells is desirable to ensure that no zones are missed. High vacuum pressure will lead to high air extraction rates and thus a rapid decrease in contaminant concentrations. Other operating practices involve continuous low extraction rates and intermittent high-rate extraction.

The following parameters should be determined to design a soil vapor extraction process:
- the type and quantities of VOC contaminants on the soil;
- the amount of dissolved VOCs (from soil moisture measurements, liquid extraction analysis, and solubility data);
- the quantity of vapor phase VOCs (from gas analysis); and
- the flow characteristics and appropriate extraction rates (from pilot tests).

Soil vapor extraction is excellent for contaminants that consist mainly of VOCs. It is also efficient for contaminant mixtures such as gasoline and for relatively compacted soils, since air is easier to move through these type of soils than water. Contaminants such as hydrogen sulfide, petroleum VOCs, ammonia, methylene chloride, and styrene can be treated in the air phase by biofiltration, as discussed in Chapter 3. For less biodegradable components such as trichloroethylene (TCE), PCE, and tetrachloroethane (TCA), catalytic oxidation or activated carbon adsorption should be used for treatment. Since the costs of treating a large area can be substantial, this type of treatment is more suited to small contaminant zones.

Co-metabolism

The injection of natural gas or methane with air can influence the metabolism of microorganisms called methanotrophs. These bacteria possess monooxygenase enzymes that oxidize methane, alkanes, alkenes, and halogenated alkanes (Thomas and Ward 1989). Some compounds that are recalcitrant under normal conditions can be degraded when methane is used as a carbon and energy source. Chlorinated solvents such as tetrachloroethylene; trichloroethylene; 1,1,1-trichloroethane; 1,1,2-trichloroethane; 1,1-dichloroethane; cis- and trans-1,2 dichloroethylene; 1,2 dichlorethene; vinyl chloride; dichloromethane; carbon tetrachloride; and chloroform. Semprini et al. (1990) showed that alternating pulses of methane and oxygen into groundwater could enhance the biodegradation of trichloroethylene, cis-dichloroethylene, trans-dichloroethylene, and vinyl chloride.

The acclimation and subsequent growth of methanotrophs can take considerable time depending largely on the breakthrough of methane. Once methane is detected, fluid injection should be stopped to allow the bacteria to grow. Fluid injection should then restart when the bacterium's population is well established. Residence time should be approximately 100 days (McCarty et al. 1991).

Degradation of TCE has been performed at field scale (Hazen 1991). Competition of methane and TCE is a problem since methane has a higher affinity for methane monooxygenase than TCE. Other difficulties include loss of enzyme activity upon TCE oxidation and the requirement for external energy sources to complete the oxidation.

In Situ Treatment of Sediments

In situ biological treatment of sediments is difficult since there is no oxygen, the temperature is low, and there is little mixing of water within the pores. The contaminants are not very bioavailable under these conditions. A combined biological/chemical treatment can work in this situation. Environment Canada performed a pilot test in Hamilton Harbor to remediate organic compounds from waste disposal, overflow from sewers, and runoff from coal piles (Anonymous 1999). Oxidants and nutrients (18.5 tonnes of calcium nitrate and 5 tonnes of organics) were injected in a 1,000 x 100 m area over 14 months. Total PAH levels decreased by 48% and hydrocarbon levels decreased by 57%. When attempts were made to inject the nutrients deeper into the sediments, methane was produced in the sediments and competed with the nitrate oxidant. More oxidant could have been added. Costs were approximately 20% that required for dredging.

Microbial Remediation of Metal-contamination

Techniques for microbial remediation of metals include bioaccumulation, biological oxidation/reduction, and biomethylation (Soesilo and Wilson 1997). Living cells can accumulate heavy metals through ion exchange, precipitation, and complexation on and within the cell surface, which contains hydroxyl, carboxyl, and phosphate groups. Bacterial oxidation/reduction could be used to alter the mobility. For example, some bacteria can reduce Cr(IV) to Cr(III) which is less toxic and mobile (Bader et al. 1996).

Sulfate-reducing bacteria (SRB) form metal (Me) sulfides that are insoluble as shown in the following reactions:

$$CH_3COOH + SO_4^{2-} \rightarrow 2HCO^-_3 + HS^- + H^+$$

$$H_2S + Me^{2+} \rightarrow MeS + 2H^+$$

As shown in the equations, sulfate, low redox conditions, and an electron donor such as methanol are required. Oxygen should not be there and nutrients must be added. Stimulating sulfate reduction can also increase pH and form metal hydroxides and oxides that precipitate and do not migrate in soils and groundwater. A project proposed by the Flemish Institute for Technological Research for NATO/CCMS 1999 and NATO/CCMS 2000 involved bioprecipitation of the heavy metals (zinc, cadmium, arsenic, lead, chromium, nickel, and copper) within biological reactive zones or biowalls. The first site contained zinc (0–150 mg/L), cadmium (0.4–4 mg/L), and arsenic (20–270 g/L) with high concentrations of sulfate (400–700 mg/L), ideal for SRB activity. At the first site, it was determined that acetate would be the best carbon source to add. A low concentration of acetate led to the reduction of zinc concentration from 10,700 g to 213 g/L. Arsenic and cadmium were also removed by precipitation. If high concentrations of acetate were used, methanogenic bacteria would dominate and no metal removal would occur. The addition of SRB bacteria decreased lag time in some cases. It was also found that redox must be below -220 mV. The second site had copper (up to 92 mg/L), chromium (up to 78 mg/L), zinc (up to 8.3 mg/L), and nickel (up to 3.5 mg/L) with levels of sulfate (up to 3,000 mg/L). Results showed that when the sulfate concentration was below 200 mg/L, the addition of extra sulfate (2,000 mg/L) or zero valent iron was necessary to reduce the redox to -200 mV. Under these conditions, complete nickel removal was obtained. Lead, zinc, chromium, and cadmium removal were also observed. Carbon addition was a necessity. Column tests are now underway to generate kinetic data for pilot tests.

Biomethylation involves the addition of a methyl ($-CH_3$) group to a metal such as arsenic, mercury, cadmium, or lead (Smith et al. 1995). The methylated forms are more mobile and can migrate into the groundwater. Although methylation increases volatility, it is not likely that methylation of metals such as arsenic will be performed for remediation since the by-products are more toxic. Methods for biomethylation are currently under development and not commercially available. Volatilization of selenium from contaminated agricultural soils has shown some promise (Thompson-Eagle and Frankenberger 1990).

Heap and *in situ* leaching has been used in the mining industry to recover copper and uranium from soil (Rawlings 1997). Copper ores usually have less than 0.5% copper contents. Because *Thiobacillus* produce sulfuric acid from reduced sulfur, they are autotrophic anaerobes that grow within the pH range of 1 to 4. *Thiobacillus* grow between 15 and 55°C, depending on the strain. Leaching can be performed by indirect means—acidification of sulfur compounds to produce sulfuric acid, which then can desorb the metals on the soil by substitution of protons. Direct leaching solubilizes metal sulfides by oxidation to metal sulfates. The acid solubilizes metals, which can then be recovered through precipitation at higher pH values. The bioleaching efficiency depends on the speciation of the metals. Leaching of other metals, including aluminum, cadmium, chromium, iron, manganese, mercury, nickel, selenium, and tin is being developed (Smith et al 1994).

Several options available for bioleaching include heap leaching, bioslurry reactors, and *in situ* methods. Anoxic sediments are more suitable for treatment since the bacteria can solubilize the metal compounds without substantially decreasing the pH. Soils require lower pH values to extract the metals since they have already been exposed to oxidizing conditions. Bacteria and sulfur compounds are added for both heap leaching and reactors. Mixing is used in the reactor, and pH can be controlled more easily. Leachate is recycled during heap leaching or *in situ* leaching. Copper, zinc, uranium, and gold have been removed by *Thiobacillus* sp. in biohydrometallurgical processes (Karavaiko et al. 1988). Fungus such as *Aspergillus niger* can produce citric and gluconic acids, which can act as chelating agents for the removal of metals, such as copper from oxide mining residues (Mulligan et al. 1999a). Several feasibility studies have indicated that contaminated soils can be remediated (Tichy et al. 1992). Sludges from anaerobic processes that contain metal sulfides could be treated in this manner (Blais et al. 1992).

The research in the area for metal removal is still quite limited (U.S. EPA 1987). Agents for metal removal include organic and inorganic acids, sodium hydroxide which can dissolve organic soil matter, water soluble solvents such as methanol, displacement of toxic cations with nontoxic cations, and complexing agents such as ethylenediaminetetraacetic (EDTA) acids in combination with complexation agents or oxidizing/reducing agents. Soil pH, soil type, cation exchange capacity (CEC), particle size, permeabilities, and contaminants all affect metal removal efficiencies. High clay and organic matter contents are particularly detrimental to metal.

A combination of hydrocarbons and metals is found at 49% of the Superfund Sites with signed Records of Decision (U.S. EPA 1997). Even though organic and metal contamination is a major concern, very few technologies are capable of dealing with both types of contaminants. It has only recently been shown that surfactants can be used to enhance metal removal (Mulligan et al. 1999b; Mulligan et al. 1999c). Biologically produced surfactants, surfactin, rhamnolipids, and sophorolipids (Mulligan et al. 1999a) are able to remove copper and zinc from a hydrocarbon-contaminated soil. This is because of the anionic character of these surfactants. In a test, a series of washings was performed with surfactin and then compared to a control. Initial concentrations of oil and grease were 12.6%, and the initial copper content of the soil was 550 mg/kg. Five consecutive washings were performed, each lasting 24 h. While the control showed a final cumulative removal of 20% of the copper, approximately 70% of the copper was removed by the surfactin. At the same time, approximately 50% of the hydrocarbons were removed by surfactin compared to 30% by the control. Therefore, these results seem very promising. Other advantages of these biosurfactants are that they are biodegradable and low in toxicity. In addition, they can potentially be produced *in situ* using the organic contaminants as substrates for their production. Larger scale studies, however, are needed.

Anaerobic Processes

Many organic compounds such as toluene, phenol, cresol, and benzoic acid can be degraded by microorganisms acting as a consortium under anaerobic conditions. Anaerobic processes may be preferred to aerobic if it is difficult to maintain a dissolved oxygen level (DO) greater than 1 mg/L. Also, nutrient requirements are less than for aerobic processes, and anaerobic processes produce less biomass; thus, there is less chance of biofouling. Types of bacteria used in anaerobic processes include nitrate-reducing, sulfate-reducing, iron-reducing, and methanogenic bacteria.

Heterocyclic compounds with oxygen, nitrogen, or sulfur substitution within the rings can be anaerobically degraded, and sulfate or carbon dioxide can be used as alternative electron acceptors. Chlorinated aliphatic and aromatic compounds that are difficult to degrade aerobically can be treated anaerobically. Since the chlorinated compound acts as an electron acceptor, the chlorine moiety is removed and replaced by hydrogen (Vogel et al. 1987). Sims et al. (1991) indicated in a review that PCE; TCE; TCA; cis- and trans-1,1-dichlorethene; 1,1 dichloroethane; 1,2-dibromoethane; tetrachloromethane; 1,2 dichloroethane; chloroform; bromoform; and vinyl chloride have been biodegraded by anaerobic subsurface-derived or pure cultures. It has also been determined in field experiments that BTEX can be anaerobically degraded (Borden et al. 1997). Carbon dioxide, ferric iron, or sulfate were used as electron acceptors in this case.

Since only small numbers of electron acceptors are necessary and oxygen is not required, anaerobic processes are potentially viable alternatives. Acetate is used as a supplemental carbon source to stimulate nitrate- and sulfate-reducing organisms for degradation of halogenated compounds (Semprini et al. 1991). Although sulfate is very soluble and inexpensive as an electron acceptor, its use will produce sulfide, a toxic product for humans and microbes. Iron (III) is not practical alone due to its poor electron-accepting capacity, but it can be used in combination with others. Nitrate has very good potential as an electron acceptor because it is very soluble in water, highly mobile, and not very reactive. The disadvantages of this process include the cost and toxicity of nitrates, the formation of nitrogen gas, and the reduction of hydraulic conductivity.

Innovations have been made in anaerobic treatment processes. Geovation developed a process called Anaerobic Bioremediation System (ABS) for the *in situ* anaerobic treatment of petroleum hydrocarbons, halogenated solvents, and organo-pesticides. Regenesis (San Clemente, CA) has developed a timed-release hydrogen compound, HRC™, to enhance *in situ* anaerobic remediation of chlorinated hydrocarbons (PCE, TCE, DCE, and VC). It releases lactic acid, a food-grade compound that provides hydrogen and is an essential nutrient. HRC favors reductive dechlorination over methanogenic activity.

Anaerobic processes are suitable for the treatment of soil and sediments and are low cost. Field data is still limited and regulatory levels may be difficult to achieve. Recently, it was shown that a pure anaerobic strain *Desulformonile tiedjei* could be added to soil to degrade 3-chlorobenzoate (El Fantroussi et al. 1999). The tests were performed at pilot scale (500 L). Microbial activity and biogas production were high. Enhanced anaerobic dechlorination is being evaluated for *in situ* remediation of PCE and TCE contaminated groundwater.

The technology is called Reductive Anaerobic *in situ* Treatment Technology (RABITT). A treatability protocol has been applied at five U.S. Department of Defense sites (Morse 1998). The selection of an electron donor to enable the efficient biodegradation of chlorinated compounds will most likely be the crucial step in the process.

Criteria for Utilization

An initial assessment of the site must be performed to determine if biodegradation of the contaminants is feasible. Hydrogeology and other factors influence the ability to get oxygen to the bacteria. Incomplete remediation can occur in the case of fractured bedrock or poorly defined aquifers. Information about the soil, groundwater, and aquifer characteristics is essential in determining the nature and extent of contamination, the potential impact on groundwater, the hydraulic conductivity and permeability of the soil, the groundwater flow rate and direction, the cation exchange capacity of the soil, and the ionic composition of the groundwater. Hydraulic conductivities greater than 10^{-5} cm/s are preferred because fine-grained soils such as clays restrict the transport of oxygen. The determination of the geology, geochemistry, and hydrogeology factors can help prevent the spreading of the contaminants into unwanted and uncontaminated areas. Hydraulic barriers such as trenches or slurry walls may need to be installed to restrict the flow. Knowledge of the ionic composition of the groundwater (levels of calcium, iron, and manganese) can eliminate unwanted interactions with nutrients. The location of underground pipe, cables, sewers, and other buried objects can influence where remediation equipment will be placed. Fractured bedrock can also cause problems since the groundwater can transport the contaminants to places that are difficult to predict. Injection and extraction conduits must be placed to avoid the fractures.

The seasonal variations in temperature and water availability are other important considerations. Usually the groundwater temperature does not vary by more than 3 to 4°C, except in the summer in the southern U.S. and in the winter in the northern states. The pH level is usually in the range of 6 to 9. Standard test methods are shown in Table 5.2. Soil samples are taken to analyze for contaminant type and levels, and to determine the bacterial count (total and/or hydrocarbon degrading population). Evaluation of the number and type of contaminants is very important in determining the feasibility of biodegradation. The important considerations in determining biodegradability are the size, shape, volatility, toxicity, and solubility in water of the contaminants and the octanol/water coefficient (K_{ow}). Microbiological detection methods are used to examine the distribution and activities of the microorganisms (Ghiorse and Balkwill 1983). The rates of oxygen uptake, carbon dioxide production, and contaminant degradation and the number of contaminant degrading organisms are used to determine bacterial activity. Typically, 10^6 to 10^9 cells per gram are required for bioremediation. In treatability tests, a 20 to 25% biodegradation is expected in four to six weeks.

Natural consortia of bacteria are more likely to treat complex mixtures better than pure cultures or laboratory studies because they do not allow metabolic by-products to accumulate. Exceptions are noted, though, when the contaminants are radioisotopes, asbestos, or metals such as cadmium, mercury, or chromium, PCBs of molecular weight greater than 1248, and PAHs of greater than six rings. Mobilization of metals can occur through changes in the organic matter or if the pH decreases. Metal sulfides can also precipitate and cause plugging during anaerobic treatment. The same tests are also applied to groundwater samples. Additional tests for groundwater include pH, suspended solids, BOD, COD, total organic carbon (TOC), and nutrient concentration tests.

Table 5.2 Standard Methods for Soil Analyses

Parameter	Method
Particle size	ASTM D-422
Moisture content	ASTM D-2216
Minimum relative density	ASTM D-4254
In situ density	ASTM D-2435
Specific gravity	ASTM D-854
Atterberg limits	ASTM D-4318
Bulk density	ASTM D-1587 or 2937
Description of soils	ASTM D-2488
Hydraulic conductivity	ASTM D-5126
Total Kjeldahl nitrogen	EPA method 351.2
Total phosphorus	EPA method 6010
Total organic carbon	American Society of Agronomy Method 90-2

Laboratory tests are then performed with soil and groundwater samples (10 to 25% solids) with various ratios of nutrients and oxygen. Since the temperature of groundwater is usually between 10 and 15°C, it is preferable to perform tests within this range to obtain more accurate kinetic information. Hydrocarbon levels are followed for three days to two weeks. Since these tests are run under highly mixed conditions with relatively homogeneous soil and water samples, "fudge factors" in the range of two to six are used to estimate field results.

Relative toxicity tests by Microtox® and measurements of oxygen consumption can also be performed. Levin and Gealt (1993) estimate that for every 1,000 kg of hydrocarbons to be degraded, 10,000 kg of oxygen and 875 kg of ammonia are required to produce 7,000 kg of bacteria. Shallow aquifers (less than 60 m) contain some oxygen while deeper ones will contain sulfate or nitrate. Due to natural degradation, oxygen levels can be low, however. Gas chromatography-mass spectrometric (GC/MS) techniques can be used to determine degradation pathways if this is required by regulatory authorities who are concerned with the fate and transport of the degraded compounds. Radiolabeling is another method for determining the breakdown of components.

Abiotic controls are used to determine the extent of air stripping and chemical oxidation. The tests are also used to evaluate orthophosphate precipitation in the form of calcium phosphate, which can clog wells. The addition of chelating agents can reduce this problem.

Hydrogen peroxide can react with iron, manganese, and copper at levels of 10 mg/kg or less, decreasing its effectiveness as an oxygen source.

Lower pumping rates are typically used as the concentrations in the water drop throughout the procedure. Aqueous contaminant concentrations depend upon the number of contaminants, the volume of the contaminant zone, and the sorption of contaminants onto the soil. The water is then recovered aboveground for further treatment and reinjection or discharge. Hydrophobic compounds cannot be easily removed by this method. Injection enhances biodegradation within the contaminated zone. Methods for surface treatment include air stripping, carbon adsorption, chemical oxidation, and biological treatment.

Based on the initial assessment tests, nutrients are chosen and added to the groundwater before injection. Nutrients are delivered at the ratio of C:N:P of 30:5:1 for balanced growth. The C:N:P ratio of 100:10:2 is frequently used in field practice (Litchfield et al. 1994). Because nutrients often do not completely dissolve, it is preferable to use double diaphragm pumps which last longer. The main disadvantage of double diaphragm pumps is that nutrient pulses are created, which causes vibration throughout the system and mechanical stress.

Two types of approaches are used depending on the assessment of the microbial population. Biostimulation involves the stimulation of indigenous target microbes. Many products exist on the market for stimulating the bacteria. However, there does not seem to be any evidence that any of them are beneficial. Bioaugmentation is used for virgin spills and in places microbes are not present or the compounds are recalcitrant. In this case, microorganisms that are stable or have a desired functionality are added because they can survive and be effective under natural conditions. For example, the white rot fungus *Phanerochaete chrysosporium* produces a variety of organic compounds and extracellular peroxidases that can degrade. Difficulties include how to send these organisms to the contaminant area and how to maintain them. It may also be difficult for these microorganisms to compete with indigenous bacteria. In general, the cost of adding bacteria is not justified (Chapelle 1999).

Operating Conditions

Installation and Start-up

Once all the equipment, including pumps, mixers, and shed, is specified and ordered, they are delivered and installed. Utilities and fences are set up to complete the installation. Although it is more costly to bury piping, it does not need to be insulated or heat traced if it is below the frost line and cannot be tampered with. Quick-connecting devices are useful since the plumbing often needs to be rearranged as the process proceeds. Check valves are particularly useful in the oxygen supply line because it is not safe for contaminated water to enter these oxidative conditions.

Start-up procedures involve the following sequence. First the recovery pumps are started and the surge tanks are filled to their desired levels. The water treatment system is started and brought to steady conditions. When the treatment system is working efficiently, the reinjection system is initiated. Nutrients are mixed with the reinjection water. To determine the flow rate, a tracer such as chloride ion can be used. If levels are low, additional amounts are added. However, if the amount is higher than 50 mg/L, sodium bromide is used instead. Oxygen delivery is only started when the nutrient supply system is working well. Oxidant suppliers often supply start-up instructions.

Regular Operating Conditions

After start-up, monitoring is initiated. This is particularly important for the first three weeks. Necessary daily readings include pH, dissolved oxygen, temperature, oxidation/reduction potential (ORP), nutrient levels, and chloride or bromide concentrations. Any problems with flow and injection systems can be determined from these measurements. Samples should be taken every week or two weeks for Microtox assay®, oxygen uptake rate, and other wastewater parameters. After the collection of a pore volume, evidence of bioremediation should be present. Hydrocarbon levels and bacterial counts are the prime parameters for determining this. It has been suggested that levels of 10^6 per ml of CFU are optimal (King et al. 1998). Bacterial levels are controlled primarily by nutrient levels since oxygen should be maintained at saturation levels, if possible. Bacterial levels that are too high can cause excessive plugging of well screens.

Regular removal of biomass from the injection and recovery wells is necessary. With hydrogen peroxide as an oxygen source, the addition of a 0.5% solution without pumping for four hours is useful. Another method is to add bleach with the same procedure. Reversal of the flow can be used to ensure that all soil is remediated. This will enable bioremediation to occur near the initial injection well.

In the case of DNAPL remediation, the initial plume removal is followed by the injection of bioactive fluids that follow the initial path of the contamination. Careful design of the extraction and injection location and pump flow rates can be accomplished.

Production of microbial surfactants in the aquifer enhances desorption of hydrocarbons from the soils, seen as a spike, and can be interpreted as contaminant levels higher than expected. Other substances, called chelants, are produced by microorganisms and bind insoluble metals. The metals can be solubilized and mobilize into the groundwater. Although this does not always happen, be aware that it can occur, particularly when target contaminants fall below regulatory levels. Shutdown should only occur when nutrients and oxygen levels are the same as at the injection wells and bacterial levels are returned to initial levels. This will ensure that all hydrocarbons have been biodegraded.

The DARAMENDTM technology (Figure 5.1) can be applied for PAHs, PCP, manufactured gas plant PAHs, phthalates, chlorinated pesticides, herbicides, and organic explosives

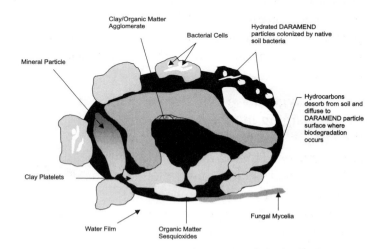

Figure 5.1 Schematic of the DARAMEND Technology

(Ferguson 1997). DARAMEND is a particle-containing native soil bacteria that works through the desorption of the contaminant from the soil into the DARAMEND particle surface. In a test, contaminated soil containing 8,700 mg total petroleum hydrocarbons (TPH)/kg decreased to 35 mg/kg in 182 days (greater than 99% removal). The treatment also decreased the level of PCP from 680 mg/kg to 6 mg/kg in 207 days. For the EPA SITE Program, 95% of PAHs and 88–96% of chlorinated phenols were desorbed. As an *ex situ* process, it costs $100-190/tonne. No excavation is required and it is unobtrusive. Its main disadvantage is that it has limited full-scale experience. DARAMEND™ II technology (U.S. Patent No. 5,618, 427) is anoxic/oxic cycling with reductive and oxidative phases. Various pesticides have been treated using this technology. In 147 days, DDT was reduced from 684 to 2 mg/kg, DDD was reduced from 141 to 64 mg/kg, Aldrin decreased from 12 to 1 mg/kg, dieldrin decreased from 128 to 38 mg/kg, and toxaphene decreased from 1,046 to 244 mg/kg.

Design

Oxygen Supply System

To obtain oxygen saturation levels (10 mg/L or higher), pure air or oxygen is pressurized into the injected water. To oxidize 1 g of gasoline, 2.5 g of oxygen are required (Verheul et al. 1988). Hydrogen peroxide can also be added to the water to enhance the oxygen content. The injection of these oxidizing agents requires special materials of construction and the suppliers' recommendations must be followed. Hydrogen peroxide breaks down into water and oxygen to yield 0.47 g of oxygen per g of hydrogen peroxide. However, concentrations of more than 1,000 mg/L should not be added because it can become toxic at these levels. Such levels can also cause precipitation in soils with a high iron and manganese content. Concentrations of 100 to 500 mg/L are usually used after stepping up from an initial concentration of 50 mg/L (Thomas and Ward 1989). Oxygen delivery by this method is also limited since hydrogen peroxide has a half-life of only 30 to 90 minutes (Hinchee and Downey 1988). In another study, Downey and Elliott (1990) determined that instability could be a major problem and that oxygen levels of 40 mg/L should be achieved for hydrogen peroxide to be economical. Hydrogen peroxide can also assist bioremediation by oxidizing some contaminants such as TCE. The oxidized compounds can be degraded more easily (Carberry 1990).

Oxygenation can also be achieved by air sparging in which air is forced into the aquifer and follows various fractures and crevices. This can be done with either vertical or horizontal well systems (Figure 5.2). A screen is placed at the lowest point of the contamination. To calculate the driving pressure, the following equation can be used:

$$P_{air} = \rho_w g h$$

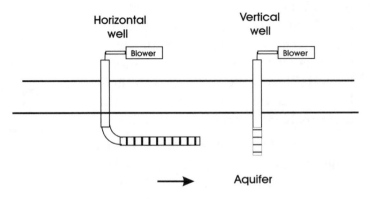

Figure 5.2 Vertical and Horizontal Well Systems for Oxygenation

where P_{air} is the air pressure to empty the well, ρ_w is the water density (kg/m^3), g is the gravitational constant ($9.81 \ m/s^2$), and h is the distance from the top of the screen to the top of the water table. Air sparging is most effective for shallow zones of 1 to 15 m in height. Air flow rates of 0.05 to 0.2 m^3/min are typical. Well diameters of 10 to 50 mm are typically used. The well casings are made of polyvinyl chloride (PVC) (Eweis et al. 1998). Spacing between the wells is a few meters to compensate for air channeling. Pilot testing to determine oxygen requirements and optimal well spacing is preferred prior to full-scale installation because biological activity in contaminated water decreases oxygen concentrations. Some remediations that have been successfully carried out are described by Brown et al. (1994a) and Billings et al. (1994). Decreased levels of contaminants near the recovery wells, increased concentrations of carbon dioxide, increased microbial numbers, and changes in the composition of the contaminants are good indications that bioremediation has taken place. Care must be taken, though, to ensure that the volatile organic compounds (VOCs) are not volatilized without treatment since these compounds are extracted by the air sparging.

Oxygen generation on site (Prosen et al. 1991) is a potentially economic alternative for use in sparging systems. However, the amount of oxygen supplied by this method is not sufficient. It has been estimated that generation costs are in the order of $0.22 to $0.24/kg of oxygen. Another novel approach is to use semi-permeable tubules that contain gaseous oxygen. The oxygen is released from the membranes as groundwater flows over the tubules (Semmens et al. 1991). With water flow rates of up to 1,000 L per minute, oxygen concentrations of 40 mg/L could be obtained (Piotrowski et al. 1994). Another method which has potential involves the injection of small bubbles of oxygen called colloidal gas aphrons (Michaelson and Lofti 1990). OHM Remediation Services Corp. (Findlay, Ohio) developed a process which injects a biodegradable surfactant at a concentration of 200 mg/L with oxygen to form bubbles of 45 to 100 microns in size. The microbubbles can enter high permeability areas. Thus, groundwater passing through the zone could be remediated with the added nutrients. OHM is looking at the technology for remediating sites of different hydrogeologic characteristics. Eftekhari and Mulligan (2000) also determined that the injection of foam enhances the remediation of PCP contaminated soil through mobilization and volatilization of the PCP.

Oxygen-releasing compounds (ORC) are another technology for supplying oxygen. They are a patented formula of magnesium peroxide that release oxygen over time when water is added; thus, the oxygen supply is continuous. ORCs are commercially available as powders or pellets from Regenesis Bioremediation Products and have been used to remediate petroleum and organic compounds, as well as some chlorinated solvents, vinyl chloride, PCP, PAHs, and MTBE. A mixture can be injected into a well as a powder or in boreholes as a slurry. In wells or trenches, they can be added to the groundwater as filter socks. Continuous mechanical operation is not required after injection. They potentially can reduce remediation costs (Fulton 1996).

Applications

Subsurface soil and groundwater can be treated together by *in situ* bioremediation, a process suitable for the treatment of low to medium subsurface contaminant levels of NAPL, chlorinated compounds, gasoline, solvents, and petroleum products. Floating free-phase NAPL, dissolved contaminants, and soils in the aquifer and groundwater can be treated simultaneously.

Costs

Operating costs for *in situ* bioremediation have been estimated at $30 to $100/m^3. Costs of *in situ* treatment are related to nutrient and oxygen delivery. The estimated cost of hydrogen peroxide is approximately $3.20 to $4.63/kg oxygen (Prosen et al. 1991). In terms of groundwater, hydrogen peroxide is estimated to cost $10 to $20 per 1,000 liters treated. Van Cauwenberghe and Roote (1998) reported that costs with ORC could be 50 to 75% of air sparging and pump and treat processes. As an *in situ* process, the DARAMEND process costs from $60 to $105/tonne. Monitoring costs can also be extensive. *In situ* bioremediation does not require removal or excavation of soil or extensive equipment.

Case Study

A demonstration was performed at the Edwards Air Force Base (Mojave Desert, California) for the treatment of groundwater contaminated with trichloroethene (TCE). The site has two homogeneous aquifers. The upper one was 8 m thick and the lower one was 5 m thick. They were separated by a 2 m aquitard. Bedrock lay below the lower aquifer. TCE concentrations varied between 500 to 1,200 μg/L with average concentrations of 680 and 750 g/L in the upper and lower aquifers, respectively. The aquifer was 9 m below the ground in a confined aquifer of sand and gravel 1.5 m in thickness. Hydraulic conductivity was on average 3.4 x 10^{-3} cm/sec and the groundwater velocity was 6.9 cm/day.

Two 20 cm diameter PVC treatment wells were installed 24 m deep and 10 m apart. Screening was placed in the wells for the upper (15 m) and lower (10 m) aquifers. The flow rate of the wells was maintained at 38 L/min in well number two and 25 mL/min in well number one to limit drawdown in the upper aquifer and changes in pressure in the lower aquifer. Toluene and oxygen (in the form of hydrogen peroxide and dissolved oxygen) were added into the wells with feed lines. Static mixers within the wells mixed the water. In one of the treatment wells, water was withdrawn from the upper aquifer and discharged into the lower aquifer; in the other, water was withdrawn from the lower aquifer and discharged into the upper one. The systems operated for 444 days, achieving steady state operation by day 142. After day 317, operation was on a balanced-flow basis. Groundwater pumping was at a rate of 25 L/min, toluene addition was 9 mg/L at a rate of 0.67 pulse per day, dissolved oxygen addition was 44 mg/L, and hydrogen peroxide addition was 47 mg/L for each well. Twenty monitoring wells were used over an area of 480 m^2. Overall TCE removal was 97.7% over the treatment period and concentrations were reduced from 1,150 μg/L to 27 μg/L. Toluene removal was almost 100%. Total capital costs were $323,453, and total annual operating and maintenance costs were $4,354. Well clogging was prevented by well redevelopment and the use of hydrogen peroxide. This contributed to the operating cost while the extensive system of monitoring wells contributed significantly to the capital costs. The dual system worked very well for the two aquifers and could potentially be used for a single aquifer where the hydraulic conductivity is layered into high and low conductivity zones.

In 1999, a system called E-DOT™, a collaboration between Compliance, Inc. (Tranverse City, Michigan) and Micro-Bac International, Inc., was installed to clean up a gasoline plume (121 m x 45 m) (Kenney and Schindler 2001). Groundwater was at a depth of .1 m. Liquid oxygen was injected into the groundwater for microbial stimulation, and a proprietary blend of microorganisms was added to enhance bioremediation. $3,500 of oxygen is added annually via pressure from the liquid oxygen tank. Replacement of the tank is done on a regular basis. Contaminant levels are expected to be below drinking water levels by the end of 2001 and closure will be in 2002.

Comparison to Other Methods

Bioremediation, in comparison to other processes, is less expensive and lower in energy requirements. It is effective, and complete mineralization is possible. Control of oxygen, moisture, and nutrient levels can become difficult. Addition of microorganisms or co-substrates may be necessary. For *in situ* remediation, the waste is treated on site without the liability associated with hauling and disposal. The risk that the contaminants will be spread further is reduced, particularly since microorganisms reduce soil permeability (U.S. DOE 1995). Bioremediation can also treat contaminants that may be inaccessible by other means. The major consideration is the time required, which can be extensive. In addition, the presence or absence of appropriate bacteria is another factor to consider when determining if biodegradation is an appropriate method. Biological treatment is able to reduce contaminant levels to the required objectives and is a cost-effective method that is less likely to produce toxic by-products. There is a growing list of compounds that can be treated biologically. The key to successful scale-up is the correct application of treatability tests (Block et al. 1997). An example of a treatability protocol is shown in Figure 5.3. Once this protocol has been performed, a decision tree similar to that in Figure 5.4 helps to select the appropriate choice of technology.

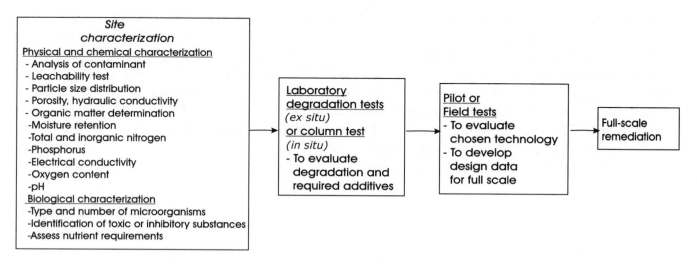

Figure 5.3 Protocol for Biological Feasibility Studies

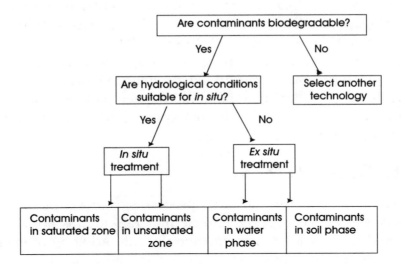

Figure 5.4 Decision Tree for Selection of Bioremediation Treatment Processes (adapted from Cookson 1995).

In the United States, biological methods are used in approximately 15 to 20% of all treatment cases (Levin and Gealt 1993). According to VISITT 6.0, bioremediation accounts for 150 out of 370 technologies. According to Record of Decisions (RODs) for 1982-1997 for Superfund remedial actions, bioremediation is the fifth most used *ex situ* technology and is being used in 33% of operational projects. For *in situ* projects, bioremediation is the third most common technology (U.S. EPA 1999). For *in situ* groundwater remediation, bioremediation is the second most used technology and has been used in 25% of the projects. Bioremediation is the second most widely used innovative technology. Bioremediation projects have increased from 1982 to 1996. Petroleum hydrocarbons, BTEX, and PAHs are treated most frequently.

Three large projects in the U.S. are treating more than 190,000 m³ by *in situ* bioremediation. The average volume of soil treated by *in situ* bioremediation was approximately 114,000 m³ more than soil vapor extraction. For *in situ* groundwater remediation, air sparging was used in 51% of the projects, *in situ* bioremediation in 29%, other technologies including dual phase extraction, permeable reactive barriers, and chemical treatment in 21%, biosparging in 3% and bioslurping in 1%. Innovative technologies for other federal programs other than Superfund and including those at Department of Defense (DoD) and Department of Energy (DOE) sites were identified as of August 1998. *In situ* bioremediation was the most common (26%) followed by SVE (25%) and *ex situ* bioremediation (19%). In this case, *in situ* bioremediation included groundwater treatment.

In Situ Bioremediation: Bioventing

Box 5-2 Overview of Bioventing

Applications	Suitable for treatment of volatile and semi-volatile chlorinated solvents and petroleum hydrocarbon in the vadose zone
Cost	Operating: $13 to $40/m^3
Advantages	Excavation not required
	Low capital and operating costs
	Numerous system variations available
Disadvantages	Difficulties with aeration
	Poorly defined aquifers can cause incomplete bioremediation
	Dependent on hydraulic conductivity and other soil characteristics
Treatment time	Six months to ten years
Other considerations:	Some of the difficulties involve passing the air through the contaminated zone. Proven at full scale

Description

Bioventing (Figure 5.5) involves the addition of air and nutrients through forced ventilation to biologically degrade semi- and nonvolatile contaminants (Henry's law coefficient less than 0.1) in the unsaturated zone with a high gas permeability. Air recovery of untreated compounds is by blower or passive means. Compared to soil vapor extraction (SVE), air is added only in sufficient quantities for biodegradation to occur. Although contaminant migration is minimized, emissions must be monitored to determine if recycling or treatment above the surface, such as by activated carbon, catalytic oxidation, thermal processes, or biofiltration, is necessary. In Europe, biofiltration has been used for 20 years and is becoming more accepted in the United States. Other alternative air treatment processes include carbon adsorption or incineration.

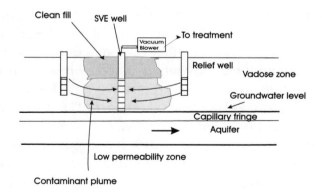

Figure 5.5 Schematic Diagram of Bioventing

Operating Conditions

Some of the difficulties with bioventing involve low moisture content, nutrient availability, and problems in passing the air through the contaminated zone. Based on soil gas pressures, oxygen concentration, and gas flow, the radius of influence is estimated and monitored. Procedures are preformed withdrawing or injecting air for 8 h at vent wells within the contaminated soil zone (U.S. Department of the Air Force 1992). Spacing of the monitoring points can then be determined. *In situ* respiration tests are used to obtain the oxygen rate uptake and can then be used to determine the biodegradation rate and, subsequently, the treatment time.

Characterization of the site for determination of the soil characteristics, moisture content, and hydraulic conductivity is essential. Addition of nutrients is difficult since saturation of the soil must be avoided because the ability to oxygenate decreases as the water content increases. Nutrient solutions can be added by allowing them to percolate through the soil. A 15% moisture content would be sufficient for a soil with porosity of 30%. Soil moisture must be low since up to 55% of water saturation can reduce gaseous permeability by more than 80% (Cookson 1995). Ridge and furrow irrigation, trenches, or spray irrigation can be used to provide the periodic addition of moisture and nutrients. Standard agricultural tests can be used to determine if nutrient availability is sufficient. Samples should be taken throughout the treatment process to determine nutrient and moisture contents. Groundwater monitoring may be necessary to ensure that nutrient and water addition do not let contaminants into the water table.

Design

Air flow is accomplished by air blowers or by the suctioning of the air through the soil. Air treatment is not required if injection is used. However, since the air must remain in the subsurface until biodegradation occurs, the flow rate may be too low. In addition, contamination might travel to other areas previously uncontaminated. Air extraction provides better control of air flow but with the added cost of surface air treatment. Wells are usually spaced according to the radius of influence with some area overlap. Pilot tests with tracer gases can be used to determine the optimal arrangement of vents and recovery wells. The calculated air permeability value can then be inserted into models that simulate air flow through the soil. Hyperventilate™ is an example of a model used for vapor extraction that has been adapted for bioventing. Another model developed by the United States Air Force Bioventing Initiative is called the Bioventing Design Tool and is based on Microsoft Excel with Visual Basic Subroutines (Lesson 1997). It can calculate oxygen utilization and carbon dioxide production. Components include site characterization, system design, process monitoring, and site information, and it has databases for case studies, equipment information, and cost estimation. This design tool estimates that closure can be achieved in two to ten years for TPH contamination and in one year for BTEX. Anaerobic bioventing can be performed by injecting gases other than oxygen.

Air injection systems include air supply compressors, pleated paper cartridge units for filtration, and air regulators to control the air flow. The frequency of injection can be adjusted in relationship to biodegradation rates. Regenerative blowers, inlet filets, a cyclonic separator, pleated filter, and silencer on the discharge are used for air extraction. If the system is shallow, the piping can be installed by horizontal drilling or by trenches. Temperature is another factor that influences biodegradation rates. Compressed air is warm and, therefore, increases the effectiveness of the bioventing process. Warming also increases the probability of contaminant transport, however, due to solubilization and volatilization.

Sampling of the off-gas is often used to monitor the biodegradation process. Increased levels of carbon dioxide and decreased levels of oxygen in the off-gas provide a good indication of microbial activity during the bioventing. The air flow may be adjusted to optimize costs and contaminant degradation. Small areas and shallow depths allow for economical installations. King et al. (1998) determined that operation at a 5% oxygen concentration is a good level for determining pump operation.

Costs

Compared to soil-vapor extraction, bioventing can take five times as long (Hinchee 1993). However, operating costs are low (in the order of $1,000 U.S. per year for a 0.2 ha site or $13 to $40/m³) because only blower costs are involved and costs are less for air treatment. A properly designed system would not require air treatment.

Case Studies

A system has been commercialized by Terra Vac Corporation (Tampa, FL) called BIOVAC®. It can be applied for vadose soil of greater than 10^{-6} cm/s hydraulic conductivity. Case studies have been performed at a manufacturing facility in Deland, FL; a former service station in Cambridge, MA; McClellan AFB Superfund Site, Sacramento, CA; and Norton AFB, San Bernadino, CA.

A combination of bioventing and biostimulation was used at another site, a former fuel storage depot in Belgium (NATO/CCMS 1999). The site was contaminated with mineral oil. Biostimulation involved mixing the soil with compost and wood flakes to add nutrients and increase the porosity of the loamy soil. The work was performed by Ecorem. The total cost of the project was $20 million U.S. The decontamination was undertaken in several stages. The first step was excavation of areas with concentration greater than 5,000 mg/kg. The second step included biodegradation by adding oxygen and nutrients and mixing with the bioactivating substrates. The last stage was soil air extraction. Structure enhancing additives were added to 0.5 m above the groundwater level to enhance air permeability. Horizontal injection and withdrawal drains were placed in layers while the soil was mixed with the structure-enhancing additives. The air system consisted of an air/water separator, air filter, vacuum pump, and air cleaning and measuring units. This system avoided the transportation of 12,000 tonnes of soil contaminated with more than 1,000 mg/kg of soil. After ten months of operation, the average concentration was less than 490 mg/kg, well below the target of 900 mg/kg.

A research program at U.S. Air Force bases in Hill, Utah, and Tyndall, Florida, proved the value of this technology. At Hill, over ten months of operation reduced the TPH level from up to 20,000 mg/kg to 80 mg/kg. Over 99% overall removal efficiency occurred after 14 months of operation. At the Tyndall site, moisture was added during the bioventing of jet fuel in sandy unsaturated soil (concentrations were up to 23,000 mg/kg). Following 200 days of aeration, initial levels of hydrocarbons were reduced by 40% and aromatics were reduced by 85%. Most of the hydrocarbons were removed by biodegradation (85%) compared to 15% by volatilization. Other tests have shown that deep vadose soils can also be remediated by bioventing as long as the air permeability in the soil is high.

Bioventing can be used when the depth to water exceeds 3 m and the surficial soils are not contaminated or are being decontaminated by another method (WASTECH 1995). If the surface is capped, soils that are more shallow or contain a shallow water table can also be biovented. Contaminants that are not biodegradable or volatile can also be removed by this process. Low-permeability clays with hydraulic conductivities below 10^{-6} cm/sec are not amenable for bioventing. In Canada, Biogenie (Ste-Foy, Quebec) has treated approximately 200,000 tonnes of soil contaminated with gasoline and diesel fuel by bioventing.

Comparison to Other Methods

Bioventing is a simple, low-cost process that is suitable for volatile and semivolatile contaminants in unsaturated soil (WASTECH 1995). It is more appropriate than land treatment, slurry bioreactors, incineration, thermal desorption, and landfill since degradable and non-degradable volatile components can be removed. Aboveground methods are more feasible only if the contaminants are difficult to treat and excavation is necessary. Addition of nutrients can be difficult, however, and many other questions remain in determining a good design. For *in situ* technologies, only SVE treats more total soil than bioremediation.

In Situ Bioremediation: Biosparging

Description

Biosparging (Figure 5.6) is a variation of bioventing and is used to promote the bioremediation of groundwater. Air is sparged below the water table to provide oxygen for biodegradation and remove VOCs by stripping. Narrow well points are used to add air gently.

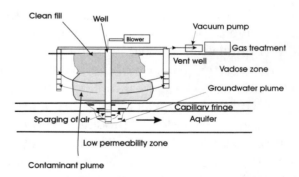

Figure 5.6 Typical Setup for Biosparging

Applications

A vapor extraction system can be used in the vadose zone with biosparging to remove the VOCs. Groundwater can be cleaned up more economically since it is not pumped. Pilot testing is required, however, as this method has not been used extensively. Geologic stratification is necessary to ensure uniform air diffusion.

Case Study

Christodoulatis et al (1996) has developed a technology called Bio-Sparge that is used to destroy hydrocarbons such as gasoline, diesel, and oil and grease at military installations, service stations, government facilities, refineries, industrial plants, and airports. Bio-Sparge uses ozone to enhance performance, in addition to proprietary remediation enhancement products. It integrates vapor extraction, sparging, oxygenation, free product recovery, injection of heat, humidity, nutrients and microorganisms, desorption of product, and biostimulation, and it can be applied to tight clays, according to AR Utility Specialists, Inc., the U.S. license holder (Anonymous 2000).

In Situ Bioremediation: Bioslurping

Description

Bioslurping is the use of soil vapor extraction for light non-aqueous phase liquid (LNAPL) from the surface of the groundwater to enhance biotreatment in the unsaturated zone and the groundwater. LNAPLs include gasoline, jet fuel, diesel fuel, or heating oils that accumulate on the water table and are difficult to remediate. Bioslurping combines vacuum extraction and bioventing. It reduces smearing of the contaminant which can occur during bioventing and increases free product recovery. The equipment called the bioslurper (AFCEE 1994) is shown in Figure 5.7.

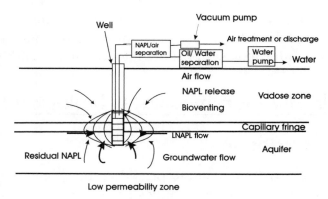

Figure 5.7 Bioslurping Process

Design

A recovery well is lowered to a depth no lower than the free product/water table level. A vacuum pump is attached to tube within the well. The free product is extracted with some water by the vacuum. Air flow in the direction of the well from the unsaturated zone is caused by the vapor going into the tube. Product and/or water are again extracted into the well. The volume is regulated by adjusting the intake depth. The bioslurper extracts vapor and water above the ground treatment while the air flow induces biodegradation in the capillary fringe zone (Baker and Bierschenk 1995).

Applications

This process is beneficial because it recovers free product while enhancing biodegradation. Since the amount of groundwater withdrawn from the aquifer is minimized, less water treatment above the surface is necessary. Bioslurping can also be applied to fractured bedrock where most other technologies cannot. Disadvantages include the potential for biological fouling of well screens and the inability to treat the residual LNAPL in the saturated soil zone.

Case Studies

To compare a pumping system with the bioslurping technology, ENSR conducted a three-week pilot study that took place at a site in South Carolina. Diesel fuel had leaked from piping into a saprolite formation. Recovery of diesel fuel by the conventional system during Phase I was negligible. During Phase II, the diesel fuel was recovered at a rate of 6.6 L/day by bioslurping. Only bioslurping was able to remove oil from the soil in the vadose zone (Baker and Bierschenk 1995).

SCG Industries, Ltd. (Saint John, New Brunswick, Canada) has developed a portable system on skids (2.5 m x 4.9 m) that requires one day to commission or decommission. It is called the Multi Phase Vacuum Extraction and Treatment system (Enviro-Access 1997a). Seven turnkey systems operate in Canada and Northeastern U.S. (Enviro-Access 1997a).

In Situ Bioremediation: Biological Treatment Barriers

Description

Barriers that are placed in the groundwater flow zone are called biobarriers. Nine bioremediation case studies of biobarriers had taken place by 1999. Systems can be passive once installed and require little maintenance, or they can be designed to extract and inject materials into the barrier. At the Lycoming Superfund Site, ARCADIS Geraghty & Miller made 20 injections of molasses into the aquifer (8 m below the surface) as pilot study. The molasses allowed the anaerobic conversion of Cr(VI) to Cr(III) and the degradation of TCE, DCE, and vinyl chloride.

An example of this technology is the Microbial Fence System by ThermoRetec Corporation (Tucson, Arizona). The system contains and degrades contaminant plumes by forming a barrier in the groundwater (Figure 5.8). A series of injection wells is used to supply oxygen and nutrients into the aquifer. Naphthalene concentrations are able to be reduced to 10 ppb from 500 ppb. Hydrocarbon content can be reduced by 80%.

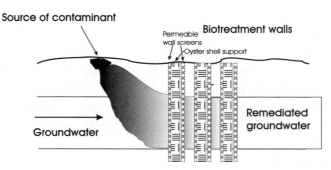

Figure 5.8 Biological Barriers for Groundwater Remediation (adapted from Pritchard et al. 1996)

Applications

Biowall and bioscreen configurations have been tested at bench, pilot and full scales by the TNO Institute (The Netherlands). Groundwater containing oil, BTEX, chlorinated solvents, and chlorinated pesticides have been treated. The setup involves a combination of funnel-and-gate™ systems with reactive trenches and biostimulated zones (NATO/CCMS 1999). Results showed that for a plume 6 to 9 m in depth, 150 m long, and 30 to 60 m in width, complete degradation of PCE to ethene and ethane could occur seven months after the installation of a 50 m semi-full-scale reactive zone (NATO/CCMS 2000). At a refinery site with a plume of more than 200 m in length, 4 m in depth, and containing 80% compounds of C6 to C12, three types of bioscreens (40 m x 0.4 m x 4 m in depth each) were installed. One was installed in a trench and backfilled with gravel. The other were vertical and horizontal fences for air sparging. After one year, 70% of the plume was biodegraded. Previously this was an anaerobic zone with insignificant natural biodegradation.

Costs

Westinghouse Savannah River Company induced methanotropic bioremediation of TCE and PCE in groundwater through horizontal wells (30 m below the surface) by injection of methane, air, nitrogen, and phosphorus. Another well at a depth of 23 m was used to extract the soil vapor untreated biologically. The TCE and PCE in the drinking water were reduced to drinking water standards in 429 days at a cost of $354,000 or $21/tonne of volatile compound removed.

Case Study

Another approach was performed as a demonstration at the Denver Federal Center by FOREMOST Solutions (Stavnes 1999). The bioremediation barrier was setup by injecting 74% porous diatomaceous earth pellets that contained liquid inoculum of selected microorganisms and nutrients into the fractures created by water jets and guar gum. The fractures provide excellent locations for the *in situ* bioremediation, whereas the pellets were excellent means for delivering the microorganisms, nutrients, and water, in addition to holding the fractures open while maintaining permeability. Wells were vented passively through the top. Benzene concentration decreased by 80% and total BTEX by more than 85%.

In Situ Bioremediation: Phytoremediation

Description

Phytoremediation is the use of plants to remove, contain, or render harmless environmental contaminants (Brar 1997). There are several mechanisms for phytoremediation (Figure 5.9). They include: phytoextraction, the uptake of contaminants through the roots and subsequent accumulation in the plants; phytodegradation, the metabolism of contaminants in the leaves, shoots, and roots release enzymes and other

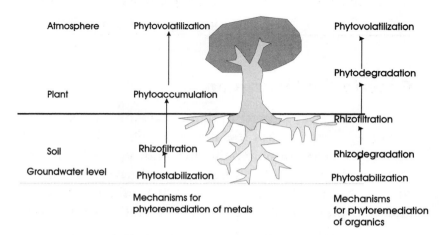

Figure 5.9 Removal Mechanisms of Organic and Inorganic Contaminants for Phytoremediation

components for stimulation of bacterial activity or biochemical conversion; and rhizodegradation, the mineralization of contaminants in the soil by microbial activity in the rhizosphere. Rhizofiltration involves the adsorption, absorption, or precipitation of contaminants in the roots. The plants are grown in green houses and then planted in the contaminated area. As the roots saturate with contaminants, the plants are removed. Another mechanism, phytovolatilization, involves volatilization of contaminants into the atmosphere by plants through uptake and transpiration. Phyto-stabilization begins in the roots of the plants where the contaminants are immobilized in the soil by absorption and accumulation, adsorption, or precipitation in the roots. Phyto-remediation is a low-cost *in situ* technology that causes minimal disturbance and is acceptable to the public and which generates low amounts of waste. This technology is being developed for a wide variety of contaminants.

Box 5-3 Overview of Phytoremediation

Applications	Suitable for treatment of organic contaminants with low K_{ow} values (less than 3) including benzene, creosote, dioxin, herbicides, insecticides, nitrotoluenes, PCBs, PCP, pesticides, phenol, TCE and toluene, and metal contaminants in soil near the surface
Cost	Operating: $0.02 to $1.00/m^3
Advantages	Excavation not required
	Low capital and operating costs
Disadvantages	Must dispose of contaminated plants
	Low removal efficiencies
	Contaminants can be toxic to plants
Treatment time	Several weeks to years
Other considerations:	Lack of field tests, removal mechanisms poorly understood since monitoring has been poor

Criteria for Utilization

Plants will interact in different manners depending on the $\log K_{ow}$ (Brar 1997). For example, $\log K_{ow}$ values less than 0.5, will enable the contaminant to be transported through the plants. Hydrophobic pollutants with values between 0.5 and three can be removed in an efficient manner. $\log K_{ow}$ values greater than three imply that the contaminant will be bound strongly to the roots. Other factors that influence phytoextraction include the plant uptake rate, transpiration rate, and the concentration of the contaminant.

Rhizodegradation involves the excretion of carbon from the plant roots that stimulates bacterial growth. Root decay can also contribute carbon to the soil. This carbon can also decrease transport the rate of contaminants to the groundwater. The bacteria and mycorrhiza fungus can thus degrade organic contaminants.

Applications

A wide range of contaminants can be degraded or volatilized by plants or the bacteria and fungus associated with them. They include benzene, creosote, dioxin, herbicides, insecticides, nitrotoluenes, PCBs, PCP, pesticides, phenol, TCE, and toluene (Brar 1997). Table 5.3 is a summary of the phytoremediation progress.

Phytoextraction can be used to extract metals from the soil. Some plants can accumulate up to 1,000 mg/kg of plant tissue (Baker and Brooks 1989; Brown et al. 1994b). The use of trees, grasses, and other herbaceous plants for remediation of low concentrations of radionuclides (^{137}Cs and ^{90}Sr) over wide areas may be cost effective for remediation of nuclear test sites and contamination due to nuclear reactor accidents (Entry et al. 1997). After harvesting, the plants can be composted, incinerated, or landfilled. The metals could also be recovered if the process is economically viable. Currently only 100 to 200 μg metal/g soil per year up to a depth of 30 cm is removed (Vangronsveld and Cunningham 1998). Because the process is so slow, it is limited to areas where the contamination is not very deep or beyond the reach of the roots. Cadmium appears to be very toxic and cannot be efficiently extracted.

Design

Phytoextraction is a still-developing process which is undergoing field demonstration. Several factors are essential in developing this process. They include high levels of metal accumulation near the ground level (Cunningham et al. 1997a), fast growing plants, plants that are tolerant to site conditions, and plants that are easy to treat and dispose of. Other considerations for selecting the most appropriate plants are presented in Table 5.4.

Currently, lead phytoextraction is at the most advanced stage. It has been determined that chelates such as EDTA, DTPA, and HEDTA, chemicals used in agriculture, can enhance the uptake by plants (Huang et al. 1997a). Shoot lead concentration can be increased to greater than 10,000 mg/kg using the chelate EDTA (0.5 g/kg soil) when a pea plant was grown in soil contaminated with 2,500 mg/kg of lead (Huang et al. 1997b). The chelates prevent precipitation of the metals and allow easier uptake into the plants. Precautions must be taken, however, to ensure that the solubilized contaminants do not leach into the groundwater. Field testing is underway and the results indicate that the process may be efficient and cost-effective.

Phytovolatilization can take place in the case of selenium in the form of dimethyl selenide (Wilber 1980). This may not be practical, since this is a less toxic form of selenium, and may not be possible with more toxic forms. This process is also in the research stage. Recently, though, a genetically modified plant was shown to reduce ionic mercury to the volatile Hg(0) (Rugh et al. 1996). The inserted genes are from bacteria that produce reducase. More research is needed in this area.

Table 5.3 Phytoremediation Progress

Plant	Contaminant	Results	Reference
Alfalfa	Toluene, phenol (0.5 m L/L) in groundwater	Microbial degradation	Davis et al. (1993)
Carrots	Octachlorodibenzo-p-dioxin (OCDD)	OCDD taken up by roots and leaves but no translocation	Brar (1997)
Poplar	TCE	Oxidation into CO_2 and H_2O,	Brar (1997)
Grasses	PCP and PAH (180 mg/kg)	Fescues and wheatgrass gave best results	Pivetz et al. (1996)
Poplar	TCE	Degradation into trichloroethanol, di- and trichloroacetic acid	Newman et al. (1997)
Indian Mustard	Cs, near nuclear disaster	Small reduction in Cs in top 15cm of soil.	Dushenkov et al. (1999)
Brassica juncea	U from contaminated soils	Concentration of U in plants increased from 5mg/kg to 5000mg/kg by adding citrate.	Huang et al (1998)
Sunflower/ mustard	Landfill leachate	In progress (rhizofiltration)	Schnoor (1997)
Thlaspi spp	Mine wastes	Uptake of Zn and Cd rapid but soil difficult to decontaminate (phytoextraction)	Brown (1995)
Brassica sp.	Refinery wastes and agricultural soil	Se is partly taken-up and volatilized but soil difficult to decontaminate	Banuelos et al. (1993)

Table 5.4 Factors for Selecting the Most Appropriate Plants for Phytoremediation (Entry et al. 1997)

Aspect	Factor
Ecological	Longevity
	Potential impact on herbivores
	Hybridization with weeds
	Native or exotic species
	Invasiveness
Agronomic	Biomass production
	Water and fertilizer requirements
	Availability of seeds or plants
	Availability of harvesting methods and soil
	Tolerance for environmental stresses, diseases, and insects
	Suitability for climate
Reproductive	Production of seeds and pollen
Economic	Remediation efficiency
	Time requirement for remediation
	Harvesting, recovery, and disposal costs
	Potential recovery of pollutants
	Reseeding or replanting costs

Phytoremediation can also be used for phytostabilization to reduce the bioavailable metal fraction in soil. This involves reclaiming mining and smelter areas by revegetation and chemical addition. Synthetic zeolites, steel shots, phosphates, liming agents, and organic materials have been evaluated for this process (Vangronsveld and Cunningham 1998). The plants decrease soil erosion by wind and rain and water percolation into the soil. Also they can accumulate, adsorb, and precipitate metals in the roots. The plants could also alter the pH and redox potential of the soil around the roots. The plants should be metal-tolerant and fast-growing and should have shallow roots; thus, grasses are ideal. Phytorestoration can be adapted to the different types of metals and soils.

Costs

Costs are low at around \$0.02 to \$1.00/m³ (Cunningham et al. 1997a). This may not apply to all cases, particularly where metal concentrations are elevated and more than one or two metals are present.

Case Study

Field scale tests have been performed at the Aberdeen Proving Grounds, Edgewood Area J-Field Site in Edgewood, Maryland, the Edward Sears site in New Gretna, New Jersey, and Carswell Air Force Base in Fort Worth, Texas (NATO/CCMS 1999, 2000, and 2001). Hybrid poplars were used to remove TCE, PCE, DCE, trimethylbenzene (TMB), xylene, and methyl chloride from the groundwater. At the Aberdeen Proving Grounds, 1122-TCA, TCE, and DCE at concentrations of up to 260 ppm were found 2 to 14 m below the ground surface in a slow-moving plume. The Edward Sears Site had soil and groundwater contaminated with methylene chloride, tetrachloroethylene (PCE), TCE, TMB, and benzene. Concentrations of TCE were up to 390 ppb. The contamination at the aquifer of the Carswell AFB site was due to the use of chlorinated solvents. Eastern Cottonwood (*Populus deltoides*) were planted in a shallow aquifer to determine if the short-rotation woody plants could remove the contamination.

In the first case, results were not conclusive due to a lack of monitoring. However in 1998, new trees (tulip, hybrid poplar, and silver maples) were planted and new monitoring was initiated. When some trees were excavated at the end of 1998, it was found that their root systems did not extend beyond the area in which they had been planted. New trees were planted with new planting methods at the site. It was determined that the trees' uptake was about 4,129 L/day. This should increase to 7,500 L/day in about 30 years. Currently, contaminant uptake is negligible, but this should improve as the trees grow. The contaminated plume has not migrated off site, and the leaves and soil are monitored to evaluate contaminant pathways. At the Edward Sears site where poplar saplings were planted (*Populus charkowiiensis x incressata*, NE 308), results suggested that evapotranspiration of the VOCs was taking place. Better sampling procedures will have to be put in place to understand the results more fully. Only toluene was detected in evapotranspiration sample bags (8 to 11 ppb). In the four samples, dichloromethane was reduced from 490,000 to 615 ppb; 12,000 ppb to not detected; 680 ppb to not detected; and 420 ppb to 1.2 ppb. Trimethylbenzene decreased from 147 to 2 ppb; 246 to not detected; 1,900 to 50 ppb; and 8 to 1 ppb at four wells. Trimethylbenzene was unaffected at another well. There was some evidence of anaerobic dechlorination since the concentration of PCE decreased from 100 to 56 ppb and TCE increased from 9 to 35 ppb. At the last site, Carswell AFB, the trees' roots grew to reach the water table. Mass flux decreased by 11% over the three-year period. Mechanisms of reductive dechlorination and hydraulic influence are expected to significantly reduce the contaminant mass. The volume of contaminated groundwater is expected to decrease by up to 30%. By January 2001, there seemed to be some evidence that the trees were stimulating microbial reductive dechlorination due to depleted oxygen levels and elevated ferrous iron and/or sulfide levels.

Comparison to Other Methods

Many challenges remain with phytoremediation, both at the technical and regulatory levels (Brar 1997). The plants must receive adequate nutrients, the soil pH must be appropriate for growth, there must be sufficient nitrogen-fixing microorganisms, and the contaminants may be toxic to the plants. Livestock and wildlife may eat the plants; therefore, the plants will have to be isolated from wildlife and agricultural lands. Pests and insects may destroy the plants. Other natural vegetation may dominate and not allow the desired plants to grow.

Because this is a new technology, there are also challenges related to public perception of the technology and the awareness and attitude of the regulators. Metal recycling is currently an expensive process and must be improved so that the metal-containing plants are not transferred to landfills where they would create further problems. The climatic conditions and bioavailability of the metals must be taken into consideration when using this method. Once contaminated, the plants will have to be disposed of in an appropriate fashion. Some techniques include drying, incineration, gasification, pyrolysis, acid extractions, anaerobic digestion, extraction of the oil, and extraction of chlorophyll fibers from the plants (Bolenz et al. 1990), or disposal, since plants are easier to dispose of than soil. Phytoremediation will be most applicable for shallow soils with low levels of contamination (2.5 to 100 mg/kg).

In Denmark, priority field tests have been targeted for bioventing, phytoremediation, and *in situ* bioremediation. Although full-scale *in situ* biological processes have not been applied in Finland, research and development is underway for phytoremediation of contaminated soil and biodegradation of dioxins and furans.

Ex Situ Processes

Ex Situ Processes: Landfarming

Box 5-4 Overview of Landfarming

Applications	Suitable for treatment of low to medium concentration of contaminants including benzene, creosote, dioxin, herbicides, insecticides, nitrotoluenes, PCBs, PCP, pesticides, phenol, TCE and toluene, and metal contaminants in soil near the surface
Cost	Operating: $0.05/m^3 per year or $20 to $60 per tonne
Advantages	Excavation may or may not be required
	Simple
	Inexpensive to operate
Disadvantages	Large areas required
	Hydrophobic compounds difficult to remediate
	Can require long periods of time
Other considerations:	Monitoring difficult, liners and caps needed to prevent leakage

Description

Landfarming is used to treat soil with organic nonhazardous contaminants. The name originates from the fact that some of the techniques common to farmers such as tilling of the soil and adding fertilizer are used. In the early 1900s, spreading of sludge from petroleum refineries onto soil was used as a disposal method (King et al. 1998). Volatilization, leaching, and reaction with the sun decreased the sludge volume. Eventually it was determined that if soil tilling, fertilizers, and pH and moisture control were added, biodegradation could be enhanced. By the 1970s, it was recognized that volatilization and leaching had to be controlled while biodegradation must be encouraged.

Landfarming is a fairly simple process which includes the following phases (Hildebrandt and Wilson 1991): 1) determination of the optimal inorganic, organic, microbiological, and geological conditions for biodegradation; 2) design and construction of the facilities including earthwork, piping, liners, drains, etc.; 3) operation and maintenance of the biotreatment process including aeration, nutrient addition, moisture control, and monitoring; and 4) closure of the system.

Methods

Biopile

A biopile is a variation of landfarming (King et al. 1998) that can be used if there is not enough space for landfarming, if the soil is a clay or heavy silt, if the contaminants are difficult to treat, or if excavation is required. If space is limited, the soil can be piled with permeable layers in between as shown in Figure 5.10. The soil, in between the layers, is called a lift and is usually more than 280 cm in thickness. Air can either be pulled via a vacuum or pushed by a blower through slotted piping in the permeable layer. Nutrients can be added by sprinklers or hoses.

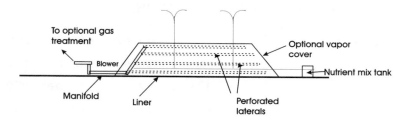

Figure 5.10 Diagram of the Biopile Process

Full-scale operation of biopiles has been accomplished in Canada (Enviro-Access 1997b). The piles are usually between 100 and 1,500 m³. Structural agents, fertilizers, and bacterial cultures are added. Costs are between $25 and $65 per tonne for gasoline-contaminated soil and $55 to $85 per tonne for heating oil contamination. Other variables which influence the cost are soil granulometry and the type and concentration of contaminants. Several companies have developed biopile technology. ADS Environonnement inc. (Montreal, Quebec) has a permanent installation in Val d'Or, Quebec. Biogenie (Ste-Foy, Quebec) has treated more than 300,000 tonnes of soil contaminated with gasoline, diesel fuel, oils, aromatic hydrocarbons, creosote, PCP, PCE, and TCE. They have seven permanent soil treatment centers. Envirosite inc. (Sherbrooke, Quebec) owns a treatment facility with a capacity of

10,000 tonnes per year. Services Environnement AES inc. has a capacity of 20,000 m³/year in Latteriere, Quebec while Recupere Sol (St. Ambroise, Quebec) has a capacity of 5,000 m³/year.

Biovault

The biovault is another variation of landfarming (Figure 5.11). A plastic liner is placed over a sand bed with low berm walls. Piping for air injection is placed in gravel at the bottom of the vault. Geofelt is then used on top of the gravel. Next, the soil is loosely piled on the base to the desired height. Another piping system is placed on the top of the soil pile either in sand or gravel. A liner covers the top of the pile. Sometimes it is welded to the bottom liner to create a seal. Air is sucked through the bottom piping systems by a vacuum or by injecting air with a blower in the top piping system. Water and nutrients are added by the top piping system or by another irrigation system. Soil moisture is monitored by moisture meters or probes. Since solubility of some compounds can be low, reducing bioavailability, surfactants in dilute concentrations can be used, particularly for contaminants of number 4 fuel oil or heavier. Biosurfactants may be more biodegradable, more tolerant to pH, salt, and temperature variation, and in some cases less expensive to use (West and Harwell 1992). Plant-based surfactant from the fruit pericarp of *Sapindus mukurossi*, a plant from the tropical regions of Asia, has shown potential for the removal of hexachlorobenzene (Roy et al. 1997). Biotreatability tests should be performed, however, to determine if the microorganisms are using the contaminant or the surfactant as substrates.

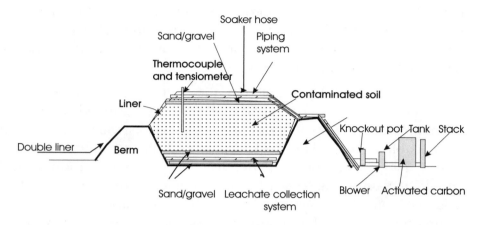

Figure 5.11 Schematic of a Biovault

An EPA Demonstration project was performed in New York (Demonstration Bulletin EPA/540/MR-95/524). The technology was developed by ENSR and Larsen Engineers to treat VOC and semivolatile VOCs. The site contained acetone, trichloroethene, tetrachloroethene, cis-1,2-dichlorethene, 2-butanone, 4-methyl-2-pentanone, and toluene. The system included two biovaults of 100 m³ with a double liner system at the bottom and a 30 mil high-density polyethylene (HDPE) liner on the sides and top. A soaker was used to supply water and a vacuum pump for air (5% oxygen). The degradation of over half of the chlorinated compounds and almost all of the remaining compounds was due to biodegradation. It was also found that the process was very sensitive to temperature decreases. Therefore, it was concluded that operation should be initiated in the spring and stopped once temperatures decreased significantly in the fall.

A variation of the biopile is called the biopit. It is a lined pit dug into the ground and covered with asphalt so it does not disturb land use. Water is added employing perforated laterates used for adding air.

Criteria for Utilization

Several tests must be performed at the beginning of the remediation work to determine if bioremediation is appropriate. Tests to obtain the moisture and nutrient content and pH levels of the soil, as well as tests for ammonia nitrogen and phosphate contents, bacterial count, hydrocarbon degrading bacterial counts, and buffer capacity, are needed to determine the proper conditions for the work.

Tests for biodegradability of the contaminants should also be performed by adding nutrients and water to obtain 50 to 80% of the saturated water capacity. Analysis of the soil for hydrocarbon content after one to four weeks will indicate if the native microorganisms are capable of degrading the hydrocarbons. Plate counts should be above 10^6 colony forming units (CFU) per gram of soil after acclimation. Hydrocarbon degrading microorganisms should account for at least 10% of the microorganisms present. If required, specific compounds should be monitored by gas chromatography-mass spectrometry (GC-MS) or other techniques. Using different nutrient ratios in different areas will indicate the optimal ratio to use at full scale. If it is determined that compounds that are difficult to degrade are present, bacteria known to degrade these compound can be added.

To obtain the field capacity, the field is saturated with water and then drained for 24 h. This also provides the moisture content of the soil. According to the U.S. EPA (1993), the moisture content of a sandy soil will be much lower than a clay soil (5% compared to 30%, respectively). Optimal moisture content for biodegradation is 60 to 80% of the field capacity. Sprinklers, trickling, or irrigation systems can be used to maintain optimal levels. The addition of water can also help in temperature control. Although anaerobic conditions will develop if too much water is added, sprinkling with water helps to reduce daily temperature variations and protects against frost. The addition of mulches such as manure, compost, wood chips, asphalt gravel, crushed stone, and sawdust can also help to maintain soil moisture and temperature (Dupont et al. 1988). However, if temperatures decrease significantly, as they do during the winter in many parts of the world, tilling and sampling may have to be discontinued until the temperature increases in the spring. The duration of this period of inactivity can be up to five months (U.S. EPA 1995a). Placing the contaminated soil in a greenhouse (Carberry et al. 1991) enables the bioremediation process to continue throughout the winter. Hot air can also be added to increase the soil temperature.

Operating Conditions

The contaminated soil is spread out in a thin layer, new material is added regularly, and tilling is performed to enhance mixing and aeration. Monitoring parameters include pH, nutrient levels, moisture content, and microbial activity. This technique could be and has been used as an *in situ* process if an impermeable layer exists underneath the contaminated zone. For example, in western Pennsylvania, a tank explosion led to the release of 3,785 m³ of number 2 fuel oil. Contamination levels in the soil averaged 10,000 to 20,000 mg TPH per kg soil. *In situ* land treatment was successfully achieved (Leavitt 1992).

In most cases, however, *ex situ* treatment is performed for soils near leaking underground storage tanks. Excavation is done to prevent further contamination problems (in the case of leaking underground storage tanks) or to prepare the land for other purposes. The treatment area can then be properly prepared with liners or other engineered controls to minimize contaminant transport. The place for treatment is called a land treatment unit (LTU). The volume of soil for treatment must be accurately estimated. The soil to be treated is spread out so that the volume of layered soil is approximately 1.25 to 1.4 times the volume of the excavated soil (King et al. 1998). The depth of the soil is minimized to ensure adequate aeration. High permeability soils can be treated at greater depths than those of low permeability. Organic material or bulking agents can be added to enhance the porosity of the soil. Depths of the soil are usually in the order of 15 to 50 cm so that regular tilling equipment can be employed. The depth of clayey soils is usually limited to 23 cm (U.S. EPA 1993). Biodegradation rates are generally slower for untilled soils; nevertheless, depths of up to 150 cm have been treated (LaGrega et al. 1994). For deeper soils, the upper layer can be treated first and then removed, allowing the next layer to be treated. These layers are often no greater than 15 cm and less than 61 cm with an average of 30 cm. Some of the removed soil can be added to the next layer since it contains already acclimated bacteria and allows dilution of contaminant concentrations. This type of procedure is often used if land area is limited.

If the ratio of C:N:P is not adequate, nutrients will have to be added. Ratios between 100 and 400:10:1 are often used. Ammonium phosphate (Genes and Cosentini 1993), ammonium nitrate (U.S. EPA 1995a), and diammonium phosphate (Flathman et al. 1995) have been used in field applications. If commercially available fertilizers are used, it should be determined if antimicrobial additives are present. These additives could cause significant difficulties for the bioremediation process. The use of slow-release nutrient pellets ensures that the nutrients will not be leached away from the treatment area; however, higher concentrations have proven to be beneficial. For example, Fogel (1994) used a C:N:P ratio of 10:3:3. Higher nutrient concentrations lead to higher costs, and they affect salinity and osmotic pressure. In this case, then, it would be better to add the nutrients in steps rather than all at the beginning. This approach also enhances long-term microbial growth. King et al. (1998) determined that adding nutrients every three months over a nine-month period gave superior results.

The pH level is another parameter that must be in the range for optimal microbial activity, usually 6.5 to 8.0. If it is not, adjustments will have to be made. In most cases, the pH must be raised to the acidity of the soil. Calcium hydroxide, calcium oxide, calcium carbonate, magnesium carbonate, and calcium silicate slag can be used. Sometimes the soil is high in carbonate due to the contamination, and acids such as sulfuric acid, liquid ammonium polysulfide or aluminum, and iron sulfates must be added to decrease the pH (Dupont et al. 1988).

Design

The components of the treatment system include an impermeable bottom layer, berms and swales, a drainage system, and a system for monitoring (Figure 5.12). A storage

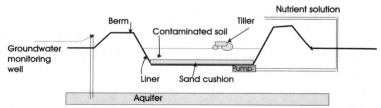

Figure 5.12 Landfarming Process Including Bottom Layer, Berms, Swales, Drainage, and Monitoring Systems

pond for water may be required for surplus rain water. The impermeable layer prevents permeation of contaminated water into the groundwater. To decrease the permeability of the soil underneath the zone, compaction is performed. On top of the compacted zone, a liner of high-density polyethylene (HDPE) with a thickness of 60 to 80 mil or compacted clay of greater than 10 cm thickness and a hydraulic permeability of less than 1×10^{-7} cm/s is used. Sometimes an already existing layer such as a paved or asphalt parking lot is used. Regardless of the type of liner used, monitoring the groundwater for potential leaks is essential by installing wells on all sides of the treatment zone at various depths.

The purpose of the drainage system is to allow the leachate from rainfall or irrigation to collect. Perforated pipes in sand or gravel beds are common. The diameter and perforation of the pipes are designed according to the flow of leachate expected. The pipes take the water into the sump. The 15 to 30 cm layer of sand makes drainage into the pipes easier, in addition to protecting the pipes from tilling equipment (U.S. EPA 1995a). A slope of 0.5 to 1.0% of the soil to be treated helps to promote drainage into the sump and reduce the formation of pools on top of the soil.

The sump is for collection of the leachate for use during the dry season or before treatment prior to discharge. The latter is not usually the case, however. The capacity of the pond depends on expected rainfall. It has been recommended that the pond should be able to hold 7.5 cm of rainfall across the soil area (King et al. 1998) and the height of the water should not be more than 15 cm above the top of the berm. According to the U.S. EPA (1993), 2.5 cm of rain corresponds to between 9 and 25 L/m^2. The sump pumping equipment is used to supply water during the dry season. The capacity of the pump is based on the drainage properties of the soil and should be sufficient to supply, within an 8-hour period, 2.5 cm of water to the entire cell. Hoses or sprinklers similar to those used in a garden are employed.

The purpose of berms and swales is to prevent uncontrolled discharges of leachate into uncontaminated soil or other treatment cells. Construction materials include uncontaminated soil with a liner, concrete, clay, or other impermeable materials. The walls should be a minimum of 30 cm above the surface of the contaminated soil and should have a slope less than 45°. The drainage base must be taken into account when determining the appropriate height for the berm walls. Although berm walls can be constructed of earth, heavy rain events have been known to wash them out, releasing large quantities of water (King et al. 1998).

Maintenance and Monitoring

If the contaminants are known to be biodegradable, a microbial count should be done to ensure that adequate numbers of bacteria are present. Monitoring throughout the treatment process will enable an evaluation as to how the process is proceeding. Usually a land treatment site is divided into cells of 4,000 m^2 with several samples being taken from a cell at one time. Analyses are performed for contaminant concentrations, nutrient levels, moisture content, and microbial activity (if required). To cut costs related to analyses, only a few cells are monitored. Once the decision has been made to close the site, all cells must be analyzed for regulatory purposes.

Tilling of the soil is done to enhance aeration and homogeneity of the soil (Fogel 1994). The selection of the equipment is based on the area of land to be tilled in one day. Equipment can vary from garden rototillers to tractors. This method is effective to a depth of approximately 30 cm. Usually this can be performed once every week or two when the soil is dry. It should be done at least 24 h after a rainfall or irrigation (U.S. EPA 1993). Optimal mixing is achieved by alternating the tilling patterns across the length, width, or diagonal of the treatment area. To reduce the amount of tilling necessary, oxygen can be added in chemical form by CaO_2 or other chemical oxidants. The economics of this method has not been proven, however, particularly since the plot should be covered and the exit air filtered.

Applications

Soils contaminated with mineral oil, polycyclic aromatic hydrocarbons, and pentachlorophenol have been treated in the Netherlands. In the United States, land treatment has been performed for petroleum refinery sites and creosote-contaminated soils and sludges (Ryan et al. 1991; U.S. EPA 1993). For example, pentachlorophenol (PCP) and polycyclic aromatic hydrocarbons (PAHs) were reduced by 95% and 50–75%, respectively, over a period of about five months (Nyer 1992). Pesticides such as 2,4 dichlorophenoxyacetic acid (2,4-D) have been reduced from 42 to 4 ppm in 77 days (Fiorenza et al. 1991). Groundwater Technology, Inc., has also developed a process for the treatment of cyclodiene insecticides that is applicable to all biodegradable compounds (U.S. EPA 1991).

Costs

Landfarming is one of the simplest methods for treatment of excavated soil. The cost of landfarming is in the order of $0.05/m^3$ per year (Cunningham et al. 1997b) or \$20 to \$60 per tonne (OMEE 1992). According to Cookson (1995), the costs for landfarming can be broken down as follows: \$40/tonne for construction, \$26 to \$53/tonne for operation, and an optional \$33/tonne for soil conditioning and \$13/tonne for soil disposal. These costs add up to \$66 to \$139 per tonne.

Comparison to Other Methods

In summary, land treatment is simple and inexpensive to operate. It can be used to treat soils with fairly high metal contents. On the other hand, land requirements and treatment times can be extensive. Time requirements are six to 36 months, depending on the type and concentration of the contaminant, soil characteristics, and weather conditions. Nutrients and moisture can be added easily. Mixing can be accomplished with conventional farming equipment, and large amounts of soil can be treated. Limitations involve difficulties with control of pH and temperature. Liners must be monitored to ensure that leakage does not occur. Hydrophobic compounds that are strongly adsorbed into the soil are particularly difficult to treat by this method. The presence of highly volatile compounds requires more enclosed structures such as greenhouses with ventilation systems to minimize volatilization. Once the soil contaminant levels are acceptable, closure may be granted. The berms are removed, and grass seed can be added to the soil or the soil may be used as fill at another site. In terms of the amount of soil treated, bioremediation (*ex situ*) has been used to treat more soil than any another process.

Ex situ bioremediation treats the largest volume per project (38,000 m³). Land treatment is the most common followed by composting. Of all biological treatment processes in the U.S., land treatment was employed in 39% of the projects, followed by bioventing (25%), *in situ* treatment (14%), and composting (9%). Other processes included, excavation with on site treatment (5%), slurry phase tank treatment (4%), *in situ* lagoon (4%), and biopile (1%).

Ex Situ Processes: Composting

Box 5-5 Overview of Composting

Applications	Suitable for treatment of nitroaromatics, sewage sludge, gasoline, high levels of surface contamination
Cost	Operating: $250 to $300 per tonne
Advantages	Rapid
	Inexpensive
	Self-heating by addition of organic amendments
Disadvantages	High exposure risk, odor
	Bulking agents can take up to 80% of volume
	Large area requirements
Treatment time	Seven days to 24 months
Other considerations:	Moisture must be maintained between 50 and 60%. Aeration is maintained by either forcing air through the pile or by use of a vacuum

Description

Composting involves biological oxidation of organic solids under aerobic conditions. This is a well-known method for the treatment of municipal solid waste, agricultural residue, yard and kitchen waste, and sewage sludge. More detail will be provided about these applications in Chapter 6. In this section, its application for the treatment of contaminated soil—a relatively new application—will be discussed (U.S. EPA 1990b). Due to the lower organic content of the contaminated material, organic material must be added to enhance the process. To date, this type of process has been used to treat petroleum hydrocarbons, chlorinated compounds, and explosives in soils, sediments, and sludges. The material to be treated should be easy to break up when mechanically turned and porous to allow aeration and should contain low levels of free liquid.

Methods

Windrow, static piles, and closed reactors are three types of composting processes. Their level of control for various parameters varies, as shown in Table 5.5. The first two types are open systems and are more common than the reactors. Windrows and static piles are placed on a concrete or asphalt surface. A polyethylene liner is frequently used to ensure that none of the leachate will pass through any cracks that may exist in the impermeable surface. The method of aeration differs between the windrow and static pile systems. In the former, the pile is turned mechanically or manually; thus, windrows are the simplest of the composting processes. The composting material is placed in long mounds, or windrows, on the platform. Pile width is usually in the range of 3 to 4 m, although some are as wide as 6 m (Cleaning Carolina Soil 1995). Pile depth is generally 1.2 to 1.5 m, which is deep enough to retain heat (Cookson 1995). If other amendments and nutrients are to be added to the composting pile, premixing is important to ensure homogeneity, which enhances biodegradation. Soluble components can be added with water. Sometimes, such as in the case at the Seymour Johnson Air Force Base, the materials can be laid out as layers and a turner is passed twice to ensure thorough mixing (Cleaning 1995). Mechanical mixing can also be done with a less expensive front-end loader, but the efficiency is not as good as with a turner, which turns and mixes the soil. During the composting period, turning is done to aerate the pile, to decrease temperatures within the middle of the pile, and to mix the contaminants. Mixing can vary from once a day (U.S. EPA 1995) to once a month (Sellers et al. 1993). If the pile is not turned, aeration would depend on the porosity of the pile and, thus, could be limited in the middle. The temperature would also be higher in the middle than on

Table 5.5 Comparison of the Levels of Control for the Various Composting Systems

Parameter	In-vessel	Static pile	Windrow
Control			
Process	High	Medium	Low
Moisture	High	Medium	Low
Air emissions	High	Medium	Low
Runoff	High	Medium	Medium
Pathogen	High	High	Medium
Requirements			
Space	Low	Medium	High
Capital costs	High	Medium	Low
Maintenance cost	High	Medium	Low
Maintenance skill level	High	Medium	Low

Adapted from Cookson (1995)

the outer surface. Oxygen limitation was noted in a case where the pile was 12 m wide by 2.5 m high by 26 m long. The soil was contaminated with ethylbenzene, styrene, and other petroleum hydrocarbons (Benazon et al. 1995). Four pipes had to be added at the bottom of the pile and another three at a depth of 1.5 m. The pile was covered with 30 cm of wood chips and a 20-mil liner of polyethylene. Within the top 0.8 m of the pile, ethylbenzene and styrene concentrations decreased from 2,190 and 365 ppm, respectively, to below 1 ppm, and petroleum hydrocarbons dropped from 30,000 to 1,000 ppm. However, below this layer, little degradation was noted due to oxygen limitations.

If the windrows are placed within warehouses or other structures, covers for the piles are not needed. However, if they are outside, covers are beneficial to maintain temperatures, reduce VOC emissions or wind erosion, and prevent drainage of contaminated leachate due to water saturation from rainfall. Common materials for the covers can be synthetic or organic such as high density polyethylene, wood chips, or compost.

In static piles, perforated pipes are placed at the bottom of the pile and air is pushed by forced aeration or pulled by vacuum through the pile. The latter is usually preferred since this reduces VOC emissions, thus reducing the need for additional treatment by biofiltration or catalytic oxidation. However, if there are cold climatic conditions, the suction of cold air could significantly decrease temperatures within the pile. Because the air from a blower is generally heated, this method is more suited for cold climates. Recycling of the offgas may be an appropriate method of dealing with emissions since the pile will act as a biofilter.

The pipes in a static pile are placed in a layer of permeable material such as gravel, compost, sand, or wood chips. Due to the aeration, static piles can be higher than windrows. Heights of 3 m are regularly used although 6 m piles or deeper are known (Cookson 1995). The deeper types are called biopiles. They may have different layers of pipes for aeration, moisture, and nutrient delivery.

To aerate a static pile, there are three types of aeration strategies which are dependent on aeration rate: fixed rate aeration, automated aeration, and variable rate (Leton and Stentiford 1990). In the first type, aeration is controlled by turning the flow on and off. This method is not optimal since the bacterial growth is low at the beginning when the air may be too high and provide too much cooling. Later, when the growth is higher, the aeration rate may not be sufficient. The automated mode involves the use of computer programming to control aeration and the resulting change in temperature. The variable rate mode, the least common strategy, employs an aeration rate that is high at the beginning and decreases with time. This method is not popular since daily monitoring is required. Due to the aeration of the piles, compensation methods must be used if an impermeable cover such as HDPE or plastic sheeting is employed. Various techniques include making slots or openings in the cover, maintaining the liner above the surface of the pile, embedding pipes in the piles to allow passive flow, or placing gravel between the pile and the cover. There are various problems with each of these methods, but they are not well documented.

In the reactor system, mixing and aeration are performed by agitation and forced aeration. Although they are more capital intensive than the other two processes, they provide optimum control of aeration and mixing. Mixing of the contaminants, microorganisms, amendments, and nutrients can be continuous or sporadic in rotating drums, chambers, or mixed tanks (Hart 1991). VOC emissions are easily controlled in the closed reactor. In addition, leaching and heat dissipation are minimized.

Operating Conditions

Temperature, aeration, pH, and moisture content are important parameters for control. To raise the temperature of the pile, readily degradable organic materials are added to serve as an energy source (LaGrega et al. 1994). The contaminants are usually not present in sufficient quantities to increase the temperature significantly. There is an initial short lag phase for acclimation followed by the exponential growth phase where the temperature rises until a maximum is reached. As the temperature reaches above 45°C, the mesophilic bacteria, which cannot survive (30 to 40°C), die or sporulate while the thermophilic bacteria begin to dominate (50 to 60°C). The choice of the energy source should be based on pilot tests because temperatures should not be allowed to increase above 55 or 60°C. Thermophilic bacteria are not able to tolerate such temperatures.

The period for the composting is complete when the temperature drops to ambient temperatures, which means the food source is depleted. The smell will change throughout the composting period from offensive to that of a garden soil. In addition, the soil texture will change from very fibrous to a more uniform texture. The mass of the pile can also be reduced up to 40%, depending on the amount of amendment used (Hart 1991).

Some examples of different additives are shown in Table 5.6. Others include tree and plant leaves, wood chips, straw, vegetation, and molasses or other food processing wastes.

It is recommended that the ratio of carbon to nitrogen within the additive should not be higher than between 20:1 and 25:1. Horse manure has a ratio of 25:1. Other additives such as nonlegume vegetables, grass clippings, poultry manure, and cow manure have ratios between 12:1 and 18:1. The manure is added to increase the microbial and nitrogen contents. Combining these with materials such as sawdust (ratio of 200:1 to 500:1) or wheat straw (ratio of 128:1 to 150:1) can increase the ratio to within the appropriate range. The choice of the material is, however, also based on price and availability.

To improve aeration, bulking agents such as wood chips are added, particularly if the soil contains substantial amounts of clay or silt. The purpose of the bulking agent is to enhance air permeability and provide better drainage in the soil. The thermal source can also serve as a bulking agent to reduce costs of other additives such as grass, wood chips, rice hulls, and hay. Inert materials such as rubber tires can be used only for bulking since they are not biodegradable. However, they do not absorb moisture and their structure is not sufficient. Organic materials such as finished compost, wood chips, and rubber tires also provide an inoculum for the pile to stimulate degradation. These materials can be recovered from the piles and used as an inoculum for other piles. Usually the soil, thermal source, and/or bulking materials provide sufficient bacterial numbers. To decrease acclimation times, sewage sludge addition can be added. If the contaminants are particularly difficult to degrade, specialized microorganisms such as the fungus *Phanerochaete chrysosporium,* which grows on wood chips, sawdust, and pine bark can be utilized to provide the appropriate enzymes to oxidize a wide variety of compounds (Holroyd and Caunt 1995). At one site the

Table 5.6 Various Additives Used in Soil Composting Projects

Process	Additive	Amount of additive	Cover	Mode of aeration	Reference
Static pile	Mix of straw, wood chips, sawdust, and pine bark inoculated with white rot fungus	5% by dry weight	HDPE liner	Positive	Holroyd and Caunt (1995)
Static pile	Straw/manure Alfalfa Horse feed	47% 38% 12%/vol.	Sawdust, wood	Negative	Williams and Myler (1990)
Windrow	Compost	10%/vol.	Three layers of 6 mil plastic	Negative	Sellers et al.(1993)
Windrow	Yard compost Turkey manure	20% 5%			Cleaning (1995)
Windrow	Dairy manure and chips, potato waste, and alfalfa mixture	70%			Cleaning (1995)

fungus was able to reduce chlorophenol concentrations from 200 ppm to 30 ppm in 24 months. This fungus has also been shown to degrade 2- and 3-ring PAHs in six to eight weeks. However, insignificant degradation of 4-ring PAHs occurred.

Design

Optimal pile design should be established in laboratory or pilot-scale tests. The percentage of amendments used is important. Higher proportions of additives give higher porosity, better air distribution, and increased water-holding ability. Stegmann et al. (1991) showed that for a diesel-contaminated soil, a ratio of 2:1 soil to compost gave better results than a soil to compost ratio of 16:1. Dooley et al. (1995) also showed that 41% amendment of a compost mix (35% wood chips with 6% cow manure) gave better biodegradation results for a herbicide-contaminated clay soil than with only a 10.8% amendment. The disadvantage of increasing the amendment content is that less soil is treated and more land may be required.

A moisture content of 60% of the water-holding capacity of the pile is considered optimal while a range of 50 to 80% is acceptable (Stegmann et al. 1991). Like landfarming, too high water contents decrease oxygen availability and too low contents reduce bioavailability of the contaminants due to strong sorption onto the soil and entrapment in pores.

Applications

Laboratory and pilot scale testing (Table 5.7) determined that composting was effective for the remediation of soils contaminated with explosives, propellant manufacturing, coal tar, and nuclear materials at a nuclear production facility in Amarillo, Texas. Within three weeks, explosives concentrations decreased to below detection limits (Doyle et al. 1992).

Costs

Composting costs for contaminated soil are in the order of $250 to $300/tonne. For example, the total costs at a Superfund site for the composting of soil contaminated with PAHs, PCP, and VOCs at the Dubose Oil Products Co., Cantonment, Florida were $5.25 million for the treatment of 17,915 tonnes of soil treated ($292/tonne). A total of 53,300 tonnes of soil was excavated in total. The concentrations of PAHs were 0.58 to 367 mg/kg, PCP were 0.058 to 51 mg/kg, and VOCs from 0.022 to 38.27 mg/kg. The cleanup objectives were 50 mg/kg PAHs, 50 mg/kg PCP, 10 mg/kg benzene, 1.5 mg/kg xylenes, 0.07 mg/kg DCE, and 0.05 mg/kg TCE. Composting took place from May to November 1993. The system consisted of a leachate collection system, aeration system with granular activated carbon (GAC) adsorbers, system for inoculum growth, and application and wastewater treatment system (US EPA 1995b).

Table 5.7 Composting Test Results

Item	Sediments and soil containing explosives and propellants (Williams et al. 1989)	Soil contaminated with coal tar (Taddeo 1989)
Contaminant	76,000 mg/kg TNT	20,000 mg/kg coal tar
Type of aeration	Static pile with vacuum	Static pile with forced air
Volume of compost pile	34 m^3	4 m^3
Bulking agent	None	Not disclosed
Additives	Alfalfa, straw, manure, horse feed, and fertilizer	Not disclosed
Ratio of waste to additives	1% volume and 24% mass	Not disclosed
Temperature	55°C	18-29°C
Test period	22 weeks	80 days
Percent removal of contaminant	99% removal of explosives	94% removal of total hydrocarbon, 95% of 2-ring PAHs, 94% of 3-ring PAH, 89% of 4-ring PAH, no removal of 5-ring PAHs

Case Studies

Pilot tests were performed by Bioremediation Systems in a 1 x 2 m pile on air ducts with forced aeration (Taddeo et al. 1997). The pile was bermed and covered and a drainage system was installed underneath the pile. Approximately 4 m³ of contaminated soil was treated. The compost mixture was created and moved to the composter. The bulking agent was placed under the duct. Moisture and temperature probes were installed in addition to the forced aeration system. Oxygen was measured by an oxygen probe. The blower was adjusted according to the oxygen level. The temperature was maintained between 18 and 30°C by the blower. Bacterial numbers were monitored. It was shown that the total microbial population increased from an initial level of 1.5 x 10^5/g to 3.7 x 10^8/g after 16 days. The PAH degrading bacteria also significantly increased. Overall, 90% of the total hydrocarbons were degraded (initial level of 6330 mg/kg TPH) after 80 days. Ninety-three percent of the alkyl benzenes and 2-ring PAHs were degraded in only nine days and 3-ring PAHs needed 23 days for the same degradation level. Eighty-nine percent removal of 4-ring PAHs was accomplished. Only 5-ring PAHs were not degraded. The final product resembled humic material, soil, and wood chips and could be used as a daily cover for landfills, industrial fill, or for parking lots (Taddeo et al. 1997).

At the Naval Surface Warfare Center (Crane, Indiana), composting is used to remediate explosives-contaminated soil (Gray 1999). The contaminants included TNT, Royal Demolition Explosives (RDX), and Her Majesty's Explosive (HMX). The center had been a bomb manufacturing operation from World War II to the mid 1970s. Composting was chosen since the costs to remediate 15,000 tonnes of soil would be $240/tonne as compared to $500/tonne for incineration. A mixture of 60% straw, 25% soil, and 15% chicken manure was used. Composting time was reduced from 20 to 30 days to a range of seven to ten days since the manure heated the mixture to 60°C within 24 h. Thus, the costs decreased to $140/tonne. The Army Corps of Engineers conducted the work. The soil was then used as cover material for the landfill. The soil is screened after excavation and then sent to the composting facility. Three buildings of 91.4 x 21.3 x 5.5 m cover two windrows each. Additives are blended with the contaminated soil in a grinder mixer. A conveyor is used to place the mixture on the floor in windrows (76 m long and 640 m³). A Scarab turns the windrows a minimum of once a day unless the temperature is lower than desired. Moisture content is monitored three times a week. Oxygen and temperature are monitored every day, and pH is monitored once a week. More than 8,660 cubic meters were remediated between April 1998 and April 1999. TNT levels decreased from 3,790 to 15 mg/kg, RDX from 15,300 to 4 mg/kg, and HMX from 10,400 to 3,300 mg/kg (Gray 1999).

Comparison to Other Processes

Composting, a method for waste treatment, has been adapted for soil treatment. Overall, composting has low energy requirements and low sludge production rates. Moisture must be maintained between 50 and 60%. Aeration is maintained by either forcing air through the pile with a blower or by pulling air through the pile with a vacuum. Metal tolerance is relatively high and composting is applicable to many organic contaminants (U.S. EPA 1983). Compared to landfarming, land requirements, water contamination problems, and air emissions are easier to control and retention times are lower (U.S. EPA 1990a). Composting is also less expensive and simpler than incineration. The major drawback of composting is that maintenance requirements to obtain good biodegradation results are higher than landfarming. Air emissions are higher due to the higher temperatures involved. Composting is ideal for treating wastes of high concentration and low permeability.

A survey of the use of biological treatment in various countries is shown in Table 5.8. Germany primarily treats soil at off-site facilities in stationary plants. Strategies and technologies in bioremediation and natural attenuation are under development. Although significant remediation has taken place in the Netherlands, as seen in Table 5.8, significant efforts have been made to develop technologies for *in situ* remediation in a program called NABIS. The government has provided $13 million, and another $7 million has been provided by the private sector. More than 50 projects took place with particular emphasis on natural attenuation for chlorinated solvents and BTEX (ten projects). A market survey in Norway showed that bioventing, biopiles, and landfarming technologies are currently available. Of these, landfarming has been the most used (NATO/CCMS 1999 and 2000). In Sweden, biological methods like composting for smaller PAHs and *in situ* processes such bioventing for petroleum stations are becoming more popular. One company is also developing a bioslurry reactor at full scale (NATO/CCMS 1998).

Although conventional processes have dominated in the United Kingdom in the past, it appears that *ex situ* bioremediation (such as biopiles) and *in situ* bioremediation is becoming more evident. There are ongoing research and development projects involving cyanide biodegradation, natural attenuation of aquifers, and white-rot fungus for xenobiotic metabolism (NATO/CCMS 2000).

Table 5.8 Use of Biological Treatment Processes for Soil Remediation in Various Countries

Country	Capacity of biological treatment processes	% of total remediation techniques
Germany	81 facilities for total of 1.2 million tonnes treated in 1996 (off-site)	56%
Netherlands	0.26 tonnes by landfarming in 1996 for organic contaminants with 60–90% effectiveness. Costs were $20–50/tonne	15%
France	In 1997, 29 sites by composting and biopiles for petroleum compounds, light and heavy oils, and PAHs Bioventing also used	13%

(NATO/CCMS 1999)

Ex Situ Processes: Slurry-Phase Reactors

Description

Slurry-phase reactors are used for the bioremediation of excavated soil. They are also called bioslurry reactors and bioreactor systems. The reactors can be open lagoons or closed tanks. The excavated soil is added to water to form a slurry. The density of the slurry depends on the amount of soil to be treated and the concentration of the contaminants. Mixing energy increases with higher slurry concentrations since mixing is required to keep the solids suspended, to break up larger soil particles, to dissolve contaminants, to assist aeration, and to increase contact between the microorganisms and the contaminants. Dilution, therefore, may be preferable if mixing becomes too difficult. The addition of nutrients and control of pH and temperature can enhance biodegradation as it does in other bioremediation processes. However, since the contact between the soil, contaminants, oxygen, water, and nutrients is superior to other bioremediation processes, the rate of biodegradation is increased. Slurry phase treatment is applicable to contaminants that are more difficult to biodegrade, such as oily or tar-like wastes (Lewis 1992). Reactors can be used on site for the treatment of excavated soils. Less land area is required than for landfarming or biopiles. Bioreactors of 3 to 15 m in diameter and 4.5 to 7.6 m in height have been used (King et al. 1998). Portable tanks with a volume of 75 m^3 have also been used.

Box 5-6 Overview of Slurry Reactors

Applications	Suitable for treatment of high concentrations of contaminants including benzene, creosote, dioxin, herbicides, insecticides, nitrotoluenes, PCBs, PCP, pesticides, phenol, TCE and toluene, and metal contaminants in soil
Cost	Operating: $100 to $250/m^3
Advantages	Good control of pH and aeration
	Good mixing of soil
	Residence times are low
Disadvantages	High capital costs
	Volatiles difficult to treat
Treatment time	Several days to ten weeks
Other considerations:	Costs increase according to sophistication. Lagoons are least expensive while reactors are the most expensive

Operating Conditions

Operation of slurry-phase reactors is usually in batch or semi-batch mode. For a batch process, the contaminated soil, microbial cultures, nutrients, and water are added to the reactor. Mixing and aeration is performed until the biodegradation process is complete. The soil is then allowed to sediment and returned to the original site. The water can be sent for wastewater treatment or recycled for the next treatment cycle. If a portion of the slurry is retained for use as an inoculum, the process is operated as a sequencing batch reactor (SBR).

Oxygen Supply

Oxygen is required for aerobic biodegradation (Figure 5.13), and can be supplied by diffused aeration, turbine spargers, or surface aerators (Metcalf and Eddy 1991). In diffused aeration, oxygen is supplied by forcing air

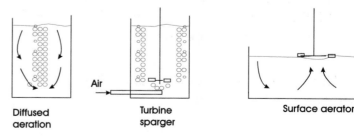

Figure 5.13 Oxygenation Setup for Slurry Reactors

through porous diffusers at the bottom of the reactor. Oxygen is transferred across the gas-liquid interface of the bubbles. Transfer rates are affected by the bubble size and contact time, which is dependent on the depth of the liquid in the tanks, usually 5 m or more. In turbine spargers, a combination of mechanical and diffused aeration occurs. The turbine enhances oxygen transfer by breaking the air into small bubbles. Deeper tanks may require multi-level turbines. Surface aeration is performed by a turbine at the surface which draws the water upward and outward, thus forming bubbles. The turbulence enables mixing to take place as well as aeration. If this method is used, the tanks cannot be too deep. Draft tubes can be added to increase the depth, however. Oxygen transfer rates for all three types of aerators can be approximated as 1.2 to 2.5 kg O_2 per kWh for dilute systems. As the solids content increases toward 10%, the rate decreases to 1 kg O_2 per kWh.

The requirements for the oxygen uptake rate depend on the rates of biodegradation (r_o) and microbial growth since the bacteria require oxygen for growth and oxidation of the contaminants. Since the biomass growth in slurry systems is difficult to estimate, an expression for wastewater treatment design is used as shown below (Metcalf and Eddy 1991):

$$r_{O2} = r_o(1 - (\frac{0.6}{1 + 0.05\tau}))$$

where r_{O2} is the oxygen uptake rate (mg/L.h) and τ is the solids retention time (days).

The solids retention time can be calculated by the equation:

$$\tau = (1 + \frac{V_R}{V})t$$

where V_R is the volume of settled solids in the tank (m³), V is the volume of the tank (m³), and t is the reaction time (days).

Mixing

Mixing in slurry reactors is performed to increase homogeneity in the reactor, maximize desorption rates, and decrease toxicity to the microorganisms. Since mixing theory is primarily applicable to reaction vessels below 50 m³, it is more appropriate to use experimentation, experience, and manufacturers' recommendations concerning mixing design for slurry-phase vessels, which are often greater than 500 m³ in volume. For a turbine mixer, the radius of influence is two to three times the diameter of the impeller. Therefore, for a square tank with sides of 60 m, approximately 16 to 36 impellers with a 5 m diameter would be required.

The Reynolds number (N_{Re}) is used to characterize the turbulence of the impeller and is determined by the equation:

$$N_{Re} = \frac{ND_i^2\rho}{\mu}$$

where N is the rotational speed of the impeller (rps), D_i is the diameter of the impeller, ρ is the density of the slurry (kg/m³), and μ is the dynamic viscosity of the slurry (kg/m.s). The value for the Reynolds number should be above 10,000 to maintain turbulent conditions. Assuming that the slurry has the same density (1,000 kg/m³) and dynamic viscosity (0.001 kg/m.s), then the rotational speed required for a 5 m diameter impeller would be 0.024 min⁻¹. In practice, however, values of 40 to 60 rpm (0.7 to 1 rps) are used to maintain Reynolds numbers in the range of 1 to 3 x 10⁷. The power requirements can be estimated from manufacturers' specifications; these are usually in the range of 20 to 50 kW/1,000 m³. For thick suspensions, these values may be five times higher. Since the radius of influence would also be significantly decreased, small-diameter deep tanks are used for thick suspensions.

Nutrient Requirements

Laboratory studies are the best way to estimate nutrient requirements because of the inorganic material present. Another method of estimating nutrient requirements is similar to that for oxygen requirements. Assuming that nitrogen comprises 10 to 14% of bacterial mass and that phosphorus accounts for 0.5 to 2% of the cell mass, the growth rate can be estimated as a function of organic removal in terms of COD or BOD removal. Thus, nitrogen requirements, r_N, can be estimated as:

$$r_N = r_o \left(\frac{0.06}{1 + 0.05\tau} \right)$$

and phosphorus requirements, r_P, can be estimated as:

$$r_P = r_o \left(\frac{0.01}{1 + 0.05\tau} \right)$$

Nitrogen can be added in the form of ammonium or nitrate while phosphorus can be added as phosphate.

Pretreatment

Pretreatment of contaminated soil by either soil fractionation or soil washing can serve to remove uncontaminated soil, soil with less contamination, or other materials that can be detrimental to the soil treatment process. Soil fractionation by screening can also reduce the amount of soil to be treated. For example, larger particle size material usually does not contain high concentrations of contaminants like fine soil and, therefore, can be removed. Soil larger than 60 mesh is also removed since it is difficult to remain suspended. Black et al. (1991) showed that smaller particles, due to their high surface area to volume ratio, absorb more contaminants on a mass basis in a typical soil. For example, although the fines accounted for 55% of the total mass, they contained 88.8% of the COD; the coarse fraction contained 29% by mass of the soil but only 2.9% of the COD. Another project showed that 37% of the dry mass contained fines which contained 94% of the PAHs (Ahlert and Kosson 1989). Other materials such as plastics, wood branches, metal parts, and so on should also be removed prior to slurry-phase treatment.

Soil washing can also remove the more contaminated soil fraction from the less contaminated one. The first such application in the United States took place in the 1980s. Some of the concepts have been derived from practices in the mining and ore processing industries. Coarse screening is usually the first step to remove large stones, plastics, and other unwanted materials. An attrition cell is used to clean the large particles. Water is added to the soil to obtain a slurry which is mixed by rotating impellers. Warmer water or addition of surfactants to the water may be used to enhance the solubilization of the contaminants. The fine and larger soil particles will separate during the scrubbing process. Surface abrasion will clean the larger particles. Countercurrent flow can be employed to enhance soil washing.

The slurry is then sent to rotary screens, vibrating screens, or hydrocyclones for partial particle classification. BioTrol (Chaska, MN) used equipment adapted from the mining industry and added equipment such as attrition scrubbers, froth flotation, mixing trommels, and pug mills (U.S. EPA 1999). The discharged water can be recycled or sent for discharge. Particles greater than 74 mm (200 mesh) in diameter can often be discarded since they will be clean (Griffiths 1995). The remaining solids are removed by settling, sometimes with the aid of coagulants or flocculants, or by air flotation. The sludge is then sent to the bioslurry reactors. Metal concentrations should be monitored to avoid toxicity problems. It may be necessary to add acids to precipitate the metals.

Soil washing is fairly expensive due to its complexity. For this process to be economical, more than 70% of the solids must be recovered as clean material (Boyle 1993). Therefore, it is advisable first to determine if the process will be economically feasible. It is usually more economical for sandy, coarse soils than for clay soils. Fractionation may be more appropriate where the concentrations are low and the contaminants are mainly associated with the finer particles.

Inoculum Development

It is extremely important to develop a good microbial population to perform the biodegradation of the contaminants in the slurry process. Sources of microorganisms include sludges from municipal wastewater treatment plants, agricultural soils, and contaminated soils. To find microorganisms that are capable of degrading the soil contaminants, add a range of contaminant concentrations to a mixture of the sources. Contaminant concentrations or COD can then be followed to determine toxicity levels for the microbial population. Biodegradation rates and stoichiometric parameters can be obtained for the best cultures. The cultures can then be further developed so that higher contaminant concentrations can be biodegraded. Once the optimal microbial population has been developed, large quantities of the microorganisms must be developed to decrease start-up times and enhance resistance to toxicity and upsets. Microbial counts greater than 10^8/mL are preferable.

Mixing

Solids concentrations of up to 50% are possible when using a steel reactor with a tapered bottom, aerators, and a good mixer to prevent solids settling (LaGrega et al. 1994). The higher the concentration, the smaller the reactor. However, this can also increase contaminants to toxic levels for the microorganisms. In addition, oxygen transfer rates are reduced. Above 40% solids content, the oxygen transfer rate has little dependence on the air flow rate (Andrews 1990). Therefore, because of the problems with mixing and aeration, most operations are performed between 10 and 40% solids (Ross 1990).

Since oxygen is usually required for slurry phase remediation, dissolved oxygen (DO) is an important operating parameter. As previously mentioned, it is desirable to maintain a DO concentration of at least 2 mg/L. The solids content is usually higher than wastewater treatment processes, thus requiring higher mixing requirements. Additional mixing in the axial direction is sometimes required to enhance solids suspension and distribute the oxygen.

Another important operating parameter is temperature. As a result of long residence times and the poor heat conductivity of water, temperatures in the slurry reactor can vary significantly. If temperature variations are severe, temperature control may be necessary. Jerger and Woodhull (1994) measured temperatures in a slurry study in Mississippi between 25 and 40°C in the spring, summer, and fall and with temperatures between 15 and 21°C in the winter.

Sometimes, other agents can be added to assist in the desorption and solubilization of the contaminants to increase biodegradation rates. Although surfactant addition has been evaluated, their role is not clear (Lewis 1992; Melcer et al. 1995). Castaldi and Ford (1992) suggest that microbially-produced surfactants called biosurfactants may be more advantageous due to their low toxicities. It has also been shown recently that biosurfactants can enhance metal removal (Mulligan et al. 1999a, b).

Oxidizing agents such as Fenton's reagent (hydrogen peroxide and iron salts) have also been evaluated to degrade high molecular weight compounds such as PAHs into smaller more easily biodegradable ones (Brown et al. 1995). As previously mentioned, hydroxylation of large PAH molecules is usually the first step in biodegradation and is often rate-limiting. Therefore, chemical assistance should enhance the biodegradation process.

Other considerations which should be taken into account are VOC emissions and foam production. High mixing and aeration rates can lead to foaming and the release of VOCs. Hoods can be installed to contain and treat the emissions. Emission concentrations will depend on the contaminant concentrations in the soil and the volatility of the compounds. If the compounds are primarily semivolatile, the cost of treatment may be high (Lewis 1992). Tests for foaming should be performed during treatability studies. The mixing speed may be reduced or an antifoaming agent may be added.

The Eimco Biolift™ bioreactor was originally developed by Eimco Process Equipment Company (Salt Lake City, UT) for the mineral processing industry and has been applied for slurry treatment. It is shown in Figure 5.14.

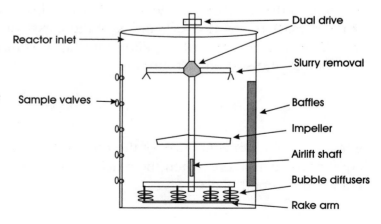

Figure 5.14 Eimco Biolift Slurry Reactor (adapted from LaGrega et al. 1994)

The reactor has a dual drive design for suspension of fine and coarse particles. Coarse particles are moved by a slow moving rake (2 rpm) to the central airlift system for their removal. Fine particles are suspended by the bubbles from rake arm diffusers. The slurry is agitated at a rate of 20 to 30 rpm by an axial flow impeller. According to the manufacturer (Brox 1988), operating costs are 20% lower than the conventional systems due to energy savings.

A biological method for remediation of metal-contaminated soil involves metal precipitation on the cell surface of bacteria in a slurry reactor. The reactor is called the Bacteria Metal Slurry Reactor (BMSR). The metals such as Cd and Zn accumulate on the biomass which is later removed. The bacterium *Alcaligenes eutrophus var. metalloterans* CH_{34} has shown the ability to accumulate metals on the proteins and polysaccharides of the cell surface (Mergeay et al. 1985). Treatment of the sandy soil by the bacteria altered the colloidal characteristics of the soil. After the bioprecipitation is complete, the slurry is sedimented and the biomass containing the metals is removed by froth flotation. The bacterial treatment allowed the soil to sediment faster, and between 30 to 70% of the metals could be removed in 12 hours (Diels 1997). More research is necessary to reduce costs, evaluate the toxicities of the remaining metals in the soil, and scale-up difficulties.

Design

To model batch processes, ideal mixing is assumed. The following mass balance can be used:

$$\frac{dM}{dt} = Vr_o$$

where M is the mass of the contaminants in the reactor (kg), t is time (days), V is the reactor volume (m^3), and r_o is the rate of contaminant biodegradation (kg/m^3-day).

The mass balance of the contaminants includes those in the solution and those sorbed on the solid phase as follows:

$$M = CV + sX_sV$$

where C is the mass of soluble contaminants (kg/m^3), s is the mass of sorbed contaminant per unit mass of soil (kg/kg), and X_s is the mass concentration of solids in the reactor (kg/m^3). If the reaction is slow, equilibrium is approximated as the coefficient, K_{SD}, that relates the soluble and sorbed distribution:

$$K_{SD} = \frac{S}{C_s}$$

where C_s is the concentration of the contaminant in the liquid phase at equilibrium. At non-equilibrium conditions when the reaction rate is fast, mass balances are required for the liquid and sorbed phases.

For the liquid phase,

$$V\frac{dC}{dt} = VK_La(C_s - C) + Vr_o$$

$$\frac{dC}{dt} = K_La(C_s - C) + r_o$$

where a is the interfacial area per unit volume (m^{-1}) and K_L is the mass transfer rate coefficient (m/day).

For the solid phase,

$$\frac{dM_{soil}}{dt} = -VK_La(C_s - C)$$

$$X_s\frac{ds}{dt} = -K_La\left(\frac{s}{K_{SD}} - C\right)$$

These equations can be approximated, substituted into the first equation, and rearranged to give:

$$\frac{dC}{dt} = \frac{r_o}{K_{SD}X_S + 1}$$

Reactor Configurations

Batch type reactors are commonly employed, particularly if the soil volume for treatment is not large and if treatment time is not a major constraint. Microbial inoculum and/or nutrient mixes can be held in holding tanks prior to addition to the slurry. If degradation kinetics is first order according to the contaminant concentration, batch operation is appropriate since the degradation will proceed at a higher rate when the contaminant concentration is also high. The disadvantage of batch operation is that lag times can be prolonged as time is required for the bacterial culture to acclimate.

In continuous reactor processes, there is no lag period since a high concentration of acclimated microorganisms is continuously fed into the slurry reactor by recycling a portion of the effluent. Continuous processes decrease the treatment residence time since there is no lag time or time required to empty and fill the reactors. Operation in this manner also dilutes high concentrations of contaminates, thus reducing toxic effects on the cultures. Volatile compounds can also be degraded more quickly, which reduces the rate of volatilization. The cost of continuous processes can be high due to continuous pumping and mixing, particularly since the residence times are typically much longer than activated sludge processes (days or weeks compared to hours).

Semibatch processes are considered to be in the middle of batch and continuous processes in terms of energy costs. For a semi-batch process, three tanks are used (Figure 5.15). The first tank is used to mix the soil with water, nutrients, and inoculum, in addition to adjusting pH levels. The mixture is then pumped to the next tank where biodegradation takes place as a result of continuous mixing and aeration. In the last tank, the liquid and solids are separated. Since the first and last tanks may be smaller than the second one, the total volume for the semi-batch process can be less than the batch in which everything is carried out in one tank. However, the semi-batch process is more complex, which may increase processing costs.

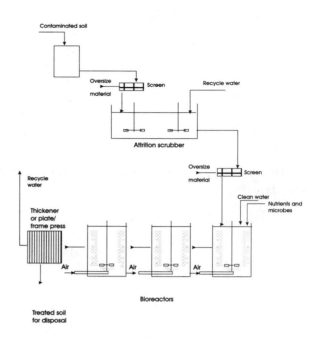

Figure 5.15 Diagram of Semibatch Sequential Slurry Reactors

The reactor size is a function of the hydraulic retention time (HRT) and is determined as follows:

$$V = Q \times HRT$$

where V is the volume of the reactor and Q is the slurry flow rate. The HRT vary according to the biodegradability of the compound, the initial contaminant concentration, and the degree of treatment desired. A simplified kinetics equation is used:

$$C = C_o e^{-kt}$$

where C is the final contaminant concentration (kg/m^3), C_o is the initial contaminant concentration (kg/m^3), k is the degradation rate (day^{-1}), and t is time in days.

Applications

There have been few large-scale demonstrations in the United States such as the demonstration through the EPA SITE program by International Technology Corporation (U.S. EPA 1992). Lagoons are more economically viable, but losses to the atmosphere via volatilization may be excessive. Full-scale slurry bioreactors are commercially available and are 3 to 15 m in diameter and 4.6 to 7.6 m in height. Slurry volumes are in the range of 56,800 to 113,600 liters. The equipment costs between $125,000 and $2 million (King et al. 1998). The tanks are usually one to three in series. Slurry concentrations are between 20 and 40%.

A 30% slurry mixture of creosote-contaminated soil from Brinard, Minnesota was treated in a bioreactor. PAH-degrading bacteria were inoculated into the reactor. Over the course of nine weeks of treatment, 77 to 97% of the PAHs were treated for an average of 89%. Liquid phase concentrations were below the detection level.

Yare (1991) demonstrated that semi-volatile compounds degraded at a much faster rate in slurry reactors compared to solid-phase bioremediation. Half-time for phenanthrene was 8.0 days in the slurry system and 32.4 days in the solid-phase system.

Costs

The main disadvantage of bioreactors is the increased capital and operating costs ($100 to $250/tonne). These costs are related to the cost of material handling, mixing, supplying oxygen, and mechanization in terms of capital, operating, and maintenance expenses. The costs for slurry preparation are in the range of $55 to $66 per tonne, $44 to $55 per tonne for biological treatment and another $22 to $33 per tonne for dewatering (Cookson 1995). Operation in semicontinuous mode could help to decrease costs.

Case Studies

Coover et al. (1994) operated a 3.8 million liter treatment at a refinery site by using a concrete clarifier (47 m in diameter) as a bioslurry reactor. A solids content of 10% was chosen and was inoculated with 83 m^3 of activated sludge from a wastewater treatment plant. The reactor was operated for eight weeks as one batch. Temperature was 20 to 25°C, the pH was 5.8, and the nutrient ratio of C:N:P was 100:2:0.2 with an oxygen uptake rate of 0.2 mg/L-min and a cell count of 10^7 to 10^8 per ml. The rate of biodegradation was the highest within the first two weeks; however, there were problems with the formation of a 10 to 25 cm thick foam layer. Another problem was that 25% of the solids settled to the bottom of the reactor.

Eko Tec AB (Sweden) performed the bioslurry remediation of 3,000 tonnes of creosote-contaminated soil and ditch sediments from a railway station area in northern Sweden (NATO/CCMS 1998). An overall total PAH level of 50 mg/kg was the objective. Benzo(a)pyrene and benzo(a)anthracene levels were to reach 10 mg/kg each. Pretreatment was by screening with a 10 mm sieve to remove large soil particles, washing, pumping to a screen to remove particles greater than 2 mm, and then pumping to 60 m^3 slurry reactor. Volume of the slurry was 25 m^3. The slurry was kept in suspension continuously, and a bacterial culture specific for PAH degradation was added with nutrients and soil activators. Parameters for this operation included dissolved oxygen, pH, temperature, and nutrient concentration. The initial total PAH concentration of 219.9 mg/kg was reduced to 26.97 ppm after 27 days of treatment, below the objective of 50 mg/kg. The levels for the two individual PAHs also reached below the objective. The treatment was closed after the 27-day period, the slurry was pumped into a concrete tank, and the slurry was allowed to sediment to separate the liquid and the solid phases. The treated soil was then used as fill material.

A process called Biolysis™ was developed by Sanexen Environmental Services (Longueuil, Quebec, Canada). A project in 1995 involved the treatment of 1,350 m^3 of soil and sediment contaminated with 13,500 mg/kg oil and grease. A settling tank was converted into a bioreactor. Dredging equipment was used for aeration and chemical and nutrient addition. Ten weeks later, the level was reduced to 1,000 mg/kg. Reactors (3 m diameter and 8 m high) can also be used when vigorous agitation of suspended solids is desired (Enviro-Access 1998). Throughput is 4 m³/day with an average retention time of three weeks. Degradation rates of 65 to 98% of heavy hydrocarbons have been achieved. The innovation in this process is the use of specialized enzymes. The first step involves increasing the bioavailability of the hydrocarbons by enzymes and surfactants. The second and third steps involve ingestion and degradation of the hydrocarbons. In reactor systems all steps occur at the same time while in the pond systems they occur sequentially.

Comparison to Other Processes

Slurry reactors provide enhanced biodegradation rates as a result of much better control of operating parameters. Pretreatment, mixing, and aeration are all combined in this process. Mueller et al. (1991) suggested that remediation rates could be an order of magnitude faster. Highly contaminated soils and sludges with concentrations in the range of 2,500 to 250,000 mg/kg were treated. Slurry reactors also take up much less space than landfarming processes. It may also be a good option for treatment of volatile compounds due to emission control.

The use of bioreactors increases the ability to control pH, oxygen, and temperature. Residence time can be significantly reduced. Recycling of the treated material can reduce contaminant levels significantly.

In determining the best method for treatment, engineering and economic considerations must be considered. The site must be examined, the amount and type of material to be degraded must be determined, large scale demonstration and cost estimation must be analyzed, regulatory issues must be considered, the time required for treatment must be estimated, and all procedures must be evaluated to determine if the procedures can be used in combination with others to reduce costs and time. The time and costs for other biological methods are compared to bioslurry processes in Table 5.9. Non-biological treatment methods include long-term storage, incineration, landfill, and air stripping (Table 5.10). Air stripping can lead to contamination of activated carbon which subsequently must be treated or disposed. A combination of physical and chemical methods may be ideal in minimizing costs and time. Table 5.11 represents a summary of the parameters that will affect operating conditions and which must be taken into consideration when characterizing the contaminated soil site.

Table 5.9 Overview of Biotreatment Processes

Type of treatment	Applications	Cost ranking	Time requirement ranking
Landfarming	PCP, oil, gasoline, PAHs, low to medium contamination	+ +	+ + +
Soil slurry reactors	Surface contamination and recalcitrant compounds	+ +	+
Subsurface soil treatment	Deep contamination, low to medium levels of contamination, oil, gasoline, chlorinated	+ +	+ + +
Bioventing	Volatile and semi-volatile chlorinated solvents and petroleum hydrocarbons in unsaturated zone	+	+ + +
Composting	Nitroaromatics, sewage sludge, gasoline, high levels of surface contamination	+ +	+ +

Cost ranking: + denotes $10 to $40/m³, ++ denotes $40 to $300/m³, and +++ denotes greater than 300/m³
Time requirements: +denotes several hours to weeks, ++ several weeks to months, +++ more than 1 year

Table 5.10 Comparison of Soil Treatment Technologies
(adapted from Levin and Gealt 1993; Cunningham et al. 1997a, b; OMEE 1992)

Process	Cost per m^3 ($) (ranking)	Months required (ranking)	Other issues
Soil vapor extraction	35–80 (++)	6–36 (+++)	Fast, unobtrusive for volatile compounds, may not obtain desired contaminant level
Soil flushing	50–150 (++)	3–24 (+++)	Unobtrusive, simple, little hydraulic control, no guarantee contaminant level reached, for soluble metals/organics
Incineration	250–2,000 (+++)	6–9 (++)	Energy requirements are high, air pollution, licensing requirements, for all contaminants
Solidification /stabilization	75–200 (++)	1–6 (+)	Long-term monitoring, leaching, mainly for metals, expensive, licensing requirements
Vitrification	1,000 (+++)	1–6 (+)	Long-term monitoring, expensive
Landfill	40–300 (++)	6–9 (++)	Long-term monitoring, leaching, disruptive, not sustainable option
Biotreatment	30–100 for *in situ* (+) 30–300 for *ex situ* (++)	3–60 (+++)	Extensive time required, intermediate products, time less for *ex situ* processes, control of leachates and emissions is superior, feasibility tests required

Cost ranking: + denotes $10 to $40/m^3, ++ denotes $40 to $300/m^3, and +++ denotes greater than 300/m^3
Time requirements: +denotes several hours to weeks, ++ several weeks to months, +++ more than 1 year

Table 5.11 Important Media Characteristics and Operating Parameters (adapted from U.S. EPA 1998)

Technology	Particle Size Distribution	Hydraulic Conductivity	Moisture Content	Air Permeability	pH	Porosity	TOC	O & G Content	Presence of NAPL	Air Flow Rate	Mix Rate	Residence Time or Throughput	Temp.
Biological													
Bioslurping	X	X	X	X	X		X	X	X	X			X
Bioventing	X		X	X	X	X	X	X	X	X			X
Bioremediation	X	X	X	X	X	X	X						X
Phyto-remediation	X	X			X		X						X
Composting	X		X		X					X	X	X	X
Slurry phase	X		X		X					X	X	X	X
Non-biological													
Vapor extraction			X	X		X	X			X			
Soil flushing	X	X			X		X	X	X			X	X
Soil washing	X				X		X					X	X
Incineration	X		X		X		X			X		X	X
Stabilization	X		X		X			X	X			X	X
Vitrification	X		X	X								X	X

Genetically Engineered Organisms

Recently, interest in genetic engineering of microorganisms to improve their performance during bioremediation processes has increased substantially. There have been concerns about the threat to our safety of manipulated bacteria. This does not seem to be the case, however. In nature, transfer of DNA fragments called plasmids is common. Zylstra et al. (1989) cloned a gene for toluene dioxygenase in *Escherichia coli*. Cloning of *Clostridium* sp. has also been accomplished to enable degradation of PCBs (Li-Hua 1998). However, there were problems with the ability of the organism to survive in the contaminated environment as well as difficulties with delivery to the contaminated zone. Social, political, and regulatory restrictions have also limited the application of engineered organisms. However, we already benefit from genetic manipulation of bacteria in industrial pharmaceutical and enzyme manufacture. Controlled release of these manipulated organisms for remediation purposes may be the main use of these organisms. Acclimation involves gene derepression and induction. In the future, manipulation of the genes involved in bioremediation will benefit this industry. To enable this technology to be developed, regulators must work with researchers during field tests. Although early investigators felt that genetically engineered microorganisms would be important (Chakrabarty et al. 1973), this has not been the case mainly because of concern by the public for the release of these organisms into the environment.

References

Air Force Center for Environmental Excellence (AFCEE). 1994. *Technology Profile: Vacuum-Mediated LNAPL Free Product Recovery/Bioremediation (Bioslurper).* AFCEE Fact Sheet. Issue 1, March. San Antonio, TX: Brooks Air Force Base.

Ahlert, R. C. and D. S. Kosson. 1989. Aerobic mineralization of organic contaminants bound on soil fines. *Third International Conference on New Frontiers for Hazardous Waste Management,* Pittsburgh, Pennsylvania. September.

Andrews, G. 1996. Large-scale bioprocessing of soils. *Biotechnology Progress,* 6:225–230.

Anonymous. 1999. Hamilton harbour cleanup makes big splash in Hamilton. Environment Canada's Green Lane. Available online http://www.on.ec.gc.ca/success-stories/co/hamilton-e.html.

Anonymous. 2000. Bio-Sparge™. ARUSI AR Utilities Specialities, Inc. Available online http://www. Arusi.net/bio-sparge.htm.

Bader, J. L., G. Gonzales, P. C. Goodell, S. D. Pillaiand, and A. S. Ali. 1996. Bioreduction of hexavalent chromium in batch cultures using indigenous soil microorganisms. *HSRC/WERC. Joint Conference on the Environment,* Albuquerque, NM. 22–24 April.

Baker, A. J. M., and R. R. Brooks. 1989. Terrestrial higher plants which hyperaccumulate metal elements: A review of their distribution, ecology, and phytochemistry. *Biorecovery,* 1:81–126.

Baker, R. S., and J. Bierschenk. 1995. Vacuum-enhanced recovery of water and NAPL: Concept and field test. *Journal of Soil Contamination.* 4(1):57–76.

Banuelos, G. S., G. E. Cardon, B. Mackey, J. Ben-Asher, L. L. Wu, P. Beuselinck, and S. Akohoue. 1993. Boron and selenium removal in boron-laden soils by four sprinkler-irrigated plant species. *Journal of Environmental Quality,* 22:786–792.

Benazon, N. D., W. Belanger, D. B. Scheulen, and M. J. Lesky. 1995. Bioremediation of ethyl-benzyne and styrene-contaminated soil using biopiles. *Biological Unit Processes for Hazardous Waste Treatment.* R. E. Hinchee, R. S. Skeen, and G. D. Sayles, eds. Columbus, OH: Batelle Press.

Billings, J. F., A. I. Cooley, and G. K. Billings. 1994. Microbial and carbon dioxide aspects of operating air-sparging sites. In *Air Sparging for Site Remediation,* ed. R. E. Hinchee. Boca Raton, FL: Lewis Publishers.

Black, W. V., R. C. Ahlert, D. S. Kosson, and J. E. Brugger. 1991. Slurry-based biotreatment of contaminants sorbed onto soil constituents. In *On-Site Bioreclamation Processes for Xenobiotic and Hydrocarbon Treatment,* eds. R. E. Hinchee and R. F. Olfenbuttel. Massachusetts: Butterworth-Heinemann.

Blais, J. F., R. S. Tyagi, and J. C. Auclair. 1992. Bioleaching of metals from sewage sludge by sulfur-oxidizing bacteria. *Journal of Environmental Engineering,* 118(5):690–707.

Block, R., H. Stroo, and G. G. Swett. 1997. Bioremediation—Why doesn't it work sometimes? In *Practical Engineering Perspectives The Environment: Air, Water and Soil,* ed. G. F. Nalven. New York: American Institute of Chemical Engineers. 277–283.

Bolenz, S., H. Omran, and K. Gierschner. 1990. Treatments of water hyacinth tissue to obtain useful products. *Biological Wastes,* 22:263–274.

Borden, R. C., M. J. Hunt, M. B. Shafer, and M. A. Barlaz. 1997. Anaerobic biodegradation of BTEX in aquifer material. *Environmental Research Brief,* EPA/600/S-97-003. National Risk Management Research Laboratory, Ada, OK. August.

Boyle, C. 1993. Soil Washing. In *Remedial Processes for Contaminated Land,* ed. M. Pratt. Warwickshire, UK: Institution of Chemical Engineers.

Brar, G. S. 1997. Phytoremediation of chlorinated solvents: Progress and challenge. *IBC's Second Annual Conference on Innovative Technologies,* Boston, MA. 21–23 July.

Brown, K. L., B. Davilla, and J. Sanseverino. 1995. Combined chemical and biological oxidation of slurry-phase polycyclic aromatic hydrocarbons. Paper 95-RA127.06. Presented at the *88th Annual Meeting and Exhibition of Air and Waste Management Association,* San Antonio, TX. June.

Brown, K. S. 1995. The Green Clean: The Emerging Field of Phytoremediation Takes Root. *Bioscience,* 45:579–582.

Brown, R. A., R. J. Hicks, and P. M. Hicks. 1994a. Use of air sparging for *in situ* bioremediation. In *Air Sparging for Site Remediation,* ed. R. E Hinchee. Boca Raton, FL: Lewis Publishers.

Brown, S. L., R. L. Chaney, J. Angle, and A. J. M. Baker. 1994b. Phytoremediation potential of *Thlaspi caerulescens* and bladder campion for zinc- and cadmium-contaminated soil. *Journal of Environmental Quality,* 23:1151–1157.

Brox, G. H. 1988. A new solid/liquid contact bioslurry reactor making bio-remediation more cost-competetive. *Proceedings of the 10ᵗʰ National Conference–Superfund '89.* Washington, D.C. 27 November.

Carberry, J. B. 1990. Enhancement of PCP and TCE biodegradation by hydrogen peroxide. *Proceedings of the 11th National Conference—Superfund '90.* Hazardous Materials Control Research Institute. Washington, D.C. November 26–28.

Carberry, J. B., J. D. Wik, and C. D. Harmon. 1991. Aerobic bioremediation of petroleum-contaminated soil using controlled landfarming. Presented at *American Chemical Society Symposium on Emergency Technologies for Hazardous Waste Management.* Atlanta, GA. 1–2 October.

Castaldi, F. J., and D. L. Ford. 1992. Slurry bioremediation of petrochemical waste sludges. *Water Science Technology,* 25(3):207–212.

Chakrabarty, A. M., G. Chou, and I. C. Gunsalus. 1973. Genetic regulation of octane dissimilation plasmid in *Pseudomonas. Proceedings of the National Academy of Sciences,* 70:1137–1140.

Chapelle, F. H. 1999. Bioremediation of petroleum hydrocarbon-contaminated ground water: The perspectives of history and hydrology. *Ground Water,* 37:122–132.

Christodoilatis, C., G. P. Korfiatis, P. N. Pal, and A. Koutsopiros. 1996. *In situ* groundwater treatment in trench Bio-sparge system. *Hazardous Waste and Hazardous Materials,* 13(2):223–236.

Cleaning Carolina Soil: Successful Bioremediation with Compost. 1995. *Biocycle,* 36(2):57–59.

Cookson, J. T. 1995. *Bioremediation Engineering Design and Application,* New York: McGraw-Hill.

Coover, M. P., R. M. Kabrick, H. F. Stroo, and D. F. Sherman. 1994. *In situ* liquid/solid treatment of petroleum impoundment sludges: Engineering aspects and field applications. In *Bioremediation Field Experience,* eds. P. L. Flathman, D. E. Jerger, and J. H. Exner. Boca Raton, FL: Lewis Publishers.

Cunningham, S. D., W. R. Berti, and J. W. Huang. 1997a. Phytoremediation of contaminated soils. *Biotechnology,* 13:393–397.

Cunningham, S. D., J. R. Shann, D. E. Crowley, and T. A. Anderson. 1997b. Phytoremediation of contaminated water and soil. In *Phytoremediation of Soil and Water Contaminants, ACS Symposium Series 664,* eds. E. L. Kruger, T. A. Anderson, and J. R. Coats. Washington D.C.: American Chemical Society. 2–19.

Davis, L. C., C. Chaffin, N. Muralidaharan, V. P. Visser, W. G. Fateley, L. E. Erickson, and R. M. Hammaker. 1993. Monitoring the beneficial effects of plants in bioremediation of volatile organic compounds. *Proceedings of Conference On Hazardous Waste Research.* 25–26 May. Kansas State University, Manhattan, KS. 236–249.

Diels, L. 1997. Heavy metal bioremediation of soil. In *Methods in Biotechnology, Vol. 2: Bioremediation Protocols,* ed. D. Sheehan. Totowa: Humana Press, Inc. 283–295.

Dooley, M. A., K. Taylor, and B. Allen. 1995. In *Composting of Herbicide-Contaminated Soil in Bioremediation of Recalcitrant Organics,* eds. R. E. Hinchee, D. B. Anderson, and R. E. Hoeppel. Columbus, OH: Battelle Press.

Downey, D. C., and M. G. Elliott. 1990. Performance of Selected *In Situ* Soil Decontamination Techniques: An Air Force Perspective. *Environmental Progress,* 9(3):169–173.

Doyle, R. C., J. F. Kitchens, and M. D. Erickson. 1992. Composting of soils contaminated with explosives. In *Proceedings of Federal Environmental Restoration Conference,* Vienna,

VA. 15–17 April. Greenbelt, MD: Hazardous Materials Control Resources Institute. 373–376.

Dupont, R. R., R. C. Sims, J. L. Sims, and D. L. Sorensen. 1988. *In Situ* Biological Treatment of Hazardous Waste-Contaminated Soils. *Bioremediation Systems, Vol. II.* ed. D. L. Wise. Florida: CRC Press.

Dushenkov, S., A. Mikheev, A. Prokhnevsky, M. Ruchko, and B. Sorochinsky. 1999. Phytoremediation of radiocesium-contaminated soil in the vicinity of Chernobyl, Ukraine. *Environmental Science and Technology,* 33(3):469–475.

Eftekhari, F., and Mulligan, C. N. 2000. Foam-surfactant technology for *in-situ* soil remediation. *6th Environmental Engineering Specialty Conference of the CSCE and 2nd Spring Conference of the Geoenvironmental Division of the Canadian Geotechnical Society.* 142–147.

EIMCO Process Equipment Company. 1994. *EIMCO biolift reactor.* Salt Lake City: EIMCO.

El Fantroussi, S., M. Belkacemi, E. M. Top, J. Mahillon, H. Naveau, and S. N. Agathos. 1999. Bioaugmentation of a soil bioreactor designed for pilot-scale anaerobic bioremediation studies. *Environmental Science & Technology.* 33:2992–3001.

Entry, J. A., L. S. Watrud, R. S. Manasse, and N. C. Vance. 1997. Phytoremediation and reclamation of soils contaminated with radionucleotides. In *Phytoremediation of Soil and Water Contaminants.* ACS Symposium Series 664, eds. E. L. Kruger, T. A. Anderson, and J. R. Coats. Washington, D.C.: American Chemical Society. 297–309.

Enviro-Access. 1997a. Bioslurping-Multiphase Vacuum Extraction & Treatment. Technological Fact Sheet, FA1-05-06.

Enviro-Access. 1997b. Pile Degradation. Technological Fact Sheet, F1-06-95. Online at http://www.enviroaccess.ca/fiches_5/F1-05-95a.html.

Enviro-Access. 1998. Biotreatment of oil sludges (Biolyses®). Technological Fact Sheet, F5-05-96. Online at http://www.enviroaccess.ca/fiches_5/F5-05-96a.html.

Eweis, J. B., S. J. Ergas, D. P. Y. Chang, and E. D. Schroeder. 1998. *Bioremediation Principles.* Boston: WCB McGraw-Hill.

Ferguson, R. 1997. DARAMEND bioremediation technology. *IBC's Second Annual Conference on Innovative Remediation Technologies.* Boston, MA. 21–23 July.

Fiorenza, S., K. L. Dusto, and C. H. Ward. 1991. Decision-making: Is bioremediation a viable option? *Journal of Hazardous Materials,* 28:171–183.

Flathman, P. E., B. J. Krupp, J. R. Trausch, J. H. Carson, R. Yao, G. J. Laird, P. M. Woodhull, D. E. Jerger, P. R. Lear, and P. Zottola. 1995. Biological solid-phase treatment of vinyl acetate-contaminated soil: An emergency response action. Paper 95-FA163.02.

Proceedings of the Air and Waste Management Association 88th Annual Meeting and Exhibition, San Antonio, TX. June.

Fogel, S. 1994. Full-scale bioremediation of No. 6 fuel oil-contaminated soil: Six months of active and three years of passive treatment. In *Bioremediation Field Experience*, eds. P. E. Flathman, D. E. Jerger, and J. H. Exner. Chelsea, MI: Lewis Publishers. 161–175.

Fulton, D. E. 1996. Selection and application of effective LNAPL recovery techniques. *Hazardous and Industrial Waste. Proceedings of the 28th Mid-Atlantic Industrial Hazardous Waste Conference*. 619.

Genes, B. R., and C. C. Cosentini. 1993. Bioremediation of polynuclear aromatic hydrocarbon-contaminated soil at three sites. In *Hydrocarbon Contaminated Soils, Vol. III*, eds. E. J. Calabreses and P. T. Kostecki. Chelsea, MI: Lewis Publishers. 323–331.

Ghiorse, W. C., and D. L. Balkwill. 1983. Estimation and morphological characterization of bacteria indigenous to subsurface environment. *Developments in Industrial Microbiology*, 24:213–224.

Gray, K. 1999. Bioremediating explosives-contaminated soil. *Biocycle*, 40(5):48–49.

Griffiths, R. A. 1995. Soil washing technology and practice. *Journal of Hazardous Materials*, 40:175–189.

Hart, S. A. 1991. Composting potentials for hazardous waste management. In *Biological Processes: Innovative Hazardous Waste Technology Series, Vol. 3*, eds. H. M. Freeman and P. R. Sferra. Lancaster, PA: Technomic.

Hazen, T. C. 1991. *Test plant for* in situ *bioremediation demonstration of the Savannah River integrated demonstration project*. DOE/OTP TTP NO.SR 0566-01, WSRC-RD-91-23. Washington, D.C.: Department of Energy.

Hildebrandt, W. W., and S. B. Wilson. 1991. On-site bioremediation systems reduce crude oil contamination. *Journal of Petroleum Technology*, 43(1):18–23.

Hinchee, R. E. 1993. Bioventing: Principles, applications and case studies. Training Program of International Network for Environmental Training, Potomac, MD.

Hinchee, R. E., and Downey, D. E. 1998. The role of hydrogen peroxide in enhanced bioreclamation. *Proceedings of Petroleum Hydrocarbons and Organic Chemicals in Groundwater: Prevention, Detections and Restoration*, Houston, TX. November 9. Dublin, OH: National Water Well Association. 715–721.

Holroyd, M. L., and Caunt, P. 1995. Large-scale soil bioremediation using white-rot fungi. *Bioaugmentation for Site Remediation*, eds. R. E. Hinchee, J. Fredickson, and B. C. Alleman. Columbus, OH: Batelle Press.

Huang, J. W., J. Chen, W. R. Berti, and S. D. Cunningham. 1997a. Phytoremediation of lead-contaminated soils: Role of synthetic chelates in lead phytoextraction. *Environmental Science & Technology.* 31:800–805.

Huang, J. W., J. Chen, and S. D. Cunningham. 1997b. Phytoextraction of lead from contaminated soils. In *Phytoremediation of Soil and Water Contaminants,* ACS Symposium Series 664, eds. E. L. Kruger, T. A. Anderson, and J. R. Coats. Washington, D.C.: American Chemical Society. 283–296.

Huang, J. W., M. J. Blaylock, Y. Kapulnik, and B. D. Ensley. 1998. Phytoremediation of uranium-contained soils: Role of organic acids in triggering uranium hyper-accumulation in plants. *Environmental Science and Technology,* 32(13):2004–2008.

Jerger, D. E., and P. M. Woodhull. 1994. Slurry-phase biological treatment of polycyclic aromatic hydrocarbons in wood preserving wastes. *Proceedings of the 87th Annual Meeting of the Air and Waste Management Association.* Cincinnati, Ohio. June.

Karavaiko, G. I., G. Rossi, A. D. Agates, S. N. Groudev, and Z. A. Avakyan. 1988. *Biogeotechnology of Metals: Manual.* Moscow, Soviet Union: Center for International Projects GKNT.

Kenney, T., and R. Schindler. 2001. Bioremediation system cleans up gasoline plume. *Water & Wastewater International,* 16(2):15.

King, R. B., G. M. Long, and J. K. Sheldon. 1998. *Practical Environmental Bioremediation: The Field Guide.* Boca Raton: Lewis Publishers.

LaGrega, M. D., P. L. Buckingham, and J. C. Evans. 1994. *Hazardous Waste Management.* New York: McGraw-Hill.

Leavitt, M. E. 1992. *In situ* bioremediation of diesel fuel in soil. *Bioremediation: The State of Practice in Hazardous Waste Remediation Operations.* Seminar sponsored by the Air & Waste Management Association and HWAC. January.

Lesson, A. 1997. Recent developments in bioventing. *IBC's Second Annual Conference on Innovative Remediation Technologies.* Boston, MA. July 21–23.

Leton, T.G., and E. I. Stentiford. 1990. Control of aeration in static pile composting. *Waste Management and Research,* 8(4):299–306.

Levin, M. A., and M. A. Gealt. 1993. Overview of biotreatment practices and promises. *Biotreatment of Industrial and Hazardous Waste.* New York: McGraw-Hill. 1–19.

Lewis, R. F. 1992. SITE Demonstration of slurry-phase biodegradation of PAH-contaminated soil. *Proceedings of the 85th Annual Meeting of the Air and Waste Management Association.* Kansas City, Missouri. June.

Li-Hua, H. 1998. Complete sequence analysis of 16s rDNA clones of *para* and *meta* anaerobic PCB dechlorinating *Clostridium* sp. Poster. *WERC/HSERC '97 Joint Conference on the Environment,* Albuquerque, New Mexico.

Litchfield, C. D., G. O. Chieruzzi, D. R. Foster, and D. L. Middleton. 1994. A biotreatment train approach to a PCP-contaminated site: *In situ* bioremediation coupled with an above the ground BIFAR system using nitrate as the electron acceptor. In *Bioremediation of Chlorinated and Polycyclic Aromatic Hydrocarbon Compounds,* eds. R. E. Hinchee, A. Lesson, and L. Semprini. Boca Raton, FL: Lewis Publishers. 155–163.

McCarty, P. L., L. Semprini, and M. E. Dolan. 1991. *In-situ* methanotrophic bioremediation for contaminated groundwater at St. Joseph, Michigan. In *On-Site Bioremediation Processes for Xenobiotic and Hydrocarbon Treatment,* eds. R. E. Hinchee and R. F. Offenbuttel. Stoneham, Mass: Butterworth-Heinemann. 16–40.

Mergeay, M., D. Niles, and H. G. Schlegel. 1985. *Alcaligenes eutrophus* CH_{34} is a facultative chemolithotroph with plasmid bound resistance to heavy metals. *Journal of Biotechnology,* 162:328–334.

Melcer, H., C. E. Aziz, and J. Anderson. 1995. Slurry Bioremediation of Coal Tar Contaminated Soil. Paper 95-FA163.01. Presented at the *88th Annual Meeting and Exhibition of Air and Waste Management,* San Antonio, TX. June.

Metcalf and Eddy, Inc. 1991. *Wastewater Engineering, 3rd Edition.* New York: McGraw-Hill.

Michaelson, D. L. and M. Lofti. 1990. Oxygen microbubbles injection for *in situ* bioremediation: Possible field scenarios. *Innovative Hazardous Waste Systems.* New York: Technotric.

Morse, J. J., B. C. Alleman, J. M. Gossett, S. H. Zinder, D. F. Fennell, G. W. Sewell, and C. M. Vogel. 1998. *Draft Technical Protocol: A Treatability Test for Evaluating the Potential Applicability of the Reductive Anaerobic Biological* In Situ *Treatment Technology (RABITT) to Remediate Chloroethenes.* DoD Environmental Security Technology Certification Program. Online at www.estcp.org.

Mueller, J. G., S. E. Lantz, B. O. Blattmann, and P. J. Chapman. 1991. Bench-scale evaluation of alternative biological treatment processes for the remediation of pentachlorophenol and creosote-contaminated materials: Slurry-phase bioremediation. *Environmental Science and Technology,* 25:1055–1061.

Mulligan, C. N., and R. Galvez-Cloutier. 2000. Bioleaching of copper mining residues by *Aspergillus niger. Water Science and Technology,* 41(12):255–262.

Mulligan, C. N., R. N. Yong, and B. F. Gibbs. 1999a. On the use of biosurfactants for the removal of heavy metals from oil-contaminated soil. *Environmental Progress,* 18(1):50–54.

————. 1999b. Removal of heavy metals from contaminated soil and sediments using the biosurfactant surfactin. *Journal of Soil Contamination,* 8:231–254.

National Research Council, NRC. 2000. *Natural Attenuation for Groundwater Remediation,* B. E. Rittmann, Chairman. Washington, D.C.: National Academy Press.

NATO/CCMS. 1999. *NATO/CCMS Pilot Study, Evaluation of Demonstrated and Emerging Technologies for the Treatment of Contaminated Land and Groundwater (Phase III), 1998.* Annual Report, Number 228, EPA/542/R-98/002. January.

————. 2000. *NATO/CCMS Pilot Study, Evaluation of Demonstrated and Emerging Technologies for the Treatment of Contaminated Land and Groundwater (Phase III), 1999.* Annual Report, Number 235, EPA/542/R-99/007. January.

————. 2001. *NATO/CCMS Pilot Study, Evaluation of Demonstrated and Emerging Technologies for the Treatment of Contaminated Land and Groundwater (Phase III), 2000.* Annual Report, Number 244, EPA/542/R-01-001. January.

Newman, L. E., S. E. Strand, and M. P. Gordon. 1997. Uptake and biotransformation of trichloroethylene by hybrid poplars. *Environmental Science & Technology,* 31(4):1062.

Nyer, E. K. 1992. Treatment for organic contaminants, physical/chemical methods. *Bioremediation: The State or Practice in Hazardous Waste Remediation Operations,* Seminar sponsored by Air and Waste Management Association and HWAC. January.

Ontario Ministry of the Environmental and Energy (OMEE). 1992. *Remediation Technologies for Contaminated Soils.* ISBN-0-7778-0343-2. Toronto.: OMEE.

Piotrowski, M. R., J. R. Doyle, D. Cosgriff, and M. C. Parsons. 1994. Bioremedial process at the Libby, Montana, Superfund site. In *Applications of Biotechnology for Site Remediation,* eds. R. E. Hinchee, D. B. Anderson, F. B. Metting, Jr., and G. D. Sayles. Boca Raton: CRC Press. 240–255.

Pivetz, B. E., J. W. Kelsey, and M. Alexander. 1996. Procedure to calculate biodegradation during preferential flow through heterogeneous soil column. *Soil Science American Journal,* 61(2):381.

Pritchard, P. H., J. E. Lin, J. G. Mueller, and M. S. Shields. 1996. Bioremediation research in EPA. In *Biotechnology in industrial waste treatment and bioremediation,* eds. R. F. Hickey and G. Smith. Boca Raton: Lewis Publishers.

Prosen, B. J., W. M. Korreck, and J. M. Armstrong. 1991. Design and preliminary performance results of a full-scale bioremediation system utilizing an on-site oxygen generator system. In *In situ bioreclamation, applications and investigations for hydrocarbon and contaminated site remediation,* eds. R. E. Hinchee and R. F. Olfenbuttel. Boston: Butterworth-Heinemann. 523–528.

Rawlings, E., ed. 1997. *Biomining: Theory, microbes and industrial processes.* Georgetown, TX: Springer-Verlag/Landes Bioscience.

Raymond, R. L. 1974. *Reclamation of hydrocarbon contaminated groundwaters.* Patent No. 3,846,290. Washington, D.C.: U.S. Patent Office. November 5.

Ross, D. 1990. Slurry-phase bioremediation: Case studies and cost comparisons. *Remediation,* 1:61–74.

Roy, D., R. R. Kommalapati, S. S. Mandava, K. T. Valsarai, and W. D. Constant. 1997. Soil washing potential of a natural surfactant. *Environmental Science & Technology,* 31:670–675.

Rugh, C. L., H. D. Wilde, and N. M. Stack. 1996. Mercuric ion reduction and resistance in transgenic *Arabidopsis thaliana* plants expressing a modified bacterial merA gene. *Proceedings of the National Academy of Science,* 93:3182–3187.

Ryan, J. R., R. C. Loehr, and E. Rucker. 1991. Bioremediation of organic contaminated soils. *Journal of Hazardous Materials,* 28:159–169.

Schafer, J. M. 1984. Determining optimum pumping rates for creation of hydraulic barriers to groundwater pollutant migration. In *Proceedings Fourth National Symposium on Aquifer Restoration and Ground Water Monitoring,* ed. D. M. Nielsen. Worthington, OH: National Water Well Association.

Schnoor, J. L. 1997. *Phytoremediation,* Technology Evaluation Report TE-98-01. GWRTAC: Pittsburg, PA.

Schwefer, H. J. 1988. Latest development of biological *in-situ* remedial action techniques portrayed by examples from Europe and U.S.A. *Contaminated Soil '88.* Boston: Kluwer Academic Publishers.

Sellers, K. L., T. A. Pederson, and C. Fan. 1993. Review of soil mound technologies for the bioremediation of hydrocarbon-contaminated soil. In *Hydrocarbon Contaminated Soils, vol. III,* eds. E. J. Calabrese and P. T. Losteck. Chelsea, MI: Lewis Publishers.

Semmens, M. L., T. Ahmed, and M. A. Voss. 1991. Field tests on a bubbleless membrane aerator. *Air-Water Mass Transfer: Selected Papers from the Second International Symposium on Gas Transfer at Water Surfaces,* 694. New York: American Society of Civil Engineers.

Semprini, L., P. V. Roberts, G. D. Hopkins, and P. L. McCarty. 1990. A field evaluation of *in-situ* biodegradation of chlorinated ethenes: Part 2, results of biostimulation and biotransformation experiments. *Groundwater,* 28:715–727.

Semprini, L., G. D. Hopkins, P. V. Roberts, and P. L. McCarty. 1991. *In situ* transformation of carbon, tetrachloride, freon-113 and 1,1,1-TCA under anoxic conditions. In *In Situ Bioreclamation, Applications and Investigations for Hydrocarbon and Contaminated Site Remediation,* eds. R. E. Hinchee and R. F. Olfenbuttel. Boston: Butterworth-Heinemann. 41–58.

Sims, J. L., J. M. Sulfita, and H. D. Russell. 1991. Reductive dehalogenation: A subsurface bioremediation process. *Remediation,* 1:75–93.

Smith, L. A., B. C. Alleman, and L. Copley-Graves. 1994. Biological treatment options. In *Emerging Technology for Bioremediation of Metals.* eds. J. L. Means and R. E. Hinchee. Boca Raton: Lewis Publishers. 1–12.

Smith, L. A., J. L. Means, A. Chen, B. Alleman, C. C. Chapman, J. S. Tixier, Jr., S. E. Brauning, A. R. Gavaskar, and M. D. Royer. 1995. *Remedial Options for Metals-Contaminated Sites.* Boca Raton, FL: Lewis Publishers.

Soesilo, J. A., and S. R. Wilson. 1997. *Site Remediation Planning and Management.* New York: Lewis Publishers.

Stavnes, S. 1999. Bioremediation barrier emplaced by hydraulic fracturing. *Groundwater Currents,* U.S. EPA Solid Waste and Emergency Response, EPA-N-099-002, No. 31. March.

Stegmann, R., S. Lotter, and J. Heerenklage. 1991. Biological treatment of oil-contaminated soils in bioreactors. In *On-Site Bioreclamation Processes for Xenobiotic and Hydrocarbon Treatment*, eds. R. E. Hinchee and R. F. Olfenbuttel. Massachusetts: Butterworth-Heinemann.

Taddeo, A. 1989. Field demonstration of a forced aeration composting treatment for coal tar. *Proceedings of the Second National Conference of Biotreatment: The Use of Microorganisms in the Treatment of Hazardous Materials and Hazardous Waste.* Washington, D.C. 27–29 November.

Taddeo, A., M. Findlay, M. Dooley-Danna, and S. Fogel. 1997. Field demonstration of a forced aeration composting treatment™ for coal tar. *IBC's Second Annual Conference on Innovative Remediation Technologies.* Boston, MA. 21–23 July.

Thomas, J. M., and C. H. Ward. 1989. *In situ* biorestoration of organic contaminants in the subsurface. *Environmental Science & Technology,* 23:760–766.

Thompson-Eagle, E. T., and W. T. Frankenberger. 1990. Volatilization of selenium from agricultural evaporation water. *Journal of Environmental Quality,* 19:125–130.

Tichy, R., J. T. C. Grotenhuis, and W. H. Rulkens. 1992. Bioleaching of zinc- contaminated soil with *Thiobacilli. Proceedings of an International Conference: Eurosol.* September. Masstricht, The Netherlands.

Tichy, R., A. Jansen, J. T. C. Grotenhuis, G. Lettinga, and W. Rulkens. 1994. Possibilities for using biologically produced sulfur for cultivation of thiobacillus with respect to bioleaching process. *Bioresource Technology,* 48:221–227.

U.S. Department of the Air Force, Air Force Center for Environmental Excellence. 1992. *Test Plan and Technical Protocol for a Field Treatability Test for Bioventing.* Environmental Services Office. May.

U.S. Department of Energy (U.S. DOE). 1995. *In situ* bioremediation of groundwater. October. Online at http://em.doe.gov/rainpluj/plum321.html.

U.S. Environmental Protection Agency (U.S. EPA). 1983. *EPA Guide for Identifying Cleanup Alternatives at Hazardous Waste Sites and Spills: Biological Treatment.* EPA 600/3-83/063. Washington, D.C.: EPA.

————. 1987. *Treatability studies under CERCLA: An Overview.* OSWER Directive 9380.3-02FS. Washington, D.C.: EPA.

————. 1990a. *Composting of Municipal Wastewater Sludges.* EPA 625/4-85-014. Cincinnati, Ohio: Office of Research and Development. August.

————. 1990b. *Available Models for Estimating Emissions Resulting from Bioremediation Processes: A Review.* EPA 600/3-90/031. Washington, D.C.: EPA.

————. 1991. *The Superfund Innovative Technology Evaluation Program: Technology Profiles.* 4th ed. EPA 54015-91/008. Washington, D.C.: Office of Research and Development. Washington, D.C.: EPA.

————. 1992. *Slurry Biodegradation.* Superfund Innovative Technology Evaluation, Demonstration Bulletin, EPA/540/M5-911/009. February. Washington, D.C.: EPA.

————. 1993. *Bioremediation Using the Land Treatment Concept,* EPA 600/R-93/164. Washington, D.C.: EPA.

————. 1995a. Member Agencies of the Federal Remediation Technologies Roundtable, *Remediation Case Studies:Bioremediation,* EPA 542-R-95-002. Washington, D.C.: EPA.

————. 1995b. Office of Solid Waste and Emergency Response, Technology Innovation Office. *Cost and Performance Report: Composting Application at the Dubose Oil Products Co. Superfund Site, Cantonment, Florida,* Contract No. 68-W3-0001. Washington, D.C.: EPA.

————. 1996. Office of Solid Waste and Emergency Response, *A Citizen's Guide to Natural Attenuation.* EPA 542-F-96-015 October. Washington, D.C.: EPA.

————. 1997. Office of Solid Waste and Emergency Response, *Recent Developments for In Situ Treatment of Metal Contaminated Soils.* Washington, D.C.: EPA.

————. 1998. Guide to documenting and managing cost and performance information for remediation projects. Revised version, EPA 542-B-98-007. Federal Remediation Technologies Roundtable. October. Available online http://www.frtr.gov.

————. 1999. Technology Innovation Office, Office of Soil Waste and Emergency Response. *The Treatment Technologies for Site Cleanup, Annual Status Report, 9th ed.* Washington, D.C.: EPA.

————. 2000. *Engineered approaches to* in situ *bioremediation of chlorinated solvents: Fundamentals and field applications.* EPA 542-R-00-008. July. Washington, D.C.: EPA.

Van Cauwenberghe, L., and D. S. Roote. 1998. *In situ* Bioremediation GWRTAC Technology Overview Report, TO-98-01. October.

Vangronsveld, J., and S. D. Cunningham. 1998. Introduction to the concepts. *Metal-contaminated Soils*. In *In situ Inactivation and Phytorestoration*, eds. J. Vangronsveld and S. D. Cunningham. Springer-Verlag Berlin Heidelberg. 1–16.

Verheul, J. H. A. M., R. Van den Berg, and D. H. Eikelboom. 1988. *In-situ* biorestoration of a subsurface contaminated soil with gasoline. *Contaminated Soil '88*. Boston: Kluwer Academic Publishers.

Vogel, T. M., C. S. Criddle, and P. L. McCarty. 1987. Transformations of halogenated aliphatic compounds. *Environmental Science & Technology*, 21:722–736.

WASTECH. 1995. *Innovative Site Remediation Technology, Bioremediation, Vol. 1*. W. C. Anderson, ed. American Academy of Environmental Engineers. New York: Springer-Verlag Berlin Heidelberg.

West, C. C., and J. H. Harwell. 1992. Surfactant and subsurface remediation. *Environmental Science and Technology*, 26:2324–2330.

Wilber, C. G. 1980. Toxicology of selenium: A review. *Clinical Toxicology*, 17:171– 230.

Williams, R. T., and C. A. Meyler. 1990. Promising research results: Bioremediation using composting. *Biocycle*. November. 78–82.

Williams, R. T., P. S. Ziegenfuss, and W. E. Sisk. 1989. Composting of explosives and propellent contaminated sediments. *Proceedings of the 21st Mid-Atlantic Industrial Waste Conference*. Pennsylvania State University. 25–27 June.

———. 1990. Composting of explosives and propellent contaminated soils under thermophilic and mesophilic conditions. *Journal of Industrial Microbiology*, 9:137–144.

Yare, B. S. 1991. A comparison of soil-phase and slurry-phase bioremediation of PNA-containing soils. In *On-Site Bioreclamation, Processes for Xenobiotic and Hydrocarbon Treatment*, eds. R. E. Hinchee and R. F. Olfenbuttel. Boston, MA: Butterworth-Heinemann. 173–195.

Zylstra, G. J., L. P. Wackett, and D. T. Gibson. 1989. Trichloroethylene degradation by *Escherichia coli* containing the cloned *Pseudomonas putida* F1 toluene dioxygenase genes. *Applied and Environmental Microbiology*, 55:3162–3166.

Solid Waste and Sludge Treatment

Introduction

Wastes are often classified as municipal, industrial, hazardous, or radioactive. Mining and agricultural industries are other sources of wastes. Some of the primary sources of wastes are shown in Table 6.1. In the mid-1990s, municipal solid waste (MSW) was generated at a rate of 208 million tonnes annually (U.S. EPA 1995–96). State and federal requirements in the U.S. control the treatment, storage, and disposal of waste. Wastes can be classified as organic, inorganic, or microbiological. To determine if the waste is suitable for biological treatment, the first step is to determine if it is organic or inorganic. Organic wastes include food, paper and cardboard, plastics, clothing, yard waste, and bone. Many of these organic wastes can be biodegraded. Inorganic wastes include silicates; sulphates; cyanides; iron; trace metals such as aluminum, cadmium, copper, lead, nickel, and zinc; and arsenic and its compounds. Metals do not degrade or decompose. When they are buried in landfills, they can remobilize and threaten the environment. Wastes such as glass, some metals, plastics, ash, and wood are relatively inert but can be recovered for reuse.

The composition of MSW delivered to a landfill can vary considerably (Kiely 1997). In general, food wastes make up 30% of domestic waste; another 35% is paper and cardboard, 8% is glass, 8% are plastics, and 5% are metals. Food wastes include meat, fat, vegetables, oils, fruit, and bones. Sources of MSW include residential (apartments and houses), commercial (restaurants, office buildings, stores, service stations), institutional (schools, hospitals, courthouses, etc.), construction and demolition sites, and municipal services (wastewater treatment, street-cleaning, garden and park landscaping). The types of commercial establishments in an area influence the composition of the waste. Offices, restaurants, schools, hospitals, and retail outlets all generate different percentages of food, paper, plastics, glass, metal, and other materials.

Table 6.1 Sources of Organic Wastes

Industrial and commercial sources	Agricultural sources	Municipal and domestic sources
Breweries	Pigs	Raw sewage
Dairies	Chickens	Organic fraction of
Food processing, packaging,	Cattle	municipal sludge
and shipping	Farmyards	Food, paper
Chemical industries	Crop residues	Activated sludge
Pharmaceutical	Rotten products	Household waste
Wineries		Grass
Hazardous wastes		Leaves
Oils		Market wastes
Paper manufacture (pulps)		
Slaughterhouses		

The biodegradable fraction (BF) of the food fraction of MSW can be estimated by the equation (Kiely 1997):

$$BF = 0.83 - 0.028 \; LC$$

where BF is the biodegradable fraction of volatile solids (VS) basis and LC is the lignin content (percent of dry weight). Based on this equation, the BF of food waste is 0.82, newsprint is 0.22, office paper is 0.82, cardboard is 0.47, and yard wastes are 0.72. Thus, food waste is highly biodegradable. Separation of these types of wastes from MSW and industrial waste at the source can facilitate biodegradation processes and eliminate expensive and difficult sorting.

Manufacturing and chemical processes produce various solid wastes, usually about four times that of municipal wastes (Tammemagi 1999). This does not include wastes produced by mining, oil and gas, and agricultural sectors. Each type of industry produces a particular waste. Some examples are distillation column bottoms from chlorobenzene production, sludges and emission control dust from coal burning, and residues from the decolorization of pharmaceuticals. The wastes are often incinerated or placed in landfills for industrial wastes.

Hazardous wastes are separated from industrial and municipal wastes. According to the U.S. Resource Conservation and Recovery Act (RCRA), waste is considered hazardous if it is ignitable, corrosive, reactive, or toxic. Approximately 15% of industrial waste and 1% of municipal waste are hazardous. Industrial wastes include organic sludges, oils and greases, solvents, heavy metal solutions, pesticides and herbicide wastes, PCBs, and contaminated soils. Varnishes, paints, turpentines, motor oil, herbicides, pesticides, batteries, and fertilizers are examples of household hazardous wastes. Annually, 40 million tonnes of hazardous wastes are produced in the U.S., 22 million tonnes in the European Union (EC, State of the Environment 1993), and 2.3 million tonnes in Canada (Tammemagi 1999). Hazardous wastes must be disposed of in specially designed landfills although liquid wastes often end up in sewage plants.

Radioactive wastes are generated by specific industries such as nuclear reactors, research labs, and medical facilities. In most countries, disposal of high-level waste is by burial in deep, stable geologic formations. Low-level wastes are disposed of in a variety of ways but mainly by burial in caverns or clay formations. They become less radioactive over time due to radioactive decay. Radioactive wastes will not be considered further in this chapter.

Various processes are available for waste treatment and disposal (Pellerin 1994). The major ones are incineration and disposal in a traditional landfill. According to the U.S. EPA (1995, 1996), 4.5 million tonnes of waste are incinerated and 24 million tonnes of industrial and household wastes are disposed of by various other processes, including thermal desorption, microwave treatment, solidification/stabilization, and vitrification (Table 6.2).

Incineration is the destruction of organic materials by combustion at high temperatures. For organic wastes, fuel is needed only for start-up. HCl is produced during the combustion of halogenated organic components. Multiple hearth, fluidized bed, and grate-type incinerators are used for municipal solid wastes since they are suitable for irregularly-shaped waste such as paper and wood. Rotary kiln types are used for 75% of hazardous waste incinerators in the U.S. (LaGrega et al. 1994). When the reaction is complete, the products are carbon dioxide and water. Some compounds such as dioxins can be formed. Although the volumes of the waste are significantly reduced, inorganic wastes remain in the ash and particulates in the exhaust gases. In the case of the biosolids, an 80% reduction of the volume occurs.

Many regulations have come into effect as a result of emissions from the incineration process. Air pollution equipment such as wet scrubbers, electrostatic precipitators, filters, and afterburners is required. In the U.S., these regulations are related to the emission of particulates, HCl, carbon monoxide, and metals. Other regulations are concerned with the destruction and removal efficiency that must be 99.99% for one or more organic components. Many facilities that incinerate biosolids have closed because other methods are more publicly favorable and less expensive. Currently 22% of biosolids are incinerated (U.S. EPA 1999a).

Landfills are necessary since waste management cannot totally reduce the wastes produced, and other treatment processes—incineration and biological treatment, for example—produce residues. Despite all design considerations, leakage is inevitable. Biodegradation rates are much higher in municipal landfills than in hazardous waste landfills because of the higher organic matter contents. However, municipal landfills are not run as bioreactors. Covers are required for emission control, and liners and leachate collection systems are needed to minimize contaminant migration. Even after closure, monitoring of emissions and groundwater quality must take place. Landfills will always be necessary, but efforts should be made to minimize their use as much as possible by recycling and converting wastes into useful products, such as compost and energy. Approximately 17% of

Table 6.2 Comparison of Various Waste Treatment/Disposal Options

Type of process	Applicability	Limitations	Time required	Cost in $/tonne (Cost rating)
Incineration	Hazardous and domestic wastes	Gaseous emissions and ash for landfill disposal	+	$1,500 to $1,800 (+++)
Traditional landfill	Hazardous and domestic types	Mainly storage with emissions	+++	$450 to $1,000(+++)
Thermal desorption	Petroleum-based wastes to obtain energy	Applicable only for volatile metals and organic materials	+	$250 to $350 (+++)
Microwave	Medical waste with no toxic emissions generated	Strictly for medical waste and no heat generation	+	$1,500 to $1,800 (+++)
Solidification/ stabilization	Conversion of waste into more stable form	Potential long-term leaching	++	$120 to $520 (+++)
Heat drying	Production of biosolids for fertilizer	Higher cost and lower nitrogen than compost	+	Not available
Vitrification	Conversion of high and low level radio-active waste into reusable glass	Expensive but glass product can be sold	++	$1,500 to $1,800 (+++)

Cost ranking: + denotes $10 to $50/tonne, ++ $50 to $200/tonne, and +++greater than $200 per tonne
Time ranking:+ denotes several hours, ++ 1-20 days, +++ more than 20 days

biosolids are landfilled (U.S. EPA 1999a). Average tipping fees increased to $34 per tonne in 1998.

One method of municipal waste disposal, land application, involves the spreading of solids wastes on or into the soil. It is particularly used for biosolids since it enriches the soil, and it can be used to enhance fertilization. Biosolids generally contain approximately 3.2% nitrogen, 2.3% phosphorus and 0.3% potassium, slightly lower than commercial fertilizers (Metcalf and Eddy 1991). Digestion, composting, and chemical or heat treatments can be used prior to land application. Although less expensive than composting, biosolids cannot be stored easily prior to application, the application is more difficult, and the product is not as high quality.

Thermal desorption is a hybrid of incineration and recycling for coal tar, heavy oil tar, and other sludges. The waste is loaded into a feed hopper, mixed, stabilized, and combined with nitrogen to reach anaerobic conditions. The system is then sealed, and the waster is fed into an externally-heated combustion chamber where the water and volatiles are removed. Then the waste goes to a higher-temperature second channel where all of the organics are volatilized. The temperature can be raised to 2,000°C, if necessary. After the separation of water and oil, the water is purified and can be returned to the soil. The oil vapors are sent to an inert-gas generator for compaction with heat. The hydrocarbons become carbon dioxide and water. There is no combustion.

Drying and dewatering can also be used for solids like biosolids, which have high moisture contents. Secondary sludges can be heat dried because heat drying does not cause odors, destroys pathogens, and reduces volume by water elimination. Although energy is required for drying, transportation costs are reduced because moisture contents are lower than compost. It can be most applicable to cities like Boston and New York City, which are far from agricultural areas. Both of these cities dry and pelletize their biosolids for transport out of state.

Dewatering is used to lower water contents before composting, heat drying, land application, or landfilling. Gravity thickening by the addition of polymers, lime, or ferric chloride is often used as a pretreatment for removal of bound water. Air drying can be used by small wastewater treatment facilities for their biosolids. Larger facilities use filter presses, centrifuges, and vacuum filters. Vacuum filters result in 12 to 22% solids, centrifuges obtain 25 to 35% solids, and belt filter presses, 20 to 32% solids content. Plate-and-frame presses produce the highest solids content (35 to 45%) and are the most expensive.

Microwave disinfection includes shredding, steam injection, and conventional micro-waves for biomedical waste treatment. Wastes include bandages, syringes, needles, blood, human and animal tissues, laboratory culture, vials, and so on. As the waste is loaded in the hopper, it is pretreated with steam. Computers control the material feed, which passes through the shredder and into the treatment chamber where it is moistened by steam. The mixture then passes to the microwave generators, which are then landfilled and sent to waste-to energy plants. However, this is an expensive process, and there is no energy generated.

Stabilization/solidification is the conversion of waste into a more stable and solid form. It is used to reduce contaminant movement and can be used where incineration or bioremediation are not effective, technically or financially. The process can be used prior to landfill or to stabilize the contaminants in place. It is particularly applicable for liquid wastes before landfill. Solidification can be used to stabilize, improve strength and compressibility, and neutralize oily sludges by the addition of cement kiln dust. Pathogens in biosolids are also reduced by stabilization. Heavy metals and radioactive materials are stabilized by a wide variety of components including lime, fly ash, Portland cement, pozzolans, thermoplastics, and organic polymers. Treatability tests are recommended and many formulations are now proprietary. Long-term stability can be an issue with some organic wastes, however, and monitoring is required.

Vitrification is the conversion, under high temperatures in a melter, of wastes into clean air, water, or glass. Metal components are converted to oxides. Radioactive wastes can leach out into environment. Water, soils, sludges, fly ash, medical waste, asbestos, and building products can all be stabilized together. The final volume for storage is reduced. The cost is about the same as incineration, but the glass produced can be sold.

In this chapter, several biological treatment processes will be discussed: anaerobic digestion, aerobic digestion, composting, landfill bioreactors, slurry reactors, landfarming, biohydrometallurgical processes, and various fermentation processes. Finally, these processes will be compared to traditional ones that have been used for many years. Biological processes are less costly, require less energy than traditional processes, and can produce highly-useful products such as methane, soil conditioners, and feeds. Extensive experience has been obtained in the design and operation of numerous biological processes such as anaerobic and aerobic digestion and composting. Public acceptance of processes such as composting is increasing. Feasibility studies are the most appropriate method for evaluating the biological treatment process. Using these biological processes can significantly decrease the amount of waste going to the landfills, thus increasing their life span.

There are several advantages of biological methods over landfilling and incineration. They are cost-effective; they require less capital expenditure, energy, and manual supervision; and they destroy toxic chemicals. Windrow composting and bioslurry processes cost half that of incineration processes.

Some of the constraints of biological processes include the length of time (days to months) required, the heavy metals and radionuclides that are not removed, the contaminants such as some chlorinated compounds that are not removed, and the strong scientific basis needed for design of successful processes. Particularly when hazardous contaminants must be treated, treatability or feasibility tests are needed.

Anaerobic Digestion

Box 6-1 Overview of Anaerobic Digestion

Applications	Suitable for treatment of organic wastes including agricultural, industrial, and municipal wastes
Cost	Operating: $20 to $70/tonne (wet basis), $60 to $150 per tonne (dry basis)
	Capital: $50,000 to $370,000/dry tonne per day
Advantages	Production of energy in the form of methane
	Stable end product
	Well-developed process
Disadvantages	Bacteria sensitive to temperature variations
	Heat often required
	Generation of high COD wastewater
Detention time	30 to 60 days (low-rate), ten to 20 days (high-rate)
Organic load	0.6 to 1.6 kg VS/m^3-day (low-rate), 2.4 to 6.4 kg VS/m^3-day (high-rate)
Other considerations:	Most common treatment for high-strength wastes

Description

Anaerobic digestion, the most common treatment used for high-strength organic industrial, agricultural, and municipal sludges, is the microbial stabilization of organic materials without oxygen to produce methane, carbon dioxide, and other inorganic products. The following reaction is typical of anaerobic digestion:

$$\text{Organic matter} + H_2O \rightarrow CH_4 + CO_2 + \text{biomass} + N_2 + H_2S + \text{heat}$$

Low-rate digesters operate at solids contents between 3 and 10%, and high-rate ones between 10 and 30%. COD removal rates range from 75 to 90%. Advantages of anaerobic digestion include reduction of waste pollution; elimination of pathogens (particularly at mesophilic and thermophilic temperatures); and production of biogas, which can be used for energy. Furthermore, the waste produced through anaerobic digestion can be used as fertilizers or fuel.

As in anaerobic wastewater treatment (Chapter 4), anaerobic digestion is performed by a consortium of bacteria in sludges containing 10^5 to 10^7 bacteria per mL. Optimal pH values are in the range of 6 to 8. The production of volatile fatty acids (VFA) can reduce the pH if the alkalinity is insufficient or if daily solids addition or removal does not exceed 3 to

5% of the total solids in the digester. A minimum alkalinity of 3,000 mg/L is usually required. Decreased pH values decrease the production of methane gas due to inhibition of methane bacteria. Bad odors, foam formation, and floating sludge are other consequences of low pH. Supernatant liquors have high organic contents (greater than 2,000 mg/L BOD) with suspended solids concentrations of 1,000 mg/L. The supernatant is usually fed into the primary clarifiers.

Methods

Low-rate Digesters

Sludge is pumped into the top of low-rate digesters where active sludge is found. Stabilized sludge is at the bottom for withdrawal. Intermittent mixing, sludge feeding, and withdrawal are utilized (Reynolds and Richards 1996). Stratification of the sludge occurs when mixing is stopped (Figure 6.1a). Digestion times are from 30 to 60 days with organic loads of 0.64 to 1.60 kg/m³-day, depending on the temperature (Reynolds and Richards 1996). Typical gas production is in the range of 0.3 to 1.0 m³/m³ reactor-day. In the mesophilic region, digestion time decreases as the temperature increases, up to 35°C. At higher temperatures, the digestion time starts to increase. The optimal temperature range is 29 to 38°C. For thermophilic digestions, the optimal temperature is 54°C; however, since thermophilic digestion is sensitive to changes, it is not often employed.

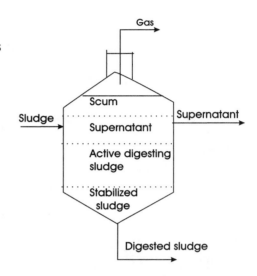

(A) Low rate digester

Figure 6.1a Stratification in a Low-rate Digester

Single-stage low-rate digesters are covered with floating or fixed covers. Diameters of the covers are from 4.6 to 38 m in 1.5 m intervals. Recycled sludge can be added to fresh sludge in digesters with floating covers. When recycling stops, the sludge stratifies. Sludge is withdrawn every two weeks and supernatant liquor every few days. Gas production raises the floating cover; the cover drops during withdrawal of sludge or supernatant.

Fixed-cover digesters are less flexible because the amount of sludge that can be added or removed is limited. The resulting gas pressure will reach a minimum of 76 mm of water when sludge is removed and a maximum of 203 mm when fresh sludge is added without withdrawing digested sludge. The amount of supernatant removal is also restricted by the fixed cover configuration. Grease buildup can also become a problem since the drying grease can accumulate and plug supernatant outlets. In some locations, fixed-cover digesters are not legal for populations greater than 10,000 persons (Reynolds and Richards 1996).

High-rate Digesters

The high-rate reactors are mixed continuously, with the exception of periods of withdrawal (Figure 6.1b). Gas production is double and retention times are half that of low-rate reactors. The mixing improves contact between the fresh and seed sludges. Typical digestion times are in the order of ten to 20 days, and organic loads are 2.4 to 6.4 kg VS/day-m³ (Reynolds and Richards 1996). Sludge feeding and removal are either continuous or intermittent. Fixed-cover digesters are used. Sludge is removed either by allowing sludge to overflow into a holding tank when fresh sludge is added or by stopping the mixing, allowing stratification to occur, and withdrawing the sludge.

First-generation reactors (batch, plug flow, continuously stirred [CSTR], and anaerobic contact) have equal HRT and solid retention times. Second generation reactors (upflow-downflow anaerobic filter, downflow stationary fixed film, fluidized bed, upflow anaerobic sludge blanket reactor [UASB], and hybrid anaerobic sludge reactors) have higher solid retention times than the HRT. Schematics of these reactors are shown in Chapter 4. Mixing can be provided by mechanical mixers, gas recirculation, and sludge recycle (Figure 6.2). Although digesters can operate in psychrophilic (0-20°C), mesophilic (25-40°C), or thermophilic (50 to 60°C) conditions, most operate under mesophilic conditions because retention times are lower than ambient temperatures, and stability and gas production is improved. Gas production is higher under thermophilic conditions, but stability is reduced and more monitoring is required.

Egg-shaped high-rate anaerobic digesters (Figure 6.3) are used extensively in Europe and Japan. The use of these digesters is increasing in North America. An external recycle pump provides mixing, particularly for volumes less than 2,800 m³. For larger volumes, a draft tube with either an impeller or jet pump in the digester is used for mixing. Tank construction is of reinforced concrete or steel. The distinct shape provides several advantages including lack of grit

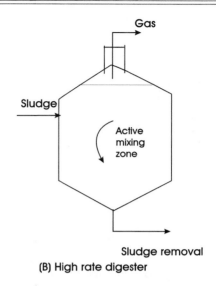

(B) High rate digester

Figure 6.1b Mixing in a High-rate Digester

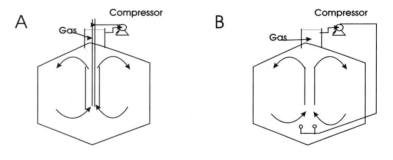

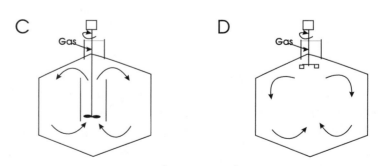

Figure 6.2 Mixing in High-rate Digesters by Gas Recycle and Draft Tube (A), Gas Recycle and Injection (B), Impeller and Draft Tube (C), and Impeller Alone (D)

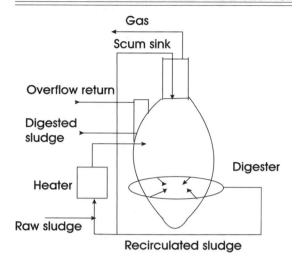

Figure 6.3 Egg-shaped High-rate Digesters Developed by CBI Walker

accumulation due to steep conical sides, enhanced mixing, increased control of scum at the top of the digester, and lower floor space requirements. They are, however, more expensive, and their height can restrict placement near residential areas.

Two-stage Digesters

Two-stage digesters are usually employed for the treatment of municipal sludge for populations between 30,000 and 50,000 persons. The first stage is a high-rate, fixed-cover digester for gasification, liquefaction of organic solids, and digestion of soluble organic materials. The second stage is a floating cover, a low-rate digester with intermittent mixing. It is employed mainly for supernatant separation, gas and digested sludge storage, and some gasification. Typical gas production rates range from 1 to 5 m^3/m^3 of digester volume, the methane content is from 50 to 70%, the COD removal rates are 70 to 90%, and loading rates range from 2 to 40 kg COD/m^3-day. The optimal operating temperature is 38°C.

Operating Conditions

Digestion technology has improved significantly since 1983. SRT and HRT are as follows:

SRT = mass of solids in reactor (kg)/rate of solids removed (kg/day)

HRT = working volume (L)/rate of sludge removed (L/day)

Municipal wastewater sludge requires an SRT of at least ten days. Mixed sludges usually require between 15 and 30 days. Volatile solids (VS) loading rate is a major parameter in designing digesters. It is defined as:

$$\text{VS loading rate} = \frac{\text{volatile solids added per day} \left(\dfrac{\text{kg VS}}{\text{day}} \right)}{\text{working volume of the reactor (m}^3)}$$

The rate is usually between 2 and 3 kg VS/m^3-day. Maximum rates are 3.2 VS/m^3-day due to toxicity from metals or ammonia (WEF 1995). Gas production in terms of VS is usually 0.5 to 1.5 m^3/kg VS destroyed. Digester tank dimensions are 5 to 50 m in diameter and 3 to 25 m in height. The material of construction is steel with insulation or reinforced concrete. Since the temperature must be raised to 38°C, heating systems are required and include jacketed pipes, water baths, internal heat exchangers, direct heat, or steam injection. Heat requirements are calculated from the following equation (WEF 1995):

$$H = WC\Delta T + UA\Delta T$$

H = heat required to heat incoming sludge (kcal/h)

W = mass flow rate of sludge (kg/h)

ΔT = difference between temperature of digester and incoming sludge (°C)

U = heat transfer through the tank walls (kcal/m²-h-°C)

A = surface area of digester (m²)

C = specific heat of feed sludge (usually 1 kcal/kg-°C)

Mixing is usually performed by mechanical means. Methane recirculation has been used but is not cost efficient. Reintroducing the effluent by recirculation pumps through an internal draft is the most common method (WEF 1995).

Design

Batch digester reactor volume can be determined by Reynolds and Richards (1996) by the equation:

$$V_s = \left[V_i - \frac{2}{3} (V_i - V_f) \right] t_d + V_2 t_s$$

where V_s = total sludge volume (m³),

V_i = initial sludge volume (m³/day),

V_f = final sludge volume (m³/day),

t_d = retention time (days),

V_2 = volume of digested sludge (m³/day), and

t_s = time for sludge storage (days).

The sludge volume takes half of the reactor, and the supernatant takes the other half of the digester. Therefore, the total reactor volume is

$$V_t = 2V_s$$

where V_t is the total digester volume (m³).

Organic loads (kg/m³-day) can also be used as a basis for design. For low-rate digesters, the organic loads can range from 0.64 to 1.6 kg/m³-day. High-rate reactors operate at 2.4 to 6.4 kg/m³-day

Reactor design can also be based on the mean cell residence time that is related to the solids production per day in the sludge and is represented as follows:

$$\theta_c = \frac{X}{\Delta X}$$

where θ_c is the mean cell residence time (days) or SRT (solids retention time), X is the mass of dry solids in the digester (kg), and ΔX is the daily production of solids in the digester (kg). In a high-rate, continuously-mixed reactor, the SRT and the hydraulic retention time (HRT) are approximately the same. Over time, the mean cell retention time will decrease until a critical minimum value is reached in which cells leave the reactor at a higher rate than they multiply. These values are dependent on temperature. McCarty (1968) suggested that θ_c^{min} would be 11 days at 18°C, eight days at 24°C, six days at 30°C, and four days at 35 to 40°C. The design mean cell residence time is often taken as 2.5 times the value of $\theta_c^{min.}$ Thus, the volume of the high-rate digester can be determined as:

$$V = Q\theta_c = Q\theta_h$$

where V = total digester volume (m³), Q is the fresh sludge flow rate (m³/day), θ_c is the design cell residence time (days), and θ_h is the design hydraulic residence time (days).

Heat Requirements

Often heat is required to increase the temperature of inlet sludge and/or to compensate for heat losses in the reactor. Outside hot water heaters and heat exchangers are often used to heat the sludge. To calculate the heat for increasing the temperature of the fresh sludge, use the following equation:

$$Q_s = P \times \frac{100}{p_s} \times (T_d - T_s) \times \frac{1}{24} \times c_p$$

where Q_s is the heat required for the sludge (J/h), P is the mass of fresh dry solids added (kg/day), p_s is the percentage dry solids in the fresh sludge, T_d is the digester (°C), T_s is the inlet sludge temperature (°C), and c_p is the specific heat capacity (4200 J/kg-°C).

Heat losses from the top, walls, and bottom of the digester can be determined by:

$$Q_d = CA\Delta T$$

where Q_d is the heat required to compensate for heat losses, C is coefficient of heat transfer (J/m²-h-°C), A is the surface area (m²), and ΔT is the temperature difference between the digester and the environment (°C). Values for C are 1634, 6128, and 5310 J/m²-h-°C for concrete against dry earth, air, and wet earth, respectively. For a floating cover, the C value is 3269 J/m²-h-°C (Imhoff et al. 1971).

The temperature of the digester should not vary more than 1°C per day since methanogenic bacteria are sensitive to temperature variations. Although digestion can take place at almost any temperature, increasing the temperature decreases reaction times. At 12°C, 90% of the digestion can be completed in 55 days. Increasing the temperature to 23°C reduces the time to 35 days; 30°C to 26 days; and 36°C to 24 days.

Methane Production

Methane production is dependent upon temperature, waste type, reactor type, and solid type. It can be determined based on stoichiometric equations. For example, if the substrate is glucose, then

$$C_6H_{12}O_6 \rightarrow 3CO_2 + 3CH_4$$

Thus, 1 kg of glucose would produce 0.27 m³ of CH_4. For continuous flow stirred tank reactors (CSTRs), the methane production rate can be determined (Tchobanoglous and Schroeder 1987). For domestic wastewater, gas production is usually 0.35 m³ per kg of COD destroyed:

$$M_{CH4} = 0.35 \left(nQC_i - 1.42 r_g V\right)$$

where M_{CH4} = methane production rate, m³/s,

n = fraction of biodegradable COD converted,

Q = flow rate, m³/s,

C_i = COD loading, kg/m³,

r_g = growth rate, kg/m³-s, and

V = volume, m³.

If the growth rate is ignored, then an estimate of the methane production can be made by removing this term and using:

$$M_{CH4} = 0.35 Q C_i$$

Digester gas is not comprised only of methane (55 to 75%). The other components are 25 to 45% carbon dioxide, water vapor, and other gases such as hydrogen sulfide, hydrogen, and nitrogen in trace amounts. If the concentration of hydrogen sulfide, a toxic and corrosive gas, is greater than 0.015%, it must be removed by scrubbing. After this step, the gas can be used in combustion engines or gas turbines. Under standard conditions, the fuel value of methane is 35,859 kJ/m³. An additional benefit is that the fuel value of the gas will also increase after removal of the hydrogen sulfide and small amounts of carbon dioxide. Digester gas can provide between 65 and 100% of the energy requirements of a municipal plant. Any excess gas can be burned in a flare or sold.

Digester Design

Thickening is often performed to reduce the reactor volume and the amount of supernatant from the sludge. For example, if a sludge contains 4% solids and then is thickened to 8%, the sludge will be half the initial volume.

Anaerobic digesters typically have dimensions in the range of 3.7 to 13.7 m in depth and 4.6 to 38.1 m in diameter. Depths of 6.1 to 10.7 m are the most common (Reynolds and Richards 1996). Digesters are constructed of reinforced concrete in a cylindrical configuration. The floor is also constructed of reinforced concrete in the form of an inverted cone, sloped to the center with a discharge pipe for removing digested sludge. Two-stage digesters include a fixed-cover digester as the first stage and a floating-cover digester as the second stage. Although sludge treatment and disposal can account for up to 40 to 45% of the capital and operating costs of a wastewater treatment plant, its role is often underestimated.

Monitoring

The digestion process can be monitored by organic matter (VS), volume and production of gases, pH , volatile acid, and alkalinity concentrations. Several of these parameters should be monitored at the same time to understand the process better. The reduction in volatile solids should be monitored weekly. If reduction rates decrease, the digester could be overloaded, mixing could be insufficient, scum could be accumulating, or the volatile solids in the feed could be lower than they were initially. The digested sludge should have a black color with no unpleasant odor, should be granular in appearance, and should settle easily to the bottom. The pH should be between 6.5 and 7.5 for optimal gas production. If the pH decreases, there could be an upset in the digester. Gas production and composition should be relatively constant. Decreases in production rates and methane concentrations could indicate toxicity. Volatile acid concentrations should also be monitored weekly. If there are any changes in the concentration (particularly acetic acid) there could be instability in the reactor. The bicarbonate alkalinity is measured to indicate the buffering capacity of the sludge, the ability to maintain a constant pH and neutralize acids. Alkalinity should normally be between 1,500 to 6,000 mg/L as calcium carbonate. In addition, if the digester is operating properly, the ratio of volatile acids to bicarbonate alkalinity should be less than 0.25.

Startup and Mixing

The fastest way to start up a reactor is to take digested sludge from another digester. This provides adequate numbers of bacteria that will become acclimatized in the shortest time. Once normal operation is initiated, adding fresh solids by mixing them into the reactor improves digestion rates. This enables good contact between solids and microorganisms, reduces the formation of scum, distributes heat and alkalinity more evenly, and disperses toxic materials. Mixing can be intermittent or continuous and performed by mechanical stirrers, forced circulation of sludge or supernatant, or discharge of compressed gas in the digester. Whereas intermittent mixing enables the supernatant to be removed after separation, continuous mixing decreases the required tank capacity due to the higher digestion rate. A secondary tank, however, is necessary in the latter case to separate the sludge and the supernatant.

Applications

Agricultural Wastes

Anaerobic digestion is frequently used to reduce agricultural wastes. These wastes can be significant since cattle produce from 10 to 14 kg per animal (10 to 14% solids content) and swine produce about 5 to 15 kg per animal (5 to 10% solids content) (Wheatley 1991). These wastes are further diluted by rain or yard runoff. Their content is similar chemically to primary sewage sludge although the fiber content is higher (Wheatley 1991). Batch, continuous stirred tank reactors, and second generation digesters have been employed. Normally, continuous stirred reactors are used if the feedstock is continuously available. SRT and HRT are the same and are in the order of 12 to 30 days. The methane-forming bacteria are limited since their doubling time is nine days. Pretreatment includes lowering solids contents to 1 to 2% and removing larger solids. Temperatures of about 36°C are used although temperatures in the thermophilic range can give higher gas yields and lower pathogenic bacteria contents. The disadvantages are higher instability and the need for more control.

Floating covers on lagoons have also been used to collect gas. At Royal Farms No. 1 in Tulane, California, hog manure was slurried and sent to a Hypalon-covered lagoon for biogas production. The gas produced fuel for a 70 kW and a 100 kW engine-generator. This generated sufficient electricity for the electric and heat requirements for the farm. The project was then adapted at other swine farms (Sharp, Resno, and Prison Farms) where floating cover lagoons were installed. A covered and lined lagoon was also installed for the treatment for organic matter from fruit crushing and wash down at Knudsen and Son (Chico, California). The biogas was subsequently burned in a boiler. Cow manure is sent to a plug flow digester in Langerwerf Dairy (Durham, California). The produced biogas is able to power an 85 kW gas engine which operates at 35 kW capacity and powers an electricity generator. It has been sufficient for the requirements of the dairy since 1982.

Resource Conservation Management specializes in anaerobic digestion systems for farm manure that produce methane and generate electricity (Feinbaum 2000). The system incorporates composting of solids and includes odor control. An example of one of their systems is a plug flow digester for 1,000 cows that was started up in December of 1996. Electricity valued at $24,000 and digester fiber valued at $30,000 were produced yearly.

Another plug flow unit operates at a Minnesota farm with 500 cows (Nelson and Lamb 2000). It was planned, designed, and constructed by the AgStar program (Environmental Protection Agency, Department of Energy, and the Department of Agriculture) as a demonstration project. Newspaper bedding is used for the cows, which adds fiber to the manure. New bedding is provided three times a day after removing and shredding the old bedding. Manure is fed into the pit and then gravity-fed through a 730 cm pipe to the mixing pit. The digester is a concrete pit with pipes for heating with hot water to 38°C. It is covered with a reinforced cloth and has dimensions of 0.9 m x 4.6 m x 0.4 m in depth. The current retention time in the digester is 35 to 40 days, but this will be reduced to 20 days. The digestate is then sent to a storage lagoon for later land application (Figure 6.4). Ninety percent of pathogens are killed during digestion. Methane gas is removed through a piping system and sent to a 150 kW engine and generator. In seven months, seven million cubic meters and 305,000 kWh of electricity were produced. This was enough to heat the digester, the entire dairy, and at least 30 homes.

In Germany, there are more than 680 agricultural biogas plants in operation. Volumes vary between 100 to 4,000 m³, and another 150 plants will be built

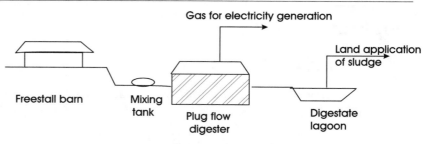

Figure 6.4 AgStar Plug Flow Anaerobic Digester

in the near future (Weiland 2000). Although 50% of the plants have volumes less than 500 m³, most in the future will be more than 1,000 m³. There are presently 14 large facilities that treat between 16,000 to 125, 000 tonnes per year. All have started up since 1995. Another trend that was initiated in 2000 is the use of agricultural crop waste for co-digestion. This waste includes maize, fodder beets, sweet sorghum, rapeseed and barley, which increase methane yield and enable the solid residue to be used as fertilizer (Table 6.3).

Table 6.3 Biogas Yield from Various Substrates (adapted from Weiland 2000)

Substrate	Biogas yield (m³ biogas/tonne substrate)
Cow manure	25
Pig manure	36
Whey	55
Fodder beet	75
Brewer grain	75
Thick stillage	80
Green wastes	110
Household biowaste	120
Flotation fat	400
Used fat	800

Industrial Wastes

Anaerobic digestion can be used for chemical, pharmaceutical, pulp and paper, and food industries. Many industries use different processes and operate in batch modes, which leads to highly variable sludge compositions. This process is most suitable for industries with continuous processes. Pretreatment consists of removing large solids and any potentially inhibiting materials. CSTRs are used with retention times of three to 12 days and organic loads of 1 to 5 kg COD/m³-day. Anaerobic filters are becoming more popular because they have high specific surface areas and void spaces and can be easily backwashed to avoid plugging. They contain lightweight synthetic packing such as PVC Flocor.

Domestic and Municipal Wastes

Domestic water consists of three streams: gray water, black water, and domestic solid waste. Gray water originates from shower/baths, kitchens, clothes washers, etc. It is usually highly degradable and generated at a rate of approximately 100 L/day per person. Black water is generated at rates of 40 L/day per person from toilet use, and it contains high levels of COD, nitrogen, and phosphorus. MSW or domestic solid waste is produced at a rate of 0.8 kg/person-day and contains 55% TS with COD load of 440 g/person-day.

A combined system has been proposed by Hammers et al. (2000) for domestic waste streams. Dry anaerobic fermentation (developed by De Baere and Verstraete 1985) is proposed for the solids with a UASB reactor for the removal of suspended solids after chemical pretreatment for the liquid streams. This system would be particularly suitable if traditional toilets are replaced by dry/vacuum toilets.

Septic tanks are used for the treatment of household waste (Figure 6.5). They usually operate at 10 to 12°C and are designed based on 200 L/person-day. HRT values are in the range of one to five days. Influent characteristics in the U.S. are 300 mg/L BOD with solids contents of 300 to 500 mg/L. Effluent values are in the order of 100 mg/L BOD and 40 mg/L TSS. COD removal rates are 25 to 50%. The tanks are usually divided into two

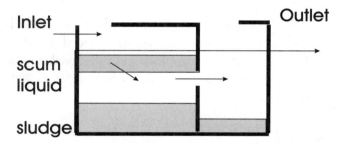

Figure 6.5 Schematic of a Septic Tank

compartments. The first, the largest, is where anaerobic digestion occurs and is where solids are allowed to settle. Scum accumulates at the top and maintains anaerobic conditions. Biogas rises to the surface as bubbles. The effluent passes into the second chamber and then is discharged. These systems have low space requirements, low costs, and no energy requirements. However, effluent qualities are low, pathogens are high, and construction must be by skilled labor.

Wastewater treatment plants are used for the treatment of domestic, commercial, and industrial wastewater. Water usually requires primary, secondary, and tertiary treatment processes. Biosolids are usually generated during secondary treatment. The type of influent wastewater influences the characteristics of the biosolids. Since industries must remove many contaminants before discharge to the water treatment facility, concentrations of metals and other pollutants such as chlorinated hydrocarbons have decreased significantly.

Municipal wastes include sewage, sewage sludge, and municipal solid waste. Individuals produces approximately 0.5 kg of feces and 1.2 kg of urine are produced every day. This waste is highly variable in physical and chemical composition. On average, it contains 35% cellulose, 34% ash, 19% crude protein, 14% lipids, and 6% hemicellulose. The carbon to nitrogen ratio is 4 to 5:1. Septic tanks for the digestion of sewage for single-family homes are common. Full-scale installations are not common but UASB systems have been used in Brazil and India where ambient temperatures are in the mesophilic range.

Anaerobic digestion of sludges from municipal treatment plants is common practice, particularly at larger facilities, to optimize gas production. Sludges from primary settling of untreated water are primary sludges. Waste activated sludge is excess activated sludge. Secondary sludges are generated from the secondary clarifier in an activated sludge or trickling filter plant.

Primary sludges can be easily digested because of their high organic contents. Activated sludges are usually digested with primary sludges. In addition to methane production, the advantages of anaerobic digestion include a 30 to 50% reduction of sludge volume, a stable odor-free sludge product, and pathogen destruction (particularly for thermophilic processes). Digestion decreases volatile solids from 32 to 48%, increases total solids from 8 to 13%, and the specific gravity increases from 1.03 to 1.05. Approximately 99.8% of coliforms are destroyed, dewatering of the sludge is easier, and the fuel value increases from 8,100 to 9,300 kJ/kg (Reynolds and Richards 1996).

Anaerobic digestion of MSW organics is performed at solids concentrations of 4 to 10%. Higher concentrations must be diluted. Gas production is 1.5 to 2.5 m^3/m^3 reactor or 0.25 to 0.45 m^3/kg of biodegradable volatile solids. Retention times are approximately 20 days. Methane concentrations are at 50 to 70%. Most reactors operate under mesophilic conditions.

Costs

Capital costs of anaerobic digesters tend to be 1.2 to 1.5 times higher than aerobic composting at large scale (10,000 to 150,000 tonnes/year) (Mata-Alvarez et al. 1999). However, costs per tonne treated are comparable with aerobic digestion when revenues from energy recovery are included. In addition, operating costs are approximately one-third those of incineration. For a plant of 55,000 tonnes/year, the capital cost has decreased from $202,000 per tonne per day in 1992 to $63,000 per tonne per day in 1999 (Jewell 1999). In Canada, a $16 million facility will be built in 2001 to treat 150,000 tonnes per year of MSW. The capital cost per tonne per day will be between $53,000 and 64,000 (N. Goldstein 2000). Tipping fees will be $42 per tonne for clean separated organic material and $55 per tonne for more contaminated material.

Clarke (2000) reviewed the capital and operating costs of anaerobic digestion systems. Capital costs were $627 per tonne per year (22,000 tonnes per year MSW) for the BTA process. For the Dranco process, capital costs were between $535 and $1,020 per tonne per year for plants between 20,000 and 50,000 tonnes per year. The least expensive at $387 per tonne per year was the Valorga which was the largest (77,000 tonnes per year). Operating costs varied from $72 per dry tonne ($23 per wet tonne) for the largest Dranco plant to $143 per dry tonne ($46 per wet tonne) for the smallest Dranco plant.

Another aspect proposed by Mata-Alvarez et al. (1999) concerned the potential benefits toward carbon dioxide sequestration. Currently, costs are between $25 and $250 per tonne in Europe. Since biowaste can sequester 25 kg of carbon per tonne, this would be $5 per tonne of organic waste, another benefit for anaerobic digestion.

Case Studies

In Europe, more than 36,000 anaerobic digestion systems are in operation, accounting for treatment of 40 to 50% of the sludge generated (Tilche and Malaspina 1998). Production of methane in biogas could exceed 15 million m³/d. Industrially, anaerobic digestion is perceived as mature (Riggle 1998). According to a study by De Baere (1999), 53 plants have capacities greater than 3,000 tonnes per year, of which 60% are operated in the mesophilic range and 40% in the thermophilic. Capacity has increased from an average of 30,000 tonnes per year from 1990 to 1995 to around 150,000 tonnes per year in 1999. This will increase anaerobic processes for waste recirculation, waste pre-treatment, and biogas production. Recently, interest in the digestion of residual refuse has increased because it allows greater flexibility and the possibility of material and energy recovery compared to landfilling and incineration (De Baere and Boelens 1999).

Since the production of biogas is promoted by the Kyoto protocol for the reduction of CO_2 emission, the government of Germany plans to increase energy production from organic wastes and biomass by a factor of four to five by the year 2010 (Weiland 2000). There are currently more than 1,230 municipal sewage treatment plants with covered digesters in Germany. Eleven percent operate with co-substrates such as grease and flotation fats, and up to 30% have excess capacity that could be used for co-substrates. Regarding biowaste from households, about 28 plants treat 5,000 to 30,000 tonnes per year. Half of these are wet digestion systems with solids contents between 8 and 12% at mesophilic temperatures, but the other half are operated as dry fermentations at thermophilic temperatures.

High Solids (Dry) Fermentation

There are three different dry fermentation processes used in Europe (Bilitewski et al. 1997). They all use solids concentrations of 25 to 35%. One of them is the Valorga process that has processed 55,000 tonnes/year of the fines from household waste in Amiens, France, since 1988. Total solids of the waste is 65% and the ratio of VS/TS = 65% (Saint-Joly et al. 2000). The wastes are combined with water to obtain a solids content of 30 to 35%. Retention times are in the order of three weeks under mesophilic conditions. The solids are then dewatered, screened, reduced in size, and composted (Figure 6.6). Average biogas

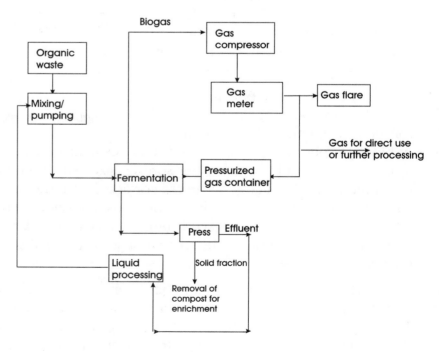

Figure 6.6 Flow Sheet of the Valorga Process

production was in the order of 145 m³/tonne. Another plant in Tilburg, The Netherlands, has been operating since 1994 and processes 40,000 tonnes per year of biowaste and 6,000 tonnes per year of paper. Total and volatile solids concentrations of incoming waste vary significantly throughout the year (between 39 and 60% TS and 36 to 64% VS) because of an increase in plant material in the summer. Average gas production was 92 m³/tonne (Saint-Joly et al. 2000). The gas production rate is lower than that of Amiens because of the lower VS content of the waste (22.5% compared to 39% at Amiens). However, the biogas production at Tillberg is higher based on VS content (224 m³ CH₄/tonne VS) compared to Amiens (205 m³/tonne VS) as a result of a higher content of kitchen waste at Tillberg. At Tillberg, productivity was also higher in the winter (320 m³ CH₄/tonne VS in winter and 170 m³ CH₄/tonne VS in summer) since the ratio of kitchen waste to garden waste increases in the winter. Therefore, biogas production depends on the composition of the waste.

Other plants have been operating since 1992 in Brecht, Belgium, using 10,500 tonnes of biowaste per year and since 1993, in Salzburg, Austria, processing 20,000 tonnes of biowaste (Bilitewski et al. 1997). These plants use a process called Dranco, which ferments the wastes at 55°C instead of the mesophilic conditions of the Valorga process. The retention time is three weeks and composting of the residue is performed for ten days (Figure 6.7). Biogas production is 5 to 8 m³/m³ reactor or 140 to 200 m³ per tonne of organic waste (40 to 60% solids).

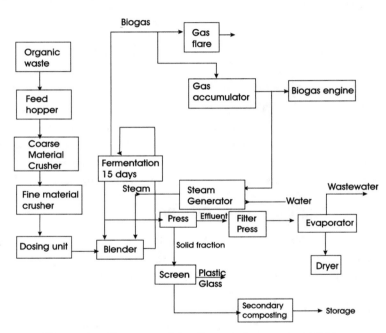

Figure 6.7 Diagram of the Dranco Process at 53°C

The Compogas process started up in 1992 in Rümlang, Switzerland, for the fermentation of 5,000 tonnes of biowaste per year. Another plant operates near Zurich, Switzerland, processing 10,000 tonnes per year of biowaste. After sorting, size reduction, and temporary storage, the material is pumped into the digester. The reactor acts as a plug flow reactor with a retention time of 15 to 20 days under thermophilic conditions. A slow-moving agitator is used to move and degas the material. The material is removed, pumped, dewatered by a worm extruder, and then cured in a shed under aerobic conditions. It remains there and is turned continuously for five to ten days.

The BIOCEL process is another type of dry mesophilic anaerobic digestion at full scale (Ten Brummeler 2000). The first unit was built in Lelystad, The Netherlands, in September of 1997. It processes 50,000 tonnes/year of organic waste from MSW in a batch high solid digestion (30 to 40% w/w). No pretreatment is necessary other than removing large objects such as tree branches. The 14 digesters are rectangular (6 x 6 m by 20 m in depth) with a volume of 720 m³ each. The effective volume is 480 m³ when loaded by shovels to the working height of 4 m. The digester floors have perforations to allow leachate to pass to the chamber below for collection. The produced biogas is converted to heat and electricity continuously in a heat/power generator. Special procedures have been designed to open and close the doors because of the explosive mixture of air and methane (Ten Brummeler 2000). Temperature is maintained at 35 to 40°C by spraying leachate previously heated by a heat exchanger on the top of the digesters. Retention time is about three weeks. Compost is also produced by a wet separation process with rotating sieve, stone separator, centrifuge, and hydrocyclones. Various fractions of 0 to 4 mm, 4 to 15 mm, and 15 to 50 mm can be produced. The flow sheet with mass balances is shown in Figure 6.8. Complete inactivation of numerous plant and animal pathogens was noted and may be due to the high volatile acid concentration in the first two weeks of the digestion.

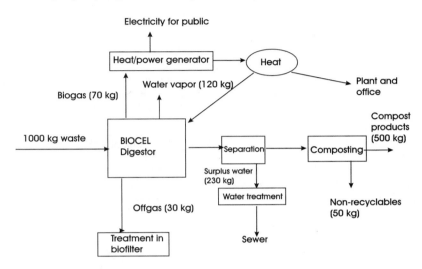

Figure 6.8 BIOCEL Process Flowsheet with Mass Balances

Low Solids (Wet) Fermentation

Several single-stage processes treat solid waste up to 15% dry solids. In Bellaria, Italy, the Solidigest Process is used for the fines from household waste. The waste is fed into the degradation towers of the municipal sewage plants. Another process was developed by BioTechnische Abfalverweertung GmbH (BTA) and Müll und Abfalltechnik GmbH (MAT) which process 5,000 tonnes of biowaste in Baden-Baden, Baden-Württemberg, Germany. The treated waste is added to the municipal sewage treatment plant. In the city of Rottweil, a full-scale plant is processing 8,500 to 9,000 tonnes of biowaste per year (Bilitewski et al. 1997). A schematic diagram of this process is shown in Figure 6.9.

Two-stage processes are also used since they can enhance methane production and stabilize the process for substances that are more difficult to degrade. In Zobes, Germany, 20,000 tonnes of chicken feces are co-processed with biowastes per year. There are five full-scale plants in Germany treating 80,000 tonnes per year, and other plants are under construction in Italy and Canada (Kubler et al. 2000). The BTA process (Figure 6.10) has also been pilot tested in Garching, Bavaria, for wet waste, biowaste, institutional food wastes, and other types of wastes. This process includes size reduction, magnetic separation, the dissolution of particles, and removal of light particles. The suspension with 8 to 10% dry solids is drained through the floor screen and heat treated at 65°C for pasteurization. After solid-liquid separation, the liquid fraction enters the methane reactor and the solid fraction enters the hydrolysis reactor (retention time of two to four days). After leaving the hydrolysis tank, the liquid fraction is sent to the methane reactor and the solids are composted. Compost quality is good

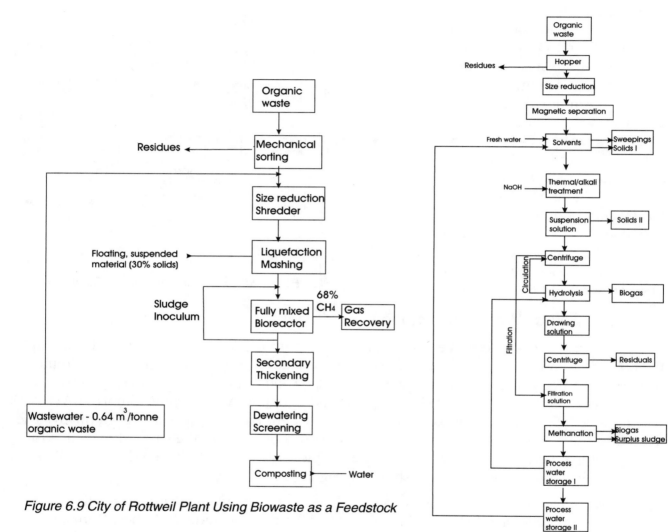

Figure 6.9 City of Rottweil Plant Using Biowaste as a Feedstock

Figure 6.10 Schematic Diagram of
the BTA Process

with low pollutant contents. Although the salinity content can increase due to co-digestion, the addition of yard waste can adjust the quality (Kubler et al. 2000). For the overall process, approximately 8% of the initial biowaste is removed by screening, 1% is removed as heavy substances, 55% is wastewater, 20% are residues, and 15% goes to biogas (Bilitewski et al. 1997). The energy yield is 260 MJ/tonne when MSW is used alone and increases to 290 MJ/tonne with 14% food waste. Wastewater is treated by activated sludge to remove carbon and nitrogen. A facility in Newmarket, Ontario, Canada is being started-up and processes 25 to 30 tonnes per day currently of restaurant and grocery wastes. At full scale, it will process 150,000 tonnes per year and produce 25 million cubic meters of gas. Detention time in the digesters is 48 h. The digested solid will then be sent for composting.

Another process was developed by AN-Maschinenbau Company that includes a fed batch hydrolysis tank with a percolation process (retention time of four days). The dissolved solids that are pumped into the methane reactor and the solids from the hydrolysis reactor are composted. A schematic diagram is shown in Figure 6.11. The compost that is produced after anaerobic treatment has similar properties to aerobic compost. Sanitization may be necessary since the pathogens may not be destroyed under mesophilic conditions.

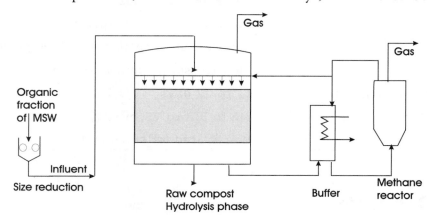

Figure 6.11 AN-Maschinenbau Process with Feed Batch Hydrolysis

Comparison to Other Processes

Anaerobic digestion is the most common biological process in the world for industrial, agricultural, commercial, and municipal sludges. It will continue to be popular since it produces methane that can be used for energy. Up to 35% solids concentrations can be processed. In addition, less carbon dioxide is produced than by aerobic processes which could become an important consideration due to the reduction requirements of the Kyoto protocol. Solids from anaerobic digestion can be composted. Supernatants from the sludge often have high organic contents and need to be treated further.

Aerobic Digestion

Box 6-2 Overview of Aerobic Digestion

Applications	Suitable for treatment of organic wastes including agricultural, industrial, and municipal wastes
Cost	Operating: $20 to $500/tonne, Capital costs 1.2 to 1.5 times less than anaerobic digestion
Advantages	High quality supernatants
	Highly stable, odorless product with nitrification
	No dangerous gases
Disadvantages	Large amounts of oxygen are required
	High operating costs due to high power requirements
	Digested sludge dewaters poorly
Detention time	12 to 22 days
Organic load	0.6 to 3.2 kg VS/m^3-day
Other considerations:	Mainly used for small and medium sized plants

Description

Aerobic digestion processes can be used for sludge produced from primary clarifiers and biological reactors, mainly for small- and medium-sized plants. The digester is aerated and operates in a stable manner. The residence time is long enough to allow nitrification to take place. The main design parameters are the sludge oxidation rate, sludge loading rate, sludge age, oxygen requirements, detention time, and efficiency desired.

The following equation can be used to represent aerobic digestion of organic matter:

$$\text{Organic matter} + O_2 \rightarrow \text{new cells} + CO_2 + \text{biomass} + H_2O + \text{energy}$$

According to Reynolds and Richards (1996), it is estimated that approximately 1.9 kg oxygen is required per kg BOD$_5$. Dissolved oxygen in the digestion tanks should not be lower than 1 to 2 mg/L. This includes oxygen required for nitrification. Initially ammonia is produced by the aerobic digestion of organic matter. Subsequently, nitrate (NO_3^-) is formed as the ammonia is oxidized.

During the digestion process, the VS levels reduce until a stable value is obtained. Solids are hydrolyzed into smaller compounds for uptake by microorganisms. Two-thirds of the cell mass is oxidized. The remaining amount is made up of compounds that are not biodegradable, such as peptidoglycan. The pH can drop to between 5 and 6, depending upon the buffer capacity of the digestion system. This drop may result from either carbon dioxide or nitrate formation, but the drop does not affect microbial activity.

Methods

Thermophilic Digestion

Thermophilic digestion is the stabilization of wastes at temperatures between 50 and 70°C. Heat is produced by the aerobic reaction when pure oxygen or air with concentrations of oxygen greater than 20% are used. Enough waste is then added for self-heating and mixing. Aeration and the reactor must be designed to maintain the temperature by minimizing heat loss. As a result of the high bacterial energy maintenance requirements, decay rates for thermophilic bacteria are higher and biomass production rates are lower than for mesophilic bacteria. This minimizes reactor volume requirements and enhances solid removal rates. Nitrogen balances must be performed and monitored. This is particularly important since temperatures above 45°C inhibit nitrification. Excess ammonia and carbon dioxide lead to the formation of ammonia bicarbonate, which is necessary for alkalinity control in the reactor (Stover 2000).

The thermophilic reaction can be represented as follows:

$$C_5H_7O_2N + 5O_2 \rightarrow 5O_2 + NH_3 + \text{energy} + \text{heat}$$

where $C_5H_7O_2N$ is the composition of the biological cell waste. From this equation, it can be seen that more than 1.42 kg of oxygen are required to destroy 1 kg of volatile solids. To maintain thermophilic conditions, the volatile solids content must be sufficient, and 18,800 to 19,200 kg of heat must be produced for every kg of volatile solids destroyed.

Design

The process can be modeled according to biological growth equations as follows:

$$r_x = -Yr_s - k_eX_v$$

where r_x is the sludge production rate,

r_s is the substrate utilization rate,

Y is the yield factor,

S is the substrate concentration,

k_e is the rate of endogenous decay coefficient, and

X_v is the VS concentration.

The equation can be simplified by eliminating the first term on the right side since the sludge was already produced during activated sludge or trickling filter processes. Even if the sludge is from a primary clarifier, the following equation can be used:

$$r_{xe} = -k_eX_v$$

where r_{xe} is the rate of endogenous decay.

Although stabilization includes both hydrolysis of the solids and metabolism of soluble products, solubilization is not specifically accounted for in the above model. It can be significant, however, but difficult to predict (Droste and Sanchez 1986). It can be determined by measuring influent and effluent total and soluble COD. VS data determined for the endogenous decay rate constant can provide an indication of solubilization.

Adams et al. (1974) presented a model taking solubilization into account by performing a mass balance around a completely mixed digester:

$$In - Out + Generation = Accumulation$$

$$QX_{vdo} - QX_{vd} + r_d V + r_{sl} V = \frac{dX_{vd}}{dt} V$$

where Q is the volumetric flow rate,

X_{vdo} is the influent degradable VS concentration,

X_{vd} is the reactor VS concentration,

r_d is the kinetic decay expression,

r_{sl} is the solubilization rate, and

V is the reactor volume.

For steady state conditions, $dX_{vd}/dt = 0$ and the VSS decay can be modeled as first-order equations for decay and solubilization.

$$r_d = -k_e X_{vd}$$
$$r_{sl} = -k_{sl} X_{vd}$$

where k_e is the endogenous decay rate coefficient, and

k_{sl} is the rate coefficient of solubilization.

If solubilization is not determined separately, an observed rate coefficient for VS, k_{obs}, can be used, which is the summation of the k values for solubilization and decay. To determine the hydraulic retention, θ_d, the following equation can be used:

$$\theta_d = \frac{X_{vdo} - X_{vd}}{(k_d + k_{sl})X_{vd}}$$

Typical solid retention times are in the order of 25 to 60 days. Values for k_e vary depending on the temperature and type of sludge. They can be as low as 0.017 days^{-1} at 15°C for primary and waste activated sludges and up to 0.71 days^{-1} for municipal waste activated sludges at 25°C. Hydraulic retention times at 20°C are from ten to 22 days, depending on the sludge type. VS in the sludge is reduced by 40 to 50%. Waste activated sludge is digested more easily than primary and waste activated sludge together. The following equation can be used to determine hydraulic retention at temperatures other than 20°C:

$$\theta_{h2} = \theta_{h20}\theta^{20-T2}$$

where θ_{h2} is the hydraulic time in days at temperature, T_2 (°C), θ_{h20} is the hydraulic retention time at 20°C, and θ is from 1.02 to 1.11.

Organic loading rates are from 0.64 to 3.2 kg VS/m^3-day. Volumetric loading rates are in the range of 0.042 and 0.113 m^3 per capita. Although solids concentrations can be up to 50,000 mg/L, they are usually between 25,000 to 35,000 mg/L. Mixing is necessary to provide adequate aeration and is in the order of 20 to 40 kW/1,000 m^3. Aerobic digestion tanks do not need to be heated or covered and, therefore, are less expensive to manufacture than anaerobic tanks. Aeration systems are similar to conventional aeration tanks and can be designed for spiral or cross roll aeration with diffused air equipment.

Applications

Aerobic digestion is traditionally used for biosolids from wastewater treatment plants of less than 38 million liters per day. According to the WEF (1995), aerobic digestion is common in wastewater treatment plants with a capacity of less than 20 million liters per day.

Costs

Cost estimates were made by Foess and Federicks (1995) for a small aerobic digestion plant treating biosolids (2.3 tonnes per day). They compared the costs of aerobic digestion to lime stabilization for operating, capital, hauling, and spreading on agricultural land. Total costs were significantly higher ($440 per tonne) compared to lime stabilization ($330 per tonne).

Costs for the thermophilic aerobic digestion have been estimated at $500 per tonne for a ten tonne dry solids per day facility without transport (Oleszkiewicz 2001). This compares to the lime stabilization ($500 per tonne) and drying ($650 per tonne) for the same size facility.

Case Studies

The Autothermal Thermophilic Aerobic Digestion (ATAD™) system, under development since the 1970s by Kruger/Fuchs, has been recognized by the U.S. EPA. More than 40 full-scale facilities operate in Europe. Others are located in Canada and the United States. Advantages include pathogen control, low space and tank requirements, and a high sludge treatment rate. Mesophilic anaerobic digesters can be converted to autothermal thermophilic digesters by covering and insulating the reactors. Sludge dewatering is about the same for thermophilic and mesophilic systems.

International Bio-Recovery Corporation (North Vancouver, BC) has developed a technology called autogenous thermophilic aerobic digestion (ATAD). After 72 h, it converts organic waste (food waste, sewage sludge, and animal manure) into single-celled protein organic fertilizer. Leafy plants can be converted in 36 to 48 h while paper products require 48 to 100 h. Up to 20% of the waste can be non-biodegradable (IBR 2001). Air flow rates of 0.5 to 2.0 volume of air per volume of reactor per hour are employed with organic loading rates of 4 to 20 kg VS/m^3-day. IBR is expanding its plant from 50 tonnes per day to 100 tonnes per day. The technology is proprietary and available for license. The process involves the removal of metals, glass, sand, and plastics (Figure 6.12a, b). The waste is macerated and inoculated with bacteria. The process is aerated with the Shearator, a submerged aeration device developed and patented by IBR. After digestion, the slurry is screened and pressed, followed by drying and pelletization of solid fertilizers or clarification, and concentration of liquid fertilizers. Odors are eliminated by a unique biological treatment and through the operation of the plant under negative pressure, allowing the plant to be situated near residential areas. Pathogens are destroyed in the process.

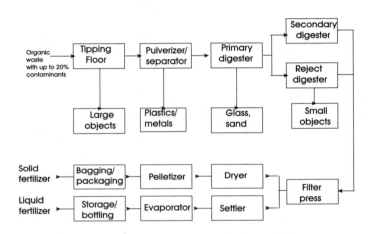

Figure 6.12a Schematic of International Bio-Recovery Process for Autogenous Thermophilic Aerobic Digestion (ATAD)

Figure 6.12b Photograph of International Bio-Recovery Process for ATAD

Other companies such as Rose & Westra (Grand Rapids, Michigan) and U.S. Filter/Jet Tech (Edwardsville, Kansas) offer variations of this process. Rose & Westra's system uses oxygen, instead of air, and an *in situ* aerator by Praxair (Praxair 2000). They have obtained solid volume reductions in the order of 40% at an installation at Rhone-Poulenc AI Company (West Virginia). The volume of the digester was 1250 m^3. Design and construction required six months for completion. U.S. Filter operates an ATAD process at the Franklin municipal wastewater treatment plant with a retention time of seven days. The primary or waste activated sludge is fed in batches after thickening into the first reactor which operates at 35 to 55°C. The second reactor operates at 55 to 65°C. Their offgas recirculation system is proprietary and minimizes heat loss. No external heat source is required. They have operated a unit treating the 95,000 L/day (4 to 5% solids) sludge from the 19,000 m^3 per day municipal wastewater facility since 1996 (U.S. Filter/Jet Tech 1999).

Comparison to Other Processes

Advantages of aerobic digestion include a stable humus end product, low capital costs, easy operation, low odor, production of non-explosive gases, operation at low pH values (down to 5.5), and higher quality supernatants than for anaerobic digestion. VS destruction rates are typically from 65% but can be between 40 and 75%. Operating costs are high because of oxygen and power requirements. Furthermore, methane is not produced, efficiency is decreased at lower temperatures, and the sludge can be difficult to dewater. From the equation, it can be seen that the volume of the reactor increases as the reaction rate decreases with temperature.

Aerobic digestion is a highly stable process that can be operated so that nitrification can also take place. Due to the aeration requirements, it is more suitable for small- and medium-sized plants. Capital costs are less than anaerobic digestion, and operation is safer because there are no explosive gases produced. Supernatants are also of higher quality. Thermophilic digestion is becoming increasingly popular due to pathogen destruction, low space and tank requirements, and high sludge treatment rates.

Composting

Box 6-3 Overview of Composting

Applications	Suitable for treatment of organic wastes including food and yard wastes, municipal and sewage sludge		
Cost	*Method*	*Operating costs ($/tonne)*	*Capital costs ($)*
	Aerated static pile	12 to 500	36,000 to 20 million
	In-vessel	18 to 540	850,000 to 33 million
	Windrow	215 to 245	50,000 to 8 million
Advantages	Product useful as soil conditioner		
	Low to high sophistication available		
Disadvantages	Processing time can be long		
	Odors can be produced from poorly maintained system		
	Sufficient moisture and aeration needed to avoid fires		
Detention time	Nine to 25 weeks for windrows, five to six weeks for aerated piles, ten to 21 days for in-vessel systems		
Other considerations:	Capital and operating costs increase with sophistication		
	Increased public acceptance		

Description

In the past, composting was a low-cost process, but recently it has developed to a high degree of sophistication, particularly in United States and Europe. Composting is "the biological decomposition of wastes consisting of organic substances of plant or animal origin under controlled conditions to a state sufficiently stable for nuisance-free storage and utilization" (Diaz et al. 1993). Bacteria, fungi, and some protozoa are involved in composting. These microorganisms are usually found in sufficient quantities in the wastes. If not, horse manure, finished compost, or rich, loamy soil can be used as an inoculum. Composting is mainly aerobic although anaerobic composting has been investigated. The process usually takes 30 days to produce a stabilized product; however, the product is not always fully stabilized and pathogens are not always eliminated.

In an integrated composting process, the sludge is dewatered until it contains 35% solids. It is then blended with bulking agents such as wood chips in a ratio of three parts chips to one part sludge. Heat generation during composting dries the sludge and kills pathogens. Mixing and turning is necessary to make sure that the sludge is totally exposed to the high temperatures. Under-floor blowers can be used to supply air in colder regions. The chips can be removed by screening at the end of the process since they remain 25 mm in size compared to a few mm for the compost.

The main parameters to consider when designing a composting process are the type of bulking agent, temperature, moisture content, and organic and nutrient contents. The criteria for the choice of bulking agents include size, cost, recoverability, carbon and nutrient content, preprocessing needs, moisture content, and porosity. The choice of bulking agents includes wood chips, sawdust, leaves, paper, and solid waste. Wood chips are used most frequently since they can be reused and are not sources of pathogens or metals. Pore volumes should be in the range of 25 to 35% (Bilitewski et al. 1997).

Methods

Systems for composting are based on the degree of mechanization and are broadly classified as static and turned windrows (piles) or mechanical enclosed systems. Capital and operating costs increase as the systems become more sophisticated. The increase in process efficiency may not offset the extra costs. The limiting factor, as in all biological processes, is the performance of the microorganisms. It is best to evaluate a system by observing an individual processing the waste. The individual should be a professional, and the observation should cover a minimal period of 8 h (Diaz et al. 1993).

Windrow Composting

Windrow composting is the oldest composting method. If no forced aeration is used, the windrows should be no higher than 2.2 m so that turning can be done. Piles can be up to 5 m, however, if forced aeration is employed. Although composting can be done if size reduction is not performed, cavities or shafts within the piles can form that will dry the windrows.

Windrow turning is accomplished by specialized windrow machines or tractors. Leachate must be collected after draining into ditches. If excessive rain occurs, a roof or covering liner is useful for preventing excessive leachate formation and water percolation. Degradation time depends on the use of turning and aeration as follows: with turning alone, nine to 12 weeks; with no turning but with forced aeration, 12 to 16 weeks; and with no turning or aeration, 20 to 25 weeks (Bilitewski et al. 1997). Triangular windrows for turning are usually 1.3 m, 1.8 m, and 2 to 2.5 m. Trapezoid windrows usually have a maximum height of 1 m.

Aerated Static Piles

In static systems, aeration is forced and the material for com-posting is not disturbed. Air is either sucked into the pile or pushed through as shown in Figure 6.13. In the former case, the air is passed through a pile of screen compost to deodorize the air.

The main steps in this method are mixing the material for composting with a bulking agent, if required, constructing the windrow, and composting. Once the com-posting is finished, the mixture is screened to remove the bulking agent, cured, and then stored.

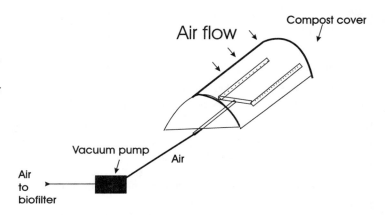

Figure 6.13 Aerated Static Piles

A perforated pipe of 10 to 15 cm in diameter is put on the compost pad and centered. The end of the pipes should be between 150 to 270 cm away from the edges of the windrow to avoid shortcutting (Diaz et al. 1993). The pipe is connected to a blower (0.25 to 3.73 kW) with a non-perforated pipe, and the bulking agent or finished compost is placed over the piping. Air flow rates of 35 to 140 m³/h are used. The material for composting is then piled on the bedding material until a windrow of 21.3 to 27.4 m in length, 150 to 270 cm in height, and 300 to 550 cm in width is reached. The pile is then covered with a 180 cm thick layer of finished compost for screened compost or 240 cm for unscreened compost. This avoids odor and temperature distribution problems. Approximately seven to 11 tonnes of material can be composted on an area of 1 ha. This includes runoff collection, administration, and storage areas.

As an example, for a pile of 120 cm high and 244 cm in diameter, sludge of 22% solids is mixed with a bulking agent of woodchips in a ratio of 1:2 (v/v). The process takes two to three weeks. The material is screened to allow the material to pass and the bulking agent to remain on the screen. Screening is easier to perform on dry days.

Naturally Aerated or Turned Piles

Aeration in naturally aerated piles is usually performed by turning the composting piles by special equipment or front-end loader (Michel 1999). The new windrow is placed behind the old one by taking the top part of the pile to make the bottom of the new windrow and then placing the bottom of the old on top of the new (Figure 6.14). The compost is allowed to fall out of the loader to ensure porosity (Richards 1998). The efficiency is highly variable and can lead to anaerobic and odorous conditions; therefore, the pile size and turning frequency are very important. The size of the facility, type of turning equipment, and volume of feedstock influence pile size.

Small windrows are 1.2 to 1.5 m high and 3 to 4 m in width. Their lengths can be up to several hundred meters in length, with volumes of 5,700 to 7,400 m³/ha. Large windrows can be 3.1 m in height and 9.2 m with volumes of 27,800 to 42,000 m³/ha. Manual turning is by four- and five-tined pitchforks while mechanical turners include rototillers or other types of machinery commonly used in landfills.

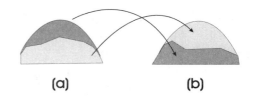

(a)   (b)

Before (a) and after (b) turning

Figure 6.14 Naturally Aerated or Turned Windrows

Continuous piles of compost up to 3.1 m in height are called slabs or briquettes. Front-end loaders or machines for shaving the slab sides are used for mixing. The capacity of this type per area is up to 170,000 m³/ha. This composting process can produce a stable product in five to six weeks with a moisture content of 30 to 40%. This is the only short process where disinfection occurs and no curing is required.

Piles for composting should be located on hard, preferably paved, surfaces to make material handling easier, to allow collection of leachate, and to prevent fly larvae from escaping. Paving (gravel, crushed stone, asphalt, or concrete) is required for heavy mechanical turners. Leachate can be odorous, so it must be collected. Shelters from wind or rain may be necessary, depending on the local climate.

Temperatures in large windrows increase up to 65°C by day compared to 57°C for small windrows (Michel 1999). Composting rates are slightly slower also, probably because of the higher temperatures. It has also been found that the number of bacteria and bacterial diversity decreased at 65°C compared to 55°C. The frequency of turning is believed to influence aeration and temperature. Turning has a significant influence on process economics and odor release, but this is poorly understood. Although air is introduced into the pile once the pile is turned, within hours the oxygen level drops to the level before the turning. Therefore, to maintain high oxygen levels, turning would have to be performed every few hours. Temperatures return to the pre-turn levels within hours and are not substantially affected by turning. Moisture, particle size, distribution, and pathogen reduction are, however, affected by turning. More frequent turning leads to higher bulk densities and lower particle size.

Aeration can be increased by maximizing the surface area to volume ratio of the windrow and using bulking agents to increase porosity. Low aeration rates increase volatile acid production, which can increase phytotoxicity and odor. Levels of ammonia also increase. Higher aeration rates liberate ammonia during composting and decreased volatile acid formation. The final compost obtained under highly aerated conditions has a pH of 8 to 8.5; an organic content of 50%; and low volatile acid, ammonia, and soluble salt contents (Michel 1999). For aerobic degradation, four liters of O_2 is required per g of dry organic solids (DOS) (Bilitewski et al. 1997). In composting, this level will vary and is highest at a temperature of 60°C.

Contained, In-vessel Systems

Contained systems have been used for small-scale systems at schools, restaurants, cafeterias, or markets (Rynk 2000). They have also been employed for food composting at large-scale facilities including prisons, hospitals, office buildings, colleges, and food processing plants. Process and odor control is good, and retention times and labor costs are low.

Passive aerated bins are simple with no moving parts or electricity requirements. The key is to employ good operating procedures. The Hot Box was developed by Open-Road, a non-profit organization (Rynk 2000). Its dimensions are 1 m x 1 m x 1 m, and it can be made of wood or recycled plastic lumber. One wall is made of planks to facilitate manual loading and unloading. To increase air movement, two rows of PVC perforated pipes are inserted through one wall and rest on the other wall. One end is open to the atmosphere. Wood ships are placed on the floor. The compost mixture, a one to one ratio of food waste and wood chips, is placed on top. If odors become a problem, biofilter mesh bags containing compost and wood chips can be placed on the end of aeration pipes. After one month, the compost is unloaded for curing.

Passive aeration works well for small-scale systems but is not usually sufficient for large volumes. One exception that seems to work well is the TEG silo-cage (Rynk 2000a). The composting materials are placed in a series of mesh cages, 4.3 m high, 6.1 m long and 1.2 m in width. An air space of 10.2 cm between the cages enables air to flow. An auger removes material from the bottom of each cage while a hopper is used to fill the top of the cage with mixed and shredded feedstock. The auger and hopper move among the cages; thus materials move vertically from top to bottom in a period between eight to 24 days, depending on the desired compost quality. The number of cages varies from six to about 20 for a full-scale system. A ten-cage system with a retention time of 18 days can have a throughput of 22.5 m³/day. To fully enclose the system, a building would be required. Insulation via a thermal jacket may also be necessary.

Aerated, Contained Systems

Although internal agitation is not used, fans are utilized to enhance aeration in aerated, contained systems. Suppliers of larger aerated containers include the CompTrainer (Green Mountain Technologies), NaturTech, Stinnes Enerco, and Eco Pod system by Ag-Bag Environmental (Rynk 2000b). These systems have been used for on-site composting at prisons, recreational parks, and military bases. Tunnel-style composters have primarily been used in the mushroom industry (Figure 6.15). The CompTrainer, NaturTech, and Stinnes Enerco systems are similar to roll-off containers which are modular and mobile. They include air

Horizontal bed reactor

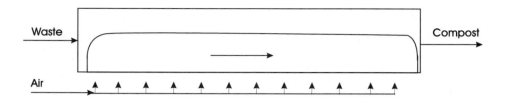

Vertical reactor

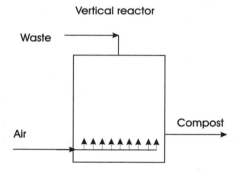

Figure 6.15 Aerated, Contained Composting Systems

distribution systems, process control, and leachate collection with the capability of leachate reuse. Biofiltration is employed for odor control. The air distribution and process control systems are highly automated with monitoring of air flow rates, direction, and recirculation. Unloading, remixing, and reloading during the process is possible.

The smallest system is the GMT CompTrainer with dimensions of 2.4 m in width, 3.2 m in length and 1.7 m in depth. The Stinnes Enerco system is 2.4 m on each side. The NaturTech and GMT supply containers are 2.4 to 2.7 m in depth and 7.6 m in length. Volumetric capacities vary from 12 to 38 m³. Retention times are usually in the order of two to three weeks for production of finished products. However, all three units can also be used in an initial composting phase (seven-day retention time) followed by windrows or aerated piles (Rynk 2000b).

The Ag-Bag EcoPod composting system is similar to an aerated static pile with the composting materials contained in a polyethylene "pod" or bag 61 m in length and 1.5 m in diameter. The bag has a capacity of 152 m³ of composting material. Larger pods of 3.1 m in diameter (380 m³) are also available. Mechanical rams or auger-like rotors with conveyors and feed tables are used for loading the premixed feedstock. The pod is unrolled by aeration. Air is exhausted through side ports on the pods. These vents can also be used for temperature probes. Aeration is done until the pod is full. Agitation and transport are not possible with this type of system. When the composting is complete, the pod is sliced open and the contents are removed by bucket loader. The pod cannot be reused; however, if the pod is punctured during use, it can be repaired with polyethylene tape. The containment is fast and flexible, and the only site requirements are a level grade and electrical outlets for the fan.

Tunnel-type systems are expensive on a small scale but feasible for on-site food residuals. These systems consist of forced aeration through a plenum in the floor, internal circulation, and a biofilter. They are operated as a plug-flow batch after loading from one end. Double-T Equipment has designed 0.9 to 1.8 tonne/day food residual systems while Kelly Green Environmental Services has developed systems for yard trimmings and food residuals with dimensions of 3.7 m in width, 3.7 m in height, and 7.3 m in depth and with capacities of 1.8 to 2.3 tonnes/day.

Agitated-aerated Containers and Rotating Drums

Currently, there are only two commercially available systems combining both aeration and agitation. They are supplied by Earth Tub and Wright Environmental Management, Inc. (WEMI). The latter has a capacity of more than one tonne per day and up to hundreds of tonnes per day. Capacity can be increased by larger units or using the units in tandem.

Rotating drums are used for small-scale backyard composting up to municipal solid waste composting (Figure 6.16). Tumbling assists in shredding the material. Retention times can range from hours to days. Small drums of 1.2 to 1.5 m in diameter and 2.4 to 4.6 m in length are available while larger ones are 2.4 to 3.1 m in diameter and 9.2 to 15.3 m in length. Suppliers include BW Organics, Augspurger Engineering, and Environmental Products and Technologies Corporation (EPTC). Systems with short retention times can be used as a pretreatment before using windrows or aerated static piles.

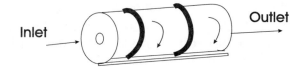

Figure 6.16 Rotating Drums for Composting

Agitated Beds

In agitated beds, composting materials are contained by long channel concrete walls with forced aeration through the floor channels. The materials to be composted are loaded at the front of the channel. Agitation is done by a turning machine that moves only the top of the beds from the discharge to the front end. The material is moved toward the back in a fixed distance each time (2.1 to 4.3 m) until discharged. Green houses or other buildings are required to totally contain the system since the top of the channels are open. The composting period depends on the turning frequency and the length of the channel but is usually in the range of ten to 28 days (Rynk 2000a).

Biomax, Farmer Automatic, Global Earth Products, Longwood Manufacturing Corporation (LMC), Resource Optimization Technologies (ROT Box), Transform Compost Systems, Ltd., and U.S. Filter (IPS) supply agitated bed systems (Rynk 2000b). The channels range from 0.9 to 2.4 m in depth, 1.8 to 3.7 m in width, and 61.0 to 91.5 m in length. Although most systems use multi channels and a turner machine, the Farmer Automatic has a one channel system which is 1 m deep. Since the channels are shallow, turning is only used for aeration. The ROT Box comprises a bed of 7.3 to 12.2 m wide and 12.2 m long with an overhead crane to support the 1.8 m wide turner that travels down the bed in strips.

Agitated bed processes were used as prototypes by the University of Maryland, the USDA, and the University of Guelph as early as the 1970s. These systems are applicable for composting of foods at large food service facilities, prisons, and schools (Rynk 2000b).

Hazardous Waste Composting

ABB Environmental Services (Wakefield, Massachusetts) has applied composting to petroleum sludges, coal tars, munitions, and wastes with high concentrations of contaminants by (U.S. EPA 2000). Wood chips or other bulking agents are added by special equipment to increase air permeability. Forced aeration or windrow methods are used. A full-scale system has been used to treat sludge at a petroleum production facility in Utah. When the soil properties are not appropriate for landfarming, the results are better with this method. Costs are in the range of $20 to $75 per cubic meter.

Anaerobic Composting

Anaerobic composting is currently being used to anaerobically digest municipal, commercial, and agricultural wastes obtained from organic recycling projects. A high-quality fertilizer is obtained after digestion. For example, at the White Street Landfill in Greensboro, North Carolina, a facility is starting up that will treat 30,000 tonnes per year of yard trimmings (Goldstein 2000a). Local food processing companies will add their high nitrogen waste. The 950,000 liter slurry preparation tank is placed on top of the 1,320,000 liter process water tank. The digesters are 176,000 m^3 in capacity each. The bacteria obtained by a proprietary process work at 35°C. The obtained compost has a high concentration of nitrogen. The biogas will be sold to a nearby industrial plant.

Anaerobic composting can be performed with up to 40% solids contents. Hydraulic retention times (HRT) are from 60 to 90 days at 55°C and 120 to 200 days at 35°C. Volatile solids can be reduced by up to 90%.

The Anaerobic Composting process has been developed by Pinnacle Biotechnologies International at the National Renewable Energy Laboratory in Golden, CO. It is applicable for organic wastes such as refuse, agricultural residues, sewage sludge, animal manure, and food processing wastes. The process is fully computer-controlled, thus enabling optimization of yields and control of emissions. A conventional screw press is used to dewater the final product. A process flow diagram is shown in Figure 6.17. The reactor is a closed unit with a low-speed agitator similar to a pug mill used in the mining industry for mixing rocks and gravel. It operates as a plug flow system at 60°C. Microorganisms are obtained from hot sources such as hot springs and thermal vents. Volume of the feedstock is reduced by 30 to 40%. Approximately 70% of the organics are converted to fuel.

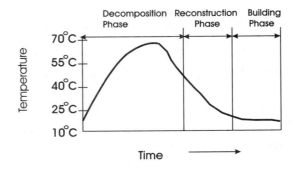

Figure 6.17 Anaerobic Composting Process Developed by Pinnacle Biotechnologies

Home Composting

To decrease the amount of waste going to landfills and to protect the environment, individual suburban homeowners have increased their interest in home compost systems. However, unless these systems are operated properly, product quality can be poor and the pile will stink, creating numerous problems with the neighbors. Telephone hotlines and volunteer organizations are available to provide expert advice on system operation.

The materials for composting must have a good composition, a minimum volume must be composted, and the conditions must be appropriate. Composting materials can include garden and vegetable cuttings, paper and cardboard, food waste, garbage and any decomposable organic matter. Human feces, plant material with pesticides, and diseased animals, however, should not be used due to health concerns. A listing of materials to include and exclude from backyard compost is shown in Table 6.4.

Even in home systems, appropriate C:N ratios must be maintained. This can be difficult and expensive for the home owner to determine; therefore, various guidelines can be used. For example, certain ratios of green garden debris to dry garden debris can be maintained. Green debris includes fresh grass, green leaves, green plant stems, and flowers whereas dry debris includes dry grass, branches, stalks, etc. Increasing the green to dry ratio decreases the amount of manure needed. Manures and green debris have high nitrogen contents (Diaz et al. 1993). One way to adjust the ratio is to form alternating green and dry brown layers of 5 to 10 cm in thickness. The dry material can also absorb moisture and maintain porosity. Dried leaves, straw, and sawdust work well for these purposes.

Table 6.4 Materials to Include or to Exclude from Backyard Compost (Martin and Gershuny 1992)

Materials to Include	Materials to Exclude
Animal manure including sheep, cow, horse and poultry	Bones
Bread	Dairy products
Coffee grounds	Diseased plants
Egg shells	Fish scraps
Fruit	Lard
Garden trimmings and clippings	Manure from humans, dogs, cats, and other non-vegetarian animals
Human and animal hair	Mayonnaise
Leaves	Meat scraps
Paper towels and napkins	Peanut butter
Shredded paper (nonrecyclable)	Salad or cooking oils or dressings
Sawdust	
Sod	
Twigs and branches	
Vegetables	
Wood chips	

To retain sufficient heat under winter conditions, the pile should have a minimum volume of 3 m³. Volumes of 1 m³ are sufficient if conditions are not as harsh. If the homeowner cannot build a large enough volume, debris from neighbours, local markets, or supermarkets can be used. Manure can also be purchased from garden centers and can serve as an absorbent and source of nitrogen. To avoid problems with odors and flies, the material should be turned occasionally.

The compost can be contained in a bin with minimal dimensions of 1 m x 1 m and 1.5 to 2 m in height (Diaz et al., 1993). The bin can be made of concrete, wood, or hardware cloth. Plastic bins or barrels of 200 L are also available. A fly-proof screen should be used to cover the bin. Usually a space of two to three times that of the bin is required in front of the bin for work space. A double bin could also be used, one for storage and one for composting. Stalks should be cut in 12.5 to 15 cm pieces to increase surface area of the material, and the pile should not be more than 2 m since higher piles can compact and decrease porosity. Lawn mowers can be used to shred leaves.

Sufficient moisture must be maintained in the compost pile and is indicated when the particle surface glistens or feels damp to the touch. If the pile is not moist enough, the pile will not heat up in an appropriate time. Too much moisture will be indicated by a bad odor or cooling of the pile. Tap water can be added, more absorbent material can be used, or turning frequency can be increased. Levels of 40 to 60% moisture are ideal.

Turning is performed by removing the front panel and taking out the material from the top. Outer particles should be moved inwards and the material should be fluffed to increase porosity. Turning once a day increases the rate of composting. Monitoring is performed with a hotbed thermometer. Approximately 30 to 40 cm inside the pile, the temperature should increase to between 43 and 49°C within a day or two and up to 54°C within the next day or two (Diaz et al. 1993). When the temperature drops back to 43°C, the process is finished. This is usually complete within six to eight weeks.

Vermicomposting

Vermicomposting is the application of worms to convert food and other organic waste into compost for use as a soil enhancer or potting soil. Bins can be made of wood or plastics (Elcock and Martens 1995). Holes are used on the bottom for drainage and aeration. Covers are required to maintain moisture and darkness for the worms. Bedding material may be composed of sawdust, shredded leaves, newspaper, peat, straw, plants, compost, and aged manure. It is here that the worms stay and that the food waste is buried. The bedding should occupy about three-fourths of the bin. *Eisenia foetida,* also called red wiggler, brandling, or manure worms, or *Lumbricus rubellis* are the two most common types of composting worms. These worms enable:

- reduction of particle size;
- elimination of old bacteria;
- stimulation of new bacterial colonies;
- excretion of nitrogen compounds, thereby enriching the compost;
- promotion of oxygen channeling;
- multiplication of carbon and nutrient exchange with bacteria and protozoa;

- regulation of pathogens; and
- production of worm casings which are beneficial for top soil.

The most appropriate ratio of worms to food is 2 kg of worms per kg of food per day. The food can include fruit and vegetable peels, broken egg shells, tea bags, and coffee grounds. Meats, dairy products, oily foods, and grains should be avoided for composting since they lead to problems with rodents, flies, and odors. Temperatures should be between 5 and 25°C. The composting process will take about two and a half months, at which time the bedding will not be visible and the contents will be brown and have an earthy texture. Dying worms should be moved to new bedding and food waste on the other side of the bin to increase their chances of survival. Control of the system must be fairly stringent. Indoor methods are also possible for apartment dwellers or those with a small yard.

Liberty Paper, a paper mill in Becker, Minnesota, participated in a pilot project involving the vermicomposting of starch biosolids (Anonymous 1998a). A module of 2.4 m x 3.0 m x 2.7 m high was used with 136 kg of *Eisenia foetida* worms commonly found in manure. These worms increased in weight to more than one tonne over the period of four months. Worms were added to the module and fed nine tonnes of material over ten days. One tonne of castings was removed and replaced by a tonne of new material every day. Separation of the worms and the castings was done after the ten days. This process was very labor intensive because it was done by a screen. Later, a blending of starch biosolids and corrugated cardboard waste was done, and the results were better. A permanent facility will be built in the future.

In Australia, a vermicomposting facility opened in March 2000 at a baked bean and soup factory (Dandenong, New South Wales). The facility includes tanks, pumps, and a worm sorter. Approximately 750 tonnes of food residuals will be processed in the first year (Anonymous 2000a). In the United States, the largest vermicomposting facility (Westley, CA) began in 1993 and is operated in windrows (Sherman-Huntoon 2000). A mechanical harvester is needed to harvest the vermicompost without the worms. Approximately 75,000 tonnes of material are processed by 220,000 kg of worms. Material includes paper pulp, tomato residuals, manure, and green waste. Beds and bins can be used by both home users and industry. A bin system was installed at the Broad River Correctional Institution in Columbia, South Carolina. About 400 to 500 kg of food waste are added in 10 to 15 cm layers. Continuous flow reactors such as the Worm Wigwam (EPM Inc. of Cottage Grove, OR) and Vermitech systems have also processed approximately 4000 kg of organics per year and 100 kg per day, respectively. Product marketing needs to be expanded to increase consumer awareness.

Operating Conditions

Temperatures for composting are between 40 and 60°C. Since low temperatures may not kill the pathogens and will increase reaction times, and too high temperatures can inhibit microbial activity, control is very important. The addition of wood chips decreases the moisture content of the sludge from between 70 and 85% to a range of 40 and 50%. Lower moisture contents increase air flow and ensure that anaerobic conditions do not result. Moisture contents should be around 55%. If the moisture contents are low, as is the case of household waste (40 to 50%), sewage sludge can be added. Oxygen levels should be above 10% (Richards 1998).

As for all microbial process, carbon to nitrogen ratios affect composting. Ratios above 30:1 inhibit composting while levels below 20:1 lead to incomplete composting. Therefore, ratios of between 20:1 and 25:1 for carbon to nitrogen are ideal. Low nitrogen levels in the compost will release nitrogen in the soil with toxic effects on the plants. The C:N ratio of dewatered sludge is 6:1 and 16:1 for digested sludge. The addition of bulking agents with a C:N ratio of 60:1 will increase the ratio. Other agents that can be added are waste paper (C:N = 170 to 300:1), grass clippings (C:N = 19:1), leaves (C:N = 40 to 80:1), manure (C:N = 25:1), sewage sludge (C:N = 15:1), sawdust (C:N = 500:1), and kitchen waste (C:N = 25:1) (Bilitewski et al. 1997).

Increased particle size can decrease the surface area available for microbial utilization. More rapid attack occurs as the surface area to volume of the particle increases. Too small particle sizes, however, can decrease porosity, making aeration more difficult. Highly rigid materials, however, can maintain the porosity even at low particle sizes. Therefore, for materials that are not very rigid, particle sizes between 1.0 to 7.5 cm should be used. Green plant material should not be less than 5 cm in diameter and not more than 12.5 cm in diameter. Generally some size reduction is required, and it is usually in the form of tumbling in a drum to break and macerate highly-organic, moist wastes

Levels of pH do not affect the composting process as much as they do anaerobic digestion. The pH will decrease initially but will increase later in the composting process. The optimal range is in the order of 6.5 to 9.5 for composting while anaerobic digestion operates mainly at 6.8 to 7.3. Other considerations include odor, pathogens, and metals. Anaerobic decomposition is usually slower, and the compost can contain undecomposed, odorous intermediates and some pathogens because of the lower temperatures.

A typical temperature curve for composting is shown in Figure 6.18. If the temperature does not increase within four days, there may be too much or too little moisture or the ratio of C:N is too high. The pH may also be too low. Once the temperature has risen, any sharp change in any parameter can indicate problems. If odors are detected, this could indicate that anaerobic conditions are developing, possibly due to high moisture contents or insufficient aeration. In the latter case, aeration rates could be increased. If the temperature declines

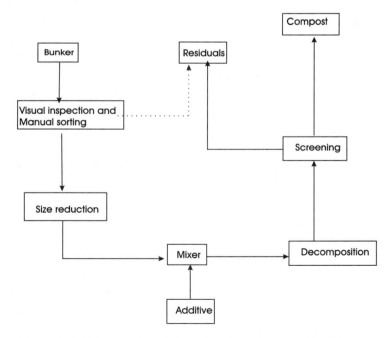

Figure 6.18 Temperature Throughout the Composting Process

with no limiting factors, the composting process is finishing. The composted material is then ready for storage. The decrease in temperature is the only reliable test for completion of the process.

Bulking Agents

Shredded tires have been used as bulking agents to provide structure and enhance porosity of the pile (Anonymous 1999). They do not decompose and can be recovered after use. Although shredded tires can be up to four times the price of wood, they can be reused many times. Wood chips can only be partially recovered, and the properties of reused wood chips are changed because the wood will have absorbed moisture in previous composting processes. Tire chips do not absorb moisture or supply carbon. If the carbon to nitrogen ratio is adequate, the tire chips can be used as the sole bulking agent. Otherwise three parts tire chips, two parts wood chips, and one part sawdust is optimal. A composting facility in Davenport, Iowa used tire chips as one-third of the bulking agent, and two-thirds were wood chips. The tire chips need to be large enough so that they can be screened out. A Rutgers study indicated that 2 cm by 2 cm up to 5 cm by 5 cm were optimal. Using these sizes, 100% of the tire chips could be recovered. One aspect that has to be monitored is the release of metals from the steel belts in the tires. There is a gradual decrease of metal release over time; however, if other metals are present, they could become a concern for the final product.

Design

A typical facility for composting includes pre-processing, degradation, and post-processing (Figure 6.19). Pre-processing removes impurities and reduces particle size by visual inspection, screening, magnetic separation, and/or manual sorting. Various types of composting units can be used, depending on the plant capacity. Odor control, loading equipment, temperature monitoring, and air distribution are required in addition to the composting vessel. Post processing further removes impurities from the final product.

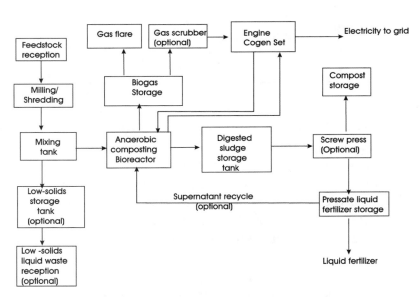

Figure 6.19 Flow Sheet of a Composting Facility

Post Treatment

Curing is done to aerobically mature the product for periods of two to eight weeks. Windrows are the most common type of curing. Trapezoid windrows are now preferred to triangular windrows because they provide faster curing and more economical use of space.

After curing, screening is performed to improved the quality of the compost. This is performed by oscillating screens, rubber mats, or trommel screens and brushes with pounding devices. Optimal screens occur when the moisture content of the compost is 30%.

Applications

The use of composting for pure sewage sludge and a mixture of household waste and sewage sludge has been well established. Recently, stabilization of municipal and industrial sewage sludge has also been demonstrated. However, composting for household waste (yard waste, fruit, vegetable, and other food waste) has recently increased substantially. The quality also is superior to MSW compost since heavy metal concentrations are lower. Heavy metal contents are shown for various types of wastes in Table 6.5.

Table 6.5 Heavy Metal Content of Various Composts Relative to Original Feedstock
(30% organic content of dry solids) (Bilitewski et al. 1997)

Element	Biosolids	Plant waste	Wet MSW compost	Mixed MSW compost
Pb	83.07	63.10	705.00	596.00
Cd	0.84	0.72	4.08	6.39
Cr	35.83	28.44	113.00	82.90
Cu	46.76	34.52	357.80	318.00
Hg	0.38	0.28	1.63	2.79
Ni	40.48	18.56	47.10	52.10
Zn	249.10	176.92	1334.00	1823.00

Alaska Earth Works is planning to construct a full-scale facility in Sitka, Alaska to compost chip woods with ground fish waste (Farrell 2000a). A pilot project in 1997 and 1998 was performed that comprised 25.8 wet tonnes of fish waste (75% coho salmon carcasses with black cod, red snapper heads, and viscera) in aerated static piles. Approximately 60% ground bark and 40% sawdust were mixed at a ratio of 2.9:1, 3.7:1, and 4.1:1 parts wood to fish in three different compost piles of 200 m^3. Mixing was performed by a front-end loader and aeration was provided by an aeration pipe. A 15-minute cycle timer was used to control the aeration blower. Exhaust air was passed through a wood biofilter to remove odors. A one-foot layer of wood debris covered the pile to limit pests, reduce odors, and retain heat. A tarp was also used to cover the piles because the area gets over 275 cm of precipitation per year. Moisture and oxygen levels, pile temperature, and odor analyses were monitored throughout the process to develop the design and cost data. Volume reductions were 58, 48 and 39%, indicating that higher levels of fish waste reduce the volume more. The finished compost will be sold as potting soil and transplant mixes in Alaska and the northwestern United States.

New systems for composting are continually being developed. For example, the Verticon is a silo reactor of approximately 12 m^3. Temperatures increase through the system until they reach higher than 70°C in the top zone. Another system is the BioMate, an enclosed, heated, and agitated fermentation tank to convert the food residuals into organic fertilizers in two days. Fermentor capacities range from 45 kg to 2.7 tonnes per day.

Costs

Costs for different composting systems are shown in Table 6.6. Lower capital costs and higher operating costs are for smaller facilities that produce only a few tonnes per year. The wide range in costs is due to the differences in facility capacities.

Table 6.6 Costs of the Various Types of Biosolids Composting (adapted from Goldstein and Block 1997)

Composting method	Operating costs ($/tonne)	Capital costs ($)
Aerated static pile	12 to 500 (dry basis)	36,000 to 20 million
In-vessel	18 to 540 (dry basis)	850,000 to 33 million
Windrow	2.15 to 245	50,000 to 8 million

In general, the development of systems has been accomplished through pilot projects funded by government grants to universities and companies. These systems are more sophisticated than they seem. Although homemade systems are available, commercial ones have been tested and are more economical because of lower production costs. Technical assistance can be supplied. Operators for the systems are necessary and must be knowledgeable in feedstock recipes, composition, and addition of amendments.

Capital costs for larger systems can be high. Depending on the sophistication and size, the cost can range from $100,000 to $500,000. Small systems such as Hot Boxes are in the order of $1,500 to $2,000 while others such as rotating drums cost from $10,000 to $20,000. In terms of operating costs, the systems are only economical if the landfill tipping fees are around $55 per tonne. Grants and the benefits of recycling increase the benefits of composting. New designs will improve costs and efficiency in the near future.

Case Studies

Sewage sludge composting is formed on mechanically dewatered sludge (Figure 6.20). Nitrogen contents of compost from sewage sludge can be lower than other types of compost; nevertheless, the resulting compost is an excellent soil conditioner, provides a long-term source of nutrients, and reduces the need for fertilizers and pesticides (Garland et al. 1995). By 1997, over 198 composting facilities were in operation in the United States (Anonymous 1998b). Dry aggregate or composted material is added to enhance aeration in the bioreactor. Retention time is 14 days. Sanitation is accomplished due to the thermophilic temperatures. Curing for four to eight weeks is then performed. Heavy metal contents determine if the compost can then be used in agricultural applications. Currently, approximately 6.8% of biosolids are composted (U.S. EPA 1999a). This could increase to 13.5% by 2010 because of decreased disposal methods (40% in 1996 to 30% in 2010).

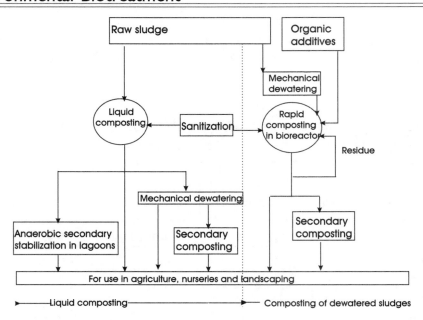

Figure 6.20 Sewage Sludge Composting

Moores Creek Advanced Wastewater Treatment Plant (Charlottesville, Virginia), run by Rivanna Water and Sewer Authority, treats 42 million liters per day of wastewater (U.S. EPA 1999a). Composting began in 1984. The plant generates 5,600 cubic meters of dewatered biosolids annually from primary, secondary, and tertiary processes. Of this amount, 4,400 cubic meters are composted and 1,200 are land-applied or used as landfill cover. It is hoped that 100% of the biosolids will be composted in the future. After filtration, the solids are anaerobically digested and dewatered by plate and filter presses before composting, Three parts wood chips (from chipped pallets) to one part biosolids are mixed and piled in static piles on an asphalt pad. The temperature during composting is maintained for five days between 55 and 77°C. For the next ten days, the temperature probes are removed and the piles are left to compost. Curing then follows for one month. Testing is performed for salmonella and heavy metals on a quarterly basis. Compost is sold once a month to the public ($16 per cubic meter for bulk weights and $20 per cubic meter for small batches).

Composting MSW is also increasing since it is more economical than incineration, and it reduces the quantity of organic waste going into landfill, decreases landfill gas, and reduces leachate concentrations. A number of pilot projects have been initiated and full-scale facilities are also running (Table 6.7). It has been estimated that approximately 35% of MSW could be used for composting (Grocery Committee 1991). Most of the uses include presorting of bulky impurities, screening, manual sorting of oversize pieces, magnetic separation, and homogenization. Presorting helps to reduce metals which are found in batteries, paints, plastics, and papers, etc. Containers, degradation drums, aerated windrows, and reactors have been used to estimate mass flows of household waste, shown in Figure 6.21 (Bilitewski et al. 1997). Odors are another concern; however, use of in-vessel systems or biofilters with aerated piles reduces odor levels considerably.

Table 6.7 A List of Full-scale MSW Composting Facilities in the United States (Anonymous 1996, 2000d)

Technology	Location	Capacity (tonne/day)	Start up
In-vessel (rotating drum)	Pinetop/Lakeside, Arizona	10	1991
Windrow/rotating drum	Sumter County, Florida	75	1988
Rotating drum/windrow	Cobb County, Georgia	300	1996
Windrow	Buena Vista County, Iowa	16	1995
Windrow	Coffeyville, Kansas	50	1989
Aerated static pile	Mackinac Island, Michigan	5	1992
In-vessel	Truman, Minnesota	50	1988
Rotating drum/windrow	Marlborough, Massachusetts	120	n/a
Rotating drum/windrow	Nantucket, Massachusetts	125	n/a
Windrow	Penington County, Minnesota	60	1991
Windrow	Filmore Country, Minnesota	11	1990
Windrow	Swift County, Minnesota	5	1987
Windrow	Lake of the Woods, Minnesota	5	1990
Rotating drum/windrow	Sevierville, Tennessee	240	1998
Rotating drum/windrow	Columbia County, Wisconsin	80	1995

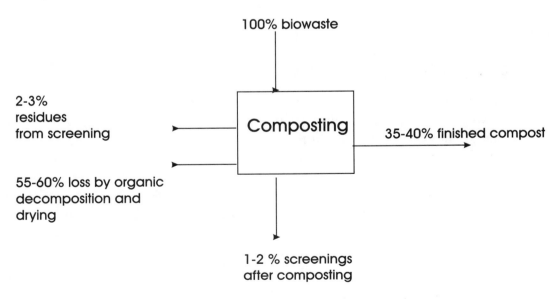

Figure 6.21 Mass Balances for MSW Composting

The largest North American co-composting facility started operation in Edmonton, Alberta, Canada in March 2000 (Goldstein 2000b). The design capacity is 180,000 tonnes of residential garbage with 22,500 tonnes of biosolids for the production of 125,000 tonnes of compost per year. The facility cost $100 million and the city of Edmonton pays the company operating the facility (TransAlta, an international electric company) $62/tonne. Rotary mixing drums with residence times of one to two days are used to combine the MSW and biosolids. A moisture content of 48%, pH of 5.5 to 6.0 and carbon to nitrogen ratios of 30:1 is obtained. Composting takes place in aeration halls in 28 days, before the screening and finishing of the final product. Fir bark and softwood biofilters are used for odor control.

Windrow and static pile composting have been used extensively in Indiana prisons (Anonymous 2000d). Forty-five hundred tonnes of food residuals, organic feedstocks, and yard trimmings were composted in 1999. All Treat Farms in Ontario, Canada, operates the largest composting facility in this province (Farrell 2000b). This facility uses windrows to process yard trimmings to produce 50,000 tonnes of compost annually. The trimmings are mixed with cow and chicken manure, seed screenings, and dirty seed from its seed factory. The materials are composted in windrows over 8.1 ha on a thick layer of gravel with limestone. The piles are turned by front loaders and bulldozers. The process takes 18 to 24 weeks, and curing takes another 12 to 18 months followed by screening with a trommel. Although low temperatures slow the process (-25°C), the windrows are still monitored and turned. Runoff is collected in five collection ponds. There have been few complaints regarding odors even though the facility is near a town of 2,000 people.

Other wastes such as the grains from a brewery have also been composted (Hundhammer 1999). Two piles were formed of 3.0 to 3.8 cubic meters. A 10 cm perforated PVC pipe was laid on the ground, hay was placed on top, and the premixed compost material was placed on top of the hay. One pile was a mixture of half grains and half horse bedding, and the other pile was half grains and half autumn leaves. Although the pile with the bedding performed the best, more experiments are needed to optimize the composting formulation.

Comparison to Other Methods

Composting is one of the simplest processes, but it can also be very sophisticated. It is mainly used for food wastes. Operating costs can range from $10 to $500 per tonne. It is, therefore, applicable for home owners, individual institutions, and even communities. It has significantly increased in popularity because of increased interest in decreasing the amount of wastes going to landfills. Moisture and oxygen levels, pile temperature, and odors must be monitored throughout the process. The carbon to nitrogen ratio is the other important factor that must be optimal to ensure the success of the process. Applications are increasing for hazardous wastes such as explosives and petroleum sludges. It can be used where soil conditions or space requirements are not appropriate for landfarming.

Composting has been used extensively in Europe—for 11% of MSW in Belgium, 16% in Spain and Portugal, and 8% in France (U.S. EPA 1995, 1996). Bioremediation can be performed on-site with little disruption, and it can also be combined with physical or chemical processes. Since waste is permanently eliminated, liability issues are also eliminated. Acceptance by the public also has increased significantly in recent years. Landfill space is running out, and no one wants to have a landfill in their backyard. There are also increasing concerns about groundwater contamination from landfill leachates. Incineration has become less favorable because of dioxin and other possible carcinogenic emissions. Anaerobic digestion also produces a highly useful product, methane. Composting methods can significantly reduce the amount of waste entering the landfill, thus extending the life of existing landfills.

Landfill Bioreactor

Box 6-4 Overview of Landfill Bioreactors

Applications	Suitable for treatment of municipal solid waste
Cost	Not fully understood. Operating costs: estimated at $9/tonne
	Capital: estimated at $5/tonne
Advantages	Increased methane production
	Increased degradation rates
	Increased landfill life
Disadvantages	Potential groundwater contamination
	Optimization of leachate recirculation and gas production required
	Lack of design criteria
Time requirements	Two to five years
Other considerations:	Reluctance by regulators to accept technology

Description

In municipal solid wastes, some wastes such as food are very biodegradable, but others such as plastics, rubber, leather, and wood are more difficult to biodegrade. Those with high lignin contents such as newsprint and cardboard have low biodegradabilities. Separation at the household level or at transfer stations can decrease the amount of waste going to landfills. Aluminum, paper, plastics, glass, and metals can be separated for recycling. Yard waste can be sent for composting and food waste can be composted, anaerobically digested, reused for animal feed, or incinerated.

There are four sequential steps in the decomposition of organic matter in traditional landfills: initial aerobic phase, first transition phase, second transition phase, and methane phase. The aerobic phase lasts only a few days to weeks until the oxygen is depleted. At the first transition phase, anaerobic conditions will begin to develop and the pH will decrease to between 4 and 6. Organic matter decomposes to organic fatty acids and then to volatile fatty acids such as acetic acid. This stage lasts from weeks to months. In the next stage, methanogenic bacteria start to grow and produce methane from acetic and formic acids. Optimal pH is between 6 and 7. This stage can take from three to five years before stability is obtained. Once stability is obtained, the last stage of methanogenesis is obtained. Methane and carbon dioxide form 45 to 65% of the gas. Gas production from MSW is in the range of 150 to 250 m^3 per tonne of wet waste.

Landfills have traditionally been operated without the addition of liquid. Increasing the moisture content of the waste, however, can increase waste degradation and methane production because most of the conversion processes are anaerobic. This is known as a landfill bioreactor, a waste treatment system increasingly used worldwide. Gas recovery can be increased up to 90% with landfill bioreactors. Leachate quality is enhanced, and wastes can be stabilized within ten years (Block 2000).

Design

At the Rabanco-owned Roosevelt Regional Landfill, adding water enhances the production of gas that is captured and used to generate electricity. By 2015, gas volumes are expected to supply 60 megawatts of electricity (Royer 1997). According to the U.S. EPA subtitle D rule, this can only be performed at landfills in the United States

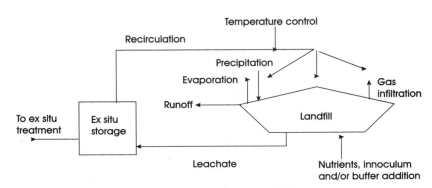

Figure 6.22 Schematic of Landfill Bioreactor

where a composite liner, such as 0.61 m of clay covered with a high density polyethylene (HDPE), is present. There are approximately ten such bioreactor projects in the U.S. Most increase the moisture content through recirculation of the leachate. Recirculation assists biodegradation by enhancing the transportation of nutrients through the waste, redistributing methane bacteria, buffering and diluting inhibitory components, and retaining the constituents of the leachate in the landfill (Figure 6.22). Recirculation can be accomplished by spray application, infiltration ponds, horizontal or vertical injection wells, or trenches. Regulations, disposal concerns, and the amount of leachate needed for each landfill compartment will determine how much leachate is allowed to accumulate. If *in situ* nitrification/denitrification are also incorporated into the landfill operation, treatment of the leachate will be minimal (Onay and Pohland 1998). Older areas in the landfill can serve as seeds for new areas. Ideally, other liquids will have to be added to increase the moisture content to 30 to 40%. A balance, however, must be made such that organic acids do not accumulate and inhibit methane production (Pohland and Kim 2000). Leachate and gas monitoring are integral to the success of the process. Disruption in the sequence of reactions that occur in the landfill (Figure 6.23) can upset the whole process. Organic waste undergoes breakdown to simpler monomers (such as sugars and amino acids) and then to alcohols, organic acids, acetic acid, and hydrogen by acid-forming bacteria. These substrates are all electron donors. For anaerobic digestion, electron acceptors include Fe(III), nitrate, or sulfate instead of oxygen for anaerobic digestion. The electron acceptors are converted to Fe(II), ammonia, and nitrogen. Hydrogen and acetic acids are also final products of the reaction and may not be converted

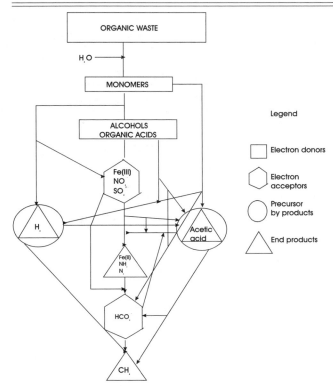

Figure 6.23 Typical Reactions at a Landfill

to methane. Ammonia produced from nitrates in the wastes and other sources usually tends to accumulate in the leachate and must be treated after discharge.

In terms of heavy metal concentration in the leachate, pH is a major factor in mobility, which increases as the pH decreases. In addition, organic and inorganic agents can serve as ligands, promoting metal transport. However, the mechanisms of precipitation, encapsulation, and sorption ensure that the heavy metals are removed in the waste. Therefore, heavy metal content must be verified in the leachate.

Solubility, volatility, hydrophobicity (K_{ow}), biodegradability, and toxicity influence the behavior of organic contaminants in the landfill. Compounds such as dibromomethane, TCE, 2- nitrophenol, nitrobenzene, pentachlorophenol (PCP), and dichlorophenol tend to be highly mobile; thus, they are found in the leachate and gas phase (Pohland and Kim 2000). They will also be biologically altered by reduction, complexation, and complete or partial degradation. Other compounds such as hexachlorobenzene, dichlorobenzene, trichlorobenzene, lindane, and dieldrin are more hydrophobic and remain in the waste for *in situ* biodegradation (Table 6.8).

Costs

Clarke (2000) compared the capital and operating costs of a landfill bioreactor with a conventional landfill with a lifetime of 20 years. Although there is uncertainty concerning the operating costs, the landfill bioreactor should have similar operating costs to the conventional landfill at $9 per tonne. Capital costs of the additional leachate distribution facilities, gas wells, and generators should add approximately $1 per tonne to a conventional landfill capital cost. A cost analysis was performed for the landfill bioreactor to estimate the economic benefit, including the sum of enhanced biogas production, saved landfill space, reduced environmental impact, and post-closure costs to implement the technology, less capital and operating costs. For a landfill life between two and five years (possible with a landfill bioreactor), the benefit compared to a conventional landfill would be $8 to $11 per tonne for an 1,800 tonne per day facility. This can be compared to anaerobic digestion (retention time of two months) which showed an $86 per tonne increase in waste treatment costs.

Table 6.8 Transport of Inorganic and Organic Components in a Landfill Bioreactor Prior to Biotransformation

Component	Transport Mechanism in Landfill
Heavy metals (Cd, Cu, Cr, Fe, Hg, Ni, Pb, Zn)	Reduction of Fe, Cr, Hg; Complexation with organic or inorganic components and mobilization; Precipitation as hydroxide (Cr) or sulfides (Cd, Cu, Fe, Hg, Ni, Pb and Zn) after sulfate reduction; Sorption and ion exchange with waste; Precipitation under alkaline conditions
Halogenated aliphatics including PCE, TCE, dibromomethane	Volatilization and mobilization in leachate due to high vapor pressure and solubility
Chlorinated benzenes such as hexachlorobenzene, trichlorobenzene and dichlorobenzene	Volatilization and sorption on waste due to low solubility and high K_{ow}
Phenols and nitroaromatics such as dichlorophenol, nitrophenol and nitrobenzene	Low volatility, vapor pressure, and K_{ow} with high solubility in leachate
PAHs and pesticides (lindane and dieldrin)	Low volatility and mobility due to low vapor pressure and high K_{ow}

(adapted from Pohland and Kim 2000)

Case Studies

At Yolo County Central Landfill near Davis, California, a 9,100 tonne bioreactor (enhanced cell) is operating alongside a conventional landfill (control cell). Each of the cells is 30 m x 30 m. The control cell has a single composite liner while the bioreactor has a double composite liner with a lower layer composed of a 0.6 m compacted clay layer, a 0.15 cm geomembrane, a drainage net, and a geotextile file. The top liner comprises 30 cm of compacted clay, a 0.15 cm geomembrane, a drainage net, and a geotextile filter (Reinhart and Townsend 1998). The leachate is distributed by a manifold at the top of the cell over 25 locations across the surface. The leachate injection pipes are buried 1.5 m below the surface of tire chips. Hydraulic head is monitored by pressure transducers. Settlement and temperature are also monitored.

The methane and gas recovery increased by a factor of ten over conventional landfill, and the stabilization time was reduced by a factor of five or more (Block 2000). For the control cell, settling was an average of 30.5 cm while for the enhanced cell, settling was an average of 104 cm from May 1996 to January 1998. Cell temperatures were also higher in the enhanced cell (43°C) compared to the control cell (35°C). During this same period, methane content was higher for the enhanced cell (57% compared to 31%) and the gas flow rate was also much higher (0.28 m³/min compared to a negligible amount for the control).

At another bioreactor site, Crow Wing Country Municipal Solid Waste Landfill (Brainerd, Minnesota), horizontal injection trenches are used to recirculate raw and pre-treated leachate (Doran 1999). Normally 5.6 to 9.5 million liters of leachate are generated annually. The leachate is collected at the base of the landfill and sent by gravity to a pump station. The leachate then goes to three aerated pretreatment ponds for primary treatment, long-term storage and settling, and secondary treatment in each of the ponds. The leachate can be applied onto the land, recirculated, or sent to a wastewater treatment plant over 250 km away. The leachate to be recirculated is sent to a series of 11 recirculation laterals that run east to west along the landfill. The laterals are made of high density polyethylene (HDPE) with a diameter of 10 cm, spaced 15.3 m apart. To avoid leachate seeping along sideslopes, the last lateral is a distance of 6.1 m away from the east and west slopes. The laterals are centered in a 0.6 x 0.6 m trench that is backfilled with 10,000 shredded waste tires. A geotextile filter and intermediate cover is used for cover. The dosage for each lateral is approximately 310 L per day per meters of trench. As the capacity of the lateral is reached, the next is opened and filled until all are filled. Velocity of the flow does not surpass 379 L/min. Approximately 26,500 L of leachate is recirculated in the landfill per day over a period of three months per year. Leachate quality and quantity, gas production, methane concentration, waste settling, and leachate head levels in the landfill are monitored. In 1998, more than a million liters of leachate were recirculated without off-site hauling. Some landfills may still need to dispose of leachate elsewhere. Since the technology is new, costs and the impact on landfill operation and closure are not fully understood. Savings can be from $96,000 to $700,000 annually. Minimal attention to leachate and gas production may be necessary during post-closure monitoring. Maintenance of a neutral pH may be difficult to maintain, increased gas production can lead to odor problems, and leachate could seep from intermediate or final cover slopes. To avoid difficulties in gas collection, the gas line should be placed above the recirculation line in the trench, and the lines should be spaced at least 3 m apart.

Waste Management designed and is operating a 600,000 m³ bioreactor landfill in Franklin, Wisconsin (Anonymous 2000c). The landfill is divided into two sections with half operating as the control and the other half as a sequential aerobic-anaerobic bioreactor. A compactor shears large objects and tears plastic bags. Leachate quality, waste mass and other characteristics, gas composition, and flow rate are monitored. High moisture contents of the waste are maintained to enhance biological activity. Aeration is also maintained by horizontal and/or vertical piping installed during construction. Various types of leachate recirculation are being tested. Methane recovery will increase and leachate quality will stabilize. Eight projects were started by the end of 2000. Other test projects have been performed in Germany, United Kingdom, Austria, Sweden, Denmark, Canada, New Zealand, and Honduras (Reinhart and Townsend 1998).

Comparison to Other Methods

The development of landfill bioreactors will continue because of their advantages over conventional landfills. Further efforts will be necessary to optimize leachate recirculation, gas generation, and removal of recalcitrant compounds. Another problem is the use of plastic bags that hides the waste from the moisture (Jones-Lee and Lee 2000). Shredding the plastic bags prior to burial of the waste should reduce this problem. Groundwater pollution could increase due to the increased hydraulic load; therefore, precautions must be taken, such as using double-composite lined landfills and following leachate recycle by clean water leaching (Jones-Lee and Lee 2000). There are still many challenges including regulator reluctance, the inability to wet the waste uniformly, and the unavailability of design criteria.

Landfarming of Sludges
Box 6-5 Overview of Landfarming

Applications	Suitable for treatment of oily and wood preservation sludges
Cost	$55 to $130/m^3 or $27 to $54 per tonne
Advantages	System simple to design and operate
	Treatment times are short
	Standard tilling equipment used
Disadvantages	Large land requirements
	Most data from petroleum industry sludges
	Leachates, volatile emissions, and dust must be monitored
Treatment time	Six to 24 months
Other considerations:	Could potentially be used for a wide variety of contaminants

Description

As described in Chapter 5, sludges can be mixed with surface soil in concentrations up to 5 to 10% in a process called landfarming. Soil containing bacteria, protozoa, fungi, mites, and earthworms serve as the inoculum for the sludge. Nutrient addition, lime, aeration, mixing, pH, and moisture adjustments may be necessary. Temperatures below 10°C can slow degradation rates significantly.

Operating Conditions

There is little operating data available, which can present problems when designing systems. In general, large rocks and debris should be removed, mixing techniques should provide at least 90% contact between soil and additives, and nutrient addition in the ratios of hydrocarbon to nitrogen to phosphorus of 100:10:1 is beneficial (Hicks 1993). Tilling provides porosity to the soils, mixes the soil and the additives, and enhances oxygen transfer. Frequency depends mainly on oxygen uptake. A redox potential of 800 mV should be maintained. Although highly permeable sludges are more easily aerated, more adsorptive soils can remove metals from the soil. The addition of bulking agents such as sand, wood chips, or sawdust (10 to 30%) can enhance remediation. The adjustment of pH (if it is not in the 6 to 8 range) can be done by adding lime, phosphoric acid, or alum. Metal buildup should be monitored to avoid toxic levels for post-closure vegetative covers. Loading rates for oily sludges are also a limiting factor.

Design

Berms and runoff channels should be incorporated to minimize runoff and washout of hazardous components. If there is potential for groundwater contamination, the operation should be performed on a liner. They are usually made of 80 mil high-density polyethylene. Greenhouses, plastic tunnels, plastic sheets, water, and planting windbreaks can also be used to control volatile emissions.

Waste application rates, frequencies, and schedules are devised based upon site and waste characteristics, feasibility studies, regulatory requirements, and climatic conditions. The best strategy to avoid toxicity to the bacteria is usually to apply low levels of wastes frequently. Sewage sludge or cow manure can be added as inoculum for substances that are difficult to degrade.

Applications

This type of treatment is often applied to petroleum sludges (diesel fuel, No. 2 and No.6 fuel oil, JP-5) and wastes containing creosote, PAHs and pentachlorophenol from the wood industry, and some pesticides (2,4-dichlorophenoxyacetic acid and 4-chloro-2-methylphenoxyacetic acid). Radioactive wastes, acids and bases, ammonia, cyanide, heavy metals, and flammable liquids are not suitable for this type of treatment. Other organic sludges can also be degraded including halogenated solvents. Because they are mobile and biodegrade slowly, leachate collection systems must be incorporated and application rates must be adjusted according to leachate concentrations. In general, more chlorinated and nitrated compounds and those of high molecular weight are difficult to degrade (Cookson 1995).

Costs

Costs for landfarming are in the range of $55 to $130/m³ ($27 to $54/tonne). This method is less expensive than slurry phase systems because cultivation equipment is often employed to a depth of 10 to 30 cm (LaGrega et al. 1994).

Comparison to Other Processes

Overall, landfarming has numerous benefits. The costs are much lower than landfilling or incinerating. Treatment times are short. Remediation periods are usually between six and 24 months. Although applications at full scale have been primarily for sludges in the petroleum industry, a wide variety of organic compounds can be degraded in this manner. The system is fairly simple to design and operate. Limitations include the following:

*   large space requirements,
*   extended remediation times,
*   nondegraded inorganic contaminants,
*   volatile solvents pretreatment,
*   characterizations of the land (topography, erosion, climate, stratigraphy, and permeability), and
*   exclusion of petroleum sludges.

Leachates and contaminants in the soil must be monitored after degradation. Landfarming seasons in colder climates can be short (seven to nine months), whereas in warmer climates, landfarming can be performed throughout the year. Initial heavy metal concentrations should be less than 2,500 mg/kg and hydrocarbons concentrations less than 50,000 mg/kg. Systems should be incorporated if mobile, halogenated solvents are to be treated. Cold climates can slow degradation rates significantly, and air emissions and leachates should be monitored.

Slurry-phase Treatment

Box 6-6 Overview of Slurry Reactors

Applications	Suitable for treatment of oil and wood preservation sludges
Cost	$130 to $200/tonne
Advantages	Accelerated biodegradation rates compared to land treatment
	Control of emissions and leachates possible
	Also applicable to metal-contaminated wastes
Disadvantages	High capital and operating costs for highly sophisticated systems
	Limited mainly to petroleum and wood preservation wastes
Detention time	Days to months
Other considerations:	Can be operated automatically or manually

Description

In slurry phase treatments, wastes are mixed with water to breakdown particle sizes, enhance contact between microorganisms and the contaminants, desorb the contaminants from the solids, and enhance aeration of the slurry. Slurry-phase treatment is faster than land application and less land is required.

Operation and Design

Process design includes pretreatment, determination of the optimal solids concentration, mixer design, and retention time. Pretreatments include particle size fractionation and/ or surfactant addition. Particle size reduction to 30 microns instead of 60 microns was found to significantly enhance biodegradation rates. The addition of surfactants can enhance contaminant desorption. Solids concentrations can be as high as 50%, but 30 to 40% is often toxic to the microorganisms. For mixing, turbine mixers or surface aerators can be used. Baffles can also be added to enhance mixing and reduce settling. Reactors can be open lagoons or closed reactors.

Applications

Full-scale systems have been performed for wood preserving wastes. Approximately 90 m³ of sludge was treated on a sequenced batch basis. Concentrations of PCP were reduced to 32 mg/L from 2,600 mg/L, and PAHs were reduced to 86 mg/L from 1,200 mg/L (La Grega et al. 1994). Oily sludges degrade at rates twice that of wood preserving wastes. Although application has been successfully applied for many years, slurry phase processes are better for controlling emissions and are faster. Degradation can occur four to 15 times faster in slurry reactors than by land treatment. Up to 95% of total PAHs can be reduced in two months.

Biohydrometallurgical Processes

Bacterial leaching of metals from mining ores, also known as bioleaching, is a full-scale process that can be performed by slurry reactors or heap leaching (Figure 6.24). Mining wastes include low-grade ores, mine tailings, and sediments from lagoons or abandoned sites. Low pH values lead to solubilization of the metals in the mining ores. Elemental sulfur or ferrous iron may be added as bacterial substrates. Reactors such as Pachuca tanks, rolling reactors, or in-propeller vessels have been used (Tyagi et al 1991). Heap leaching is more common since it allows the large volume wastes to be treated in place (Boon 2000). To enhance this process, aeration can be forced through

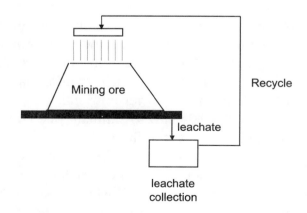

Figure 6.24 Heap Leaching of Mining Residues

the pile or hydrophilic sulfur compounds can be added (Tichý 2000). *Thiobacilli* bacteria are responsible for the oxidation of inorganic sulfur compounds. Applications include metal dissolution in low-grade sulfide ores, generation of acidic ferric sulfate leachate for hydrometallurgical purposes, and extraction of gold by oxidation of pyrite through bacterial sulfide production. The extraction of metals from low-grade metals ores and refractory gold ores is a multibillion dollar business worldwide (Rawlings 1997). Bacterial solubilization by oxidation of the sulfide minerals, pyrite, and arsenopyrite enhances gold extraction by the traditional method of cyanidation. The solubilization mechanisms have been debated extensively, however.

Biohydrometallurgical processes are highly efficient and cause fewer environmental problems then chemical methods (Torma and Bosecker 1982). For slurry processes, the oxidation rate per reactor volume, pH, temperature, particle size, bacterial strain, slurry density, and ferric and ferrous iron concentrations need to be optimized. Bioleaching is very effective for the recovery of gold from refractory gold pyrite and copper from chalcopyrite. Mulligan and Galvez-Cloutier (2000) have also shown that copper can be recovered from oxidized mineral ores by the production of organic acids by the fungus *Aspergillus niger*.

Sewage sludge, soils, and sediments have also been treated by bioleaching (Tichý 2000). Sewage sludge and other organic wastes contain metals. As these wastes are anaerobically digested, the metals can be more easily desorbed as a result of changes during the anaerobic digestion pro-cess. Bioleaching can then be used to remove the metals from the sludge (Tyagi et al. 1993). Ferrous iron is added as a substrate which is subsequently oxidized (Figure 6.25). Acidification follows, which solubilizes the metals. A completely stirred or airlift reactor can be used. The sludge can then be disposed, the processed water is neutralized, and metals are removed by precipitation.

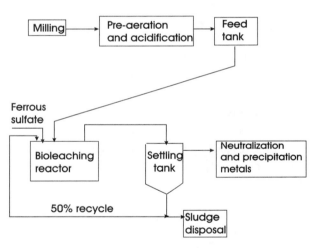

Figure 6.25 Bioleaching in a Reactor Using Ferrous Iron as the Substrate

Pintail Systems, Inc., (Aurora, Colorado) has developed a biological process (U.S. EPA 1999b) for the removal of cyanide in a heap leach process (Figure 6.26). Native microorganisms are extracted from the ore and tested for cyanide detoxification potential. Those identified are kept for bioaugmenta-tion purposes. The bacteria are then grown in spent ore infusion broths to adapt them to field conditions. The cyanide is the carbon and/or nitrogen source for the bacteria. Other nutrients are added to the broth. Tests are performed in a pilot column (15 cm x 305 cm) to simulate pile conditions and to determine detoxification rates, process parameters, and effluent characteristics. Once the test is complete, sufficient quantities of bacteria cultures are applied to the heap. The concentrations of cyanide and metals in the leachate are then determined. Metals are measured since they are also biomineralized during the leaching process. This process can also be used for spent ore heaps, waste rock dumps, mine tailings, and processed water from silver and gold mining operations. The technology has been evaluated in the SITE Demonstration Program. Two full-scale cyanide detoxification projects have been completed by Pintail Systems.

An anaerobic treatment process was developed by Geo-Microbial Technologies (Ochelata, OK) called anaerobic metals release (AMR). This acidification can contaminate streams and lakes. In contrast, the AMR technology uses *Thiobacillus* with denitrifying culture at neutral pH instead of aerobic acidophilic bacteria that form acids and solubilize metal sulfides. The anaerobic conditions are controlled by the nitrate levels in the leaching solutions. This allows solubilization of the metals. The nitrate levels are adjusted to ensure that there are no nitrates in the leachate. The metals are removed from the leachate by standard methods

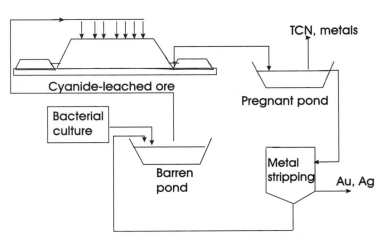

Figure 6.26 Pintail Bioleaching Process for Metal and Sulfur Removal from Anaerobically Digested Sludge (adapted from Couillard and Mercier 1994)

and the effluent is recycled. Levels of sulfide reducing bacteria and sulfides are kept to a minimum. The technology was demonstrated in 1994 in the SITE Emerging Technology Program.

Costs

Costs can range from $130 to $200/m³. If volatile compounds are present, off-gases may need to be further treated. For the Biolysis™ system, costs are in the order of $75/tonne, including $5/tonne for energy and $1/tonne for chemicals.

Case Studies

The Canton, Mississippi Superfund site was the location of a full-scale creosote waste treatment by batch airlift bioreactors from 1991–1994 (FRTR 1997). Four slurry phase bioreactors of 11.6 m diameter and 7.3 m high were used with a blower for aeration and impellers for mixing and maintaining the slurry in suspension (10% solids). Sixty batches of 140 to 160 m³ of material were treated. They were operated batchwise. Total PAH concentrations decreased from 8,545 to 634 mg/kg (93% average removal rate). The 2- to 4-ring PAHs were removed more efficiently than the larger 5- to 6-ring PAHs. Problems were encountered with foam production. Desanding had to be incorporated as a pretreatment. Treatment costs were approximately $150/tonne.

An example of a lagoon containing sludge was a Superfund site at French Ltd. (Crosby, Texas). It consisted of a 3.4 ha industrial waste management site lagoon containing liquid, sludge, subsoil, and groundwater zones (Figure 6.27). The contaminants included benzene, vinyl chloride, and benzo(a)pyrene with up to 400 mg/kg of VOCs, 750 mg/kg

PCP, 5,000 mg/kg of semivolatiles, up to 5,000 mg/kg of metals, and 616 mg/kg PCBs. The quantity of soil and sludge was approximately 300,000 tonnes with soil varying from fine grain to coarse sand and tar-like sludges (petrochemical sludges, kiln dust and styrene, and tar oils). A Mix flo system was used between January 1992 and November 1993 to minimize gaseous emissions while supplying oxygen. Pure oxygen and eductors

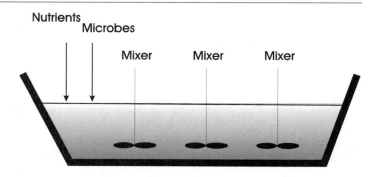

Figure 6.27 Lagoon Treatment of Oily Sludges

in a series were employed to maintain oxygen levels of 2.0 mg/L. The objectives were benzo(a)pyrene, 9 mg/kg; total PCBs, 23 mg/kg; vinyl chloride 43 mg/kg; and benzene 14 mg/kg. These objectives were met within 11 months for both pits (one of subsoil and the other of tarry sludge). Air emissions were not surpassed. This was the first application of a slurry-phase bioremediation at a Superfund site.

A system has been developed by Sanexen Environmental Services (Longueuil, Quebec, Canada) called Biolysis™ (Sanexen 2000). The reactors can treat up to 4 m³/day with 65 to 98% reductions in heavy hydrocarbons. The reactors are transportable with dimensions of 3 m in diameter and 8 m in height. Treatment can also be performed at the site in an existing basin as shown in Figure 6.28. Retention times are from five to ten weeks. The minimum volume treated in this manner is 1,000 tonnes. The hydrocarbons

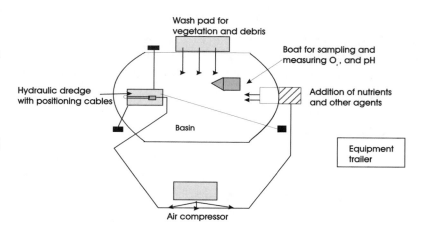

Figure 6.28 Biolysis System Developed by Sanexen for Oily Sludge Treatment

are degraded in three steps. The first stage consists of mechanical agitation, and application of enzymes and surfactants to enhance biodegradability. The latter agents are the unique aspects of the system. Then degradation occurs and the microorganisms cause uptake of the hydrocarbons The steps are simultaneous in the reactor system and sequential in the ponds. The heat generated by the process allows it to take place through all seasons. The process can be operated automatically or manually by a part- or full-time employee. Approximately 7,000 tonnes of material have been treated including oily sludges from refineries and petrochemical plants.

Comparison to Other Methods

Slurry phase systems in the form of reactors or lagoons incorporate continuous mixing, thus decreasing the degradation times and reducing land requirements. Emissions and leachates can be easily controlled. Operating costs are approximately twice those of landfarming. With the appropriate conditions, bacterial leaching of mining ores, tailings, and other metal-containing residues such as sewage sludge can be performed in slurry reactors. All biological processes are compared for their costs and treatment times in Table 6.9.

Table 6.9 Summary of Biological Processes

Process	Applicability	Cost ranking	Treatment time
Anaerobic digestion	Industrial, agricultural and municipal sludges	++	++
Aerobic digestion	Industrial, agricultural and municipal sludges from small- to medium-sized plants	++	++
Composting	Food and yard wastes, municipal and sewage sludge	++	+++
Landfill bioreactor	Municipal solid waste	+	+++
Landfarming	Oily and wood preservation sludges	+	+++
Slurry reactors	Oily and wood preservation sludges	++	++
Fermentation	Agricultural, forest, yard, food, paper wastes	++	++

Cost ranking: + denotes $10 to $50/tonne, ++ $50 to $200/tonne, and +++ greater than $200 per tonne
Time ranking: + denotes several hours, ++ 1–20 days, +++ more than 20 days

Conversion of Wastes Via Fermentation

Box 6-7 Overview of Fermentation

Applications	Suitable for treatment of cellulosic wastes including food, paper, agricultural and yard wastes
Cost	Dependent on the product and waste used
Advantages	Variety of products such as ethanol, animal feed, and fertilizer
	High temperatures used in fermentation capable of pasteurizing sludges
Disadvantages	Low operating costs and space requirements
	Cellulose must be converted to sugars
	Heat required for many processes
Detention time	Hours to days
Other considerations:	Few installations available, success depends on product market

Description

Ethanol production

Agricultural wastes, forest thinnings, wood residuals, yard debris, and low-grade paper can be used as feedstocks for the production of ethanol. These wastes contain high contents of cellulose that can be broken down into sugars by chemicals or enzymes. Fermentation is then performed to produce ethanol. Several plants have already been designed for ethanol production. They include a rice straw to ethanol and acetic acid plant by Arkenol (Sacramento, California), a bagasse to ethanol plant by BC International (Jennings, Louisiana), and a materials recovery and municipal waste to ethanol plant by Masada Resource Group (Middletown, New York) (Martin 2000). The development of an integrated approach for energy, waste, and resource issue is the ultimate goal, and biotechnology will greatly assist in reduction of environmental pollutants.

At the Middletown facility, the process uses food, paper products and biosolids for ethanol production (Gray 1999). The cellulose is hydrolyzed into glucose and other simple sugars that are fermented into ethanol. The alcohol is then distilled to market-grade ethanol. The acid is recovered for reuse, and the lignin is used for energy. Carbon dioxide is sold after recovery. The stillage from ethanol production is anaerobically treated. Methane is recovered for energy use. The capacity of the plant is 36 million liters of ethanol per year from 230,000 tonnes per year of MSW and 73,000 dry tonnes per year of biosolids. The capital cost is $105 million ($0.95 per tonne per day) with an anticipated payback of 20 years (Waste 2001).

BC International Corporation (BCI) will construct a facility that will utilize 306,000 tonnes per year sugar cane waste and rice hulls for the production of ethanol by genetically engineered bacteria. It will be open in Louisiana in February 2002. BCI will develop, own, and operate another facility in Gridley, California, for the production of 76 million liters/year of ethanol from rice hulls. They have also made an agreement with Collins Company, a timber firm, to produce 3.8 million liter/year of ethanol and to co-generate electricity from forest thinnings (Anonymous 2000b).

Other Products

Thermo Tech Technologies (B.C., Canada) developed the Thermo Master™ Process that converts organic waste into animal feed and fertilizer at high temperatures (Figure 6.29). Bakery, fruit and vegetable, food processing, dairy products, brewery, supermarket, and restaurant wastes are converted to feed supplements while animal manures, slaughterhouse wastes, and wastewater treatment sludges are converted to fertilizers. The waste is fed into the fermentor as slurries for the thermophilic aerobic bacteria. Temperatures of 70°C are reached by the exothermic reactions and maintained by external heat exchange. The heat pasteurizes the product. At the end of

Figure 6.29 Thermo Master Process for Feed Production from Organic Wastes

the fermentation, the product is centrifuged, dried, and pelleted. The product can be sold for $200 to $220/tonne (Siriu 2001). Ninety-nine percent of odors are destroyed by a regenerative thermal oxidizer. Plants located in Ontario, Canada and New York have been designed to handle from 200 to 1,200 tonnes of organic waste per day. A new facility is planned for Detroit, Michigan.

Nu Environ Technology has a process for the conversion of animal manure and other organic materials into fertilizers or animals feeds (Nu Environ Tech Inc. 2001). The process includes the receiving the organic waste at the plant and depositing it into a tank or pit. Here a herbal odor eliminator is sprayed to reduce odors. Wastes include animal manure, food and market waste, seafood waste, chicken and animal intestines, animal hair and blood, sewage sludge, chicken feathers, dead animals, and other organic wastes. The waste is then sent to a storage tank where an agent called SRM-2 Organic Fermentation Accelerator and Expander is introduced and mixed with the waste to enhance fermentation. The waste is then sent by conveyor to a Double Disk Fermentor and Dryer (Figure 6.30) where it remains for an hour. The material is then sent to the dryer where it remains for 30 minutes and then to a screen and electromagnetic separator for the removal of plastic, metals, and other debris. The product is next sent to a hopper for packaging. The equipment includes collecting tanks, input storage tanks, vacuum fermentation and drying machines, continuous drying and cooling machines, finished product hoppers and bagging equipment. The antipollution equipment includes a wastewater treatment facility, boiler and condenser, scrubber, and dust collector.

Each module can handle 90 tonnes per day of organic waste and will take up 1,900 m² of floor space. The process has been fully developed and is ready for commercialization.

Figure 6.30 Double Disk Fermentor and Dryer of Nu Environ Technology (courtesy of Nu Environ Technology)

Kankyo Systems Engineering Co. Ltd. of Japan has developed a vacuum fermentation unit called the Jey-System. A wide variety of wastes can be used. Some examples include cow or pig waste, raw garbage, bean curd waste, alcohol production waste, human manure waste, and fishery waste. Between 300 kg and 50 tonnes per day of organic waste can be taken into the system, regardless of the moisture content and shape of the material particles. Feeds with moisture contents up to 80% can be accommodated. The waste materials are sent to the fermentor with a residence time of 3 h. The fermentor is kept under vacuum at 50 to 70°C to easily reduce the moisture content and shape of the material. Oxygen is supplied to the microorganisms from the waste material by the force of the vacuum. The final product is a dried powder as a result of the high evaporation rate in the fermentor. Volume is reduced by 30 to 40%. One-stage fermentation is sufficient for some applications such as feed material or bed materials for raising cows, pigs, or chickens. The lower temperatures used to not destroy the nutritive value of the feed. A second fermentation step is used if the final product is to be used as a fertilizer. Temperatures of 55 to 60°C are used for this purpose. The system is simple with low operating costs, small space requirements, and a closed unit so there are not offensive odors. Fish bones can be taken into the fermentor due to the powerful agitator.

Agricultural wastes, wood residuals, yard debris, low-grade paper, and forest wastes can be used as feedstocks for ethanol production. The economic key to the process is the conversion of cellulose to sugars by enzymes or chemicals. In some cases, municipal wastes have been converted to ethanol. Other processes have also been developed for the conversion of wastes to fertilizer or animal feed. Their success will depend on the market for the products and the quality of the final product.

References

Adams, C. E., W. W. Eckenfelder, Jr., and W. Stein. 1974. Modifications to aerobic digester design. *Water Research*, 8(4):213–218.

Anonymous. 1996. Composting the next step. *Waste Age,* 27(3):47–58.

———. 1998a. Paper mill ventures in vermicomposting. *Biocycle,* 39:52.

———. 1998b. Biosolids management update. *Biocycle,* 39(1).

———. 1999. Reader's Q & A. *Biocycle.* 40(3):25.

———. 2000a. Vermicomposting at a food processing factory (Dandenong, New South Wales). *Biocycle,* 41(6):69.

———. 2000b. Waste-to-ethanol plant gets bond approval. *Biocycle,* 41(4):103.

———. 2000c. Testing landfill bioreactor technology. *Biocycle,* 41(5):24–25.

———. 2000d. Solid waste composting trends in the U.S. *Biocycle,* 41(11):31–38.

Bilitewski, B., G. Härdtle, and K. Marek. 1997. *Waste Management.* Berlin: Springer-Verlag.

Block, D. 2000. Reducing greenhouse gases at landfills. *Biocycle,* 41(4):40–46.

Boon, M. 2000. Bioleaching of sulfide minerals. Environmental Technologies to Treat Sulfur Pollution. In *Principles and Engineering,* eds. P. Lens and L. H. Pol. London: IWA Publishing. 105–130.

Clarke, W. P. 2000. Cost-benefit analysis of introducing technology to rapidly degrade municipal solid waste. *Waste Management and Research.* 18:510–524.

Cookson, J. T. 1995. *Bioremediation Engineering Design and Application.* New York: McGraw-Hill.

De Baere, L. 1999. Anaerobic digestion of solid waste: State of the art. *Proceedings of the Second International Symposium on Anaerobic Digestion of Solid Wastes,* Barcelona, Spain. 15–18 June. In *Graphiques 92,* eds. J. Mata-Alvarez, A. Tilche, and F. Cecchi. 1:290–299.

De Baere, L., and J. Boelens. 1999. The treatment of grey and mixed solid waste by means of anaerobic digestion: future development. *Proceedings of the Second International Symposium on Anaerobic Digestion of Solid Wastes,* Barcelona, Spain. 15–18 June. In *Graphiques 92,* eds. J. Mata-Alvarez, A. Tilche, and F. Cecchi. 1:302–305.

De Baere, L., and W. Verstraete. 1985. High rate dry anaerobic composting process for the organic fraction of solid wastes. *Biotechnology and Bioengineering,* 15:321–330.

Diaz, L. F., G. M. Savage, L. L. Eggerth, and C. G. Golueke. 1993. *Composting and Recycling Municipal Solid Waste.* Boca Raton: Lewis Publishers.

Doran, F. 1999. Is it possible there's a way to use landfill leachate to save money? Some researchers and landfill operators think recirculation is just the ticket. *Waste Age,* 30(4):74–79.

Droste, R. L., and W. A. Sanchez. 1986. Modeling active mass in aerobic sludge digestion. *Biotechnology and Bioengineering,* 28(11):1699–1706.

EC. 1993. The state of the environment in the european community. EC Official Publication. Brussels: EC.

Elcock, G., and J. Martens. 1995. Composting with red wiggler worms. *City Farmer.* Canada's Office of Urban Agriculture.

Farrell, M. 2000a. Composting fish and wood residuals in Alaska. *Biocycle,* 41(5):81–83.

———. 2000b. Processing with the end market in mind. *Biocycle,* 41(5):39–40.

Feinbaum, R. 2000. Pioneering in anaerobic digestion. *Biocycle,* 41(5):78.

Federal Remediation Technologies Roundtable (FRTR). 1997. Slurry Phase Biological Treatment. October. Available online: http://www.frtr.gov.

Foess, G. W., and D. Federicks. 1995. Comparison of Class A and Class B private biosolids stabilization technologies. *Florida Water Resources Journal.* May. 28–32.

Garland, G. A., T. A. Gist, and R. E. Green. 1995. The compost story: From soil enrichment to pollution remediation. *Biocycle,* 36(10):53.

Gloyna, E. F., and W. W. Eckenfelder, Jr., eds. 1968. *Advances in Water Quality.* Austin, Tex: University of Texas Press.

Goldstein, J. 2000a. Anaerobic digestion advances. *Biocycle,* 41(2):30–32.

———. 2000b. Edmonton gives a new dimension to MSW composting. *Biocycle,* 41(11):26–29.

Goldstein, N. 2000. Anaerobic digester on line to process MSW in Ontario. *Biocycle,* 41(11):30.

Goldstein, N., and D. Block. 1997. Biosolids composting holds its own. *Biocycle,* 38(12):64.

Gray, K. 1999. MSW and biosolids become feedstocks for ethanol. *Biocycle,* 40(8):37–38.

Grocery Committee on Solid Waste of the Food Marketing Institute. 1991. *Composting Task Force Report.*

Hammers, F., Y. Kalogo, and W. Verstraete. 2000. Anaerobic digestion technologies for closing the domestic water, carbon and nutrient cycles. *Water Science Technology,* 41:203–211.

Hicks, R. J. 1993. Aboveground bioremediation: Practical approaches and field experiences. In *Proceedings of Applied Bioremediation.* Fairfield, New Jersey. 25–26 October.

Hundhammer, M. 1999. Setting up a composting trial. Have a residual that you'd like to compost? A brewery in Maine provides an example of the steps needed to test the waters. *Biocycle,* 40(9):56–57.

Imhoff, K., W. J. Mueller, and D. K. B. Thistlethwaythe. 1971. *Disposal of Sewage and Other Waterborn Wastes.* Ann Arbor, Mich: Ann Arbor Science Publishers.

International Bio-Recovery Corporation. (IBR). 2001. *From waste to resource... Closing the loop.* IBR Technology Brochure. Downloadable. Available online http://www.ibrcorp.com/technology.html.

Jewell, W. J. 1999. Latest progress in anaerobic digestion. *Biocycle,* 40(8):64–55.

Jones-Lee, A., and G. F. Lee. 2000. Appropriate use of MSW leachate recycling in municipal solid waste landfilling. *Proceedings of the Air & Waste Association Conference and Exhibition.* Salt Lake City, Utah. 18–22 June.

Kiely, G. 1997. *Environmental Engineering.* London: McGraw-Hill.

Kubler, H., K. Hoppenheidt, P. Hirsch, A. Kottmair, R. Nimmrichter, H. Nordsieck, W. Muche, and M. Swerev. 2000. Full-scale co-digestion of organic waste. *Water Science and Technology,* 41:195–202.

LaGrega, M. D., P. L. Buckingham, and J. C. Evans. 1994. *Hazardous Waste Management.* New York: McGraw-Hill.

Martin, D., and G. Gershuny, eds. 1992. *The Rodale Book of Composting.* Emamaus, PA: Rodale Press.

Martin, K. 2000. Why solid waste managers should look at the ethanol option. *Biocycle,* 41(2):35–36.

Mata-Alvarez, J., W. Verstraete, J. Van Lier, and F. Pohland. 1999. Digesting one million tonnes of organic residuals. *Biocycle,* 40(12):68–69.

McCarty, P. L. 1968. Anaerobic treatment of soluble wastes. In *Advances in Water Quality Improvement,* eds. E. F. Gloyna and W. W. Eckenfelder, Jr. Austin, TX: University of Texas Press.

Metcalf and Eddy. 1991. *Wastewater Engineering: Treatment, Disposal and Reuse, 3rd ed.* T. Tchobanaglous and F. Burton, eds. New York: McGraw-Hill.

Michel, F. 1999. Managing compost pile to maximize natural aeration. *Biocycle,* 40(3):56–58.

Mulligan, C. N., and R. Galvez-Cloutier. 2000. Bioleaching of copper mining residues by *Aspergillus niger*. *Water Science and Technology*, 41(12):255–262.

Nelson, C., and J. Lamb. 2000. Digestors bring energy and fertilizer to dairy farms. *Biocycle*, 41(5):76–79.

Nu Environ Tech Inc. 2001. An 'in vessel process,' very environment-friendly. Available online http://www.nuenviron.com/technology.htm.

Oleszkiewicz, J. 2001. Biosolids Management. *CSCE National Lecture Tour*. Montreal, Canada. 27 April.

Onay, T. T., and F. G. Pohland. 1998. *In situ* nitrogen management in controlled bioreactor landfill. *Water Science and Technology*, 32(5):1383–1392.

Pellerin, C. 1994. Alternatives to incineration: There's more than one way to remediate. *Environmental Health Perspectives*, 12(10):1–11.

Pohland, F. G., and J. C. Kim. 2000 Microbially mediated attenuation potential of landfill bioreactor systems. *Water Science and Technology*, 41(3):247–254.

Praxair. 2000. Praxair I-SO system for industrial applications. Downloadable brochure. Available online http://www.praxair.com/praxair.nsf/

Rawlings, E., ed. 1997. *Biomining: Theory, Microbes and Industrial Processes*. Georgetown, TX: Landes and Springer-Verlag.

Reinhart, D. R., and T. G. Townsend. 1998. *Landfill Bioreactor Design and Operation*. Boca Raton: Lewis Publishers.

Reynolds, T. D., and P. A. Richards. 1996. *Unit Operations and Processes in Environmental Engineering, 2nd ed*. Boston: PWS Publishing Company.

Richards, T. 1998. *Cornell Composting, Resources: Operator's Fact Sheets*. Dept. of Agricultural and Biological Engineering, Cornell University. May.

Riggle, D. 1998. Acceptance improves for large-scale anaerobic digestion. *Biocycle*, 39(6):51–55.

Royer, B. 1997. Landfill bioreactor electrifies the northwest. *Waste Age*, 28:12.

Rynk, R. 2000a. Contained composting systems review. *Biocycle*, 41(3):30–36.

———. 2000b. Large-scale contained composting systems. *Biocycle*, 41(4):67–72.

Saint-Joly, C., S. Desbois, and J-P. Lotti. 2000. Determinant impact of waste collection and composition on anaerobic digestion performance: Industrial waste. *Water Science and Technology*, 41:291–297.

Sanexen Environmental Services. 2000. Biotreatment of oily sludges (Biolysis). Enviro-Access Technological Fact Sheet, F5-05-96.

Sherman-Huntoon, R. 2000. Latest developments in mid- to large-scale vermicomposting. *Biocycle,* 41(11):51–54.

Siriu, B. 2001. Heating Up Food Waste. *Waste Age,* 32(2):16.

Stover, E. L. 2000. Thermophilic treatment process is an effective alternative to mesophilic treatment for high-strength industrial residuals. *Industrial Wastewater,* 8:31–34.

Tammemagi, H. 1999. *The Waste Crisis: Landfills, Incinerators, and the Search for a Sustainable Future.* New York: Oxford University Press.

Tchobanoglous, G., and E. Schroeder. 1987. *Water Quality.* Reading, MA: Addison Wesley.

Ten Brummeler, E. 2000. Full scale experience with the BIOCEL process. *Water Science and Technology,* 41(3):299–304.

Tichý, R. 2000. Treatment of solid materials containing inorganic sulfur compounds. In *Environmental Technologies to Treat Sulfur Pollution: Principles and Engineering,* eds. P. Lens and L. H. Pol. London: IWA Publishing. 329–354.

Tilche, A., and F. Malaspina. 1998. Biogas production in Europe. Paper presented at the *10ᵗʰ European Conference Biomass for Energy and Industry.* Warburg, Germany. 8–11 June.

Torma, A. E., and K. Bosecker. 1982. Bacterial leaching. *Progress in Industrial Microbiology,* 16: 77–118.

Tyagi, R., D. Couillard, and F. T. Than. 1991. Comparative study of bacterial leaching of metal from sewage sludge in continuous stirred tank and air-lift reactors. *Process Biochemistry,* 26:47–54.

Tyagi, R. D., T. R. Sreekrishnan, P. G. C. Campbell, and J. F. Blais. 1993. Kinetics of heavy metal bioleaching from sewage sludge: Mathematical model. *Water Research,* 27:1653–1661.

U.S. Environmental Protection Agency (U.S. EPA). 1995–96. Office of Solid Waste. Available online: www.epa.gov/epaoswer/osw/tsd.htm.

———. 1999a. *Biosolids generation, use, and disposal in the United States.* United States Environmental Protection Agency, Solid Waste and Emergency Response (5306W), EPA530-R-99-009. September.

———. 1999b. *Superfund technology evaluation program: Technology profiles, 10ᵗʰ ed.,* Volume 1. Demonstration Program, EPA/540/R-99/500a. February.

———. 2000. EPA Reachit database. Available online: http://epareachit.org.

US Filter/Jet Tech. 2001. ATAD process treats municipal wastewater facility. Case Study. Downloadable brochure. Available online http://www.usfilter.com.

Waste treatment plant will yield ethanol and other products. 2001. *Chemical Engineering*. March. 19.

Water Environment Federation (WEF). 1995. Wastewater Residuals Stabilization. MOP FD-9, Manual of Practice (MOP). Water Environment Federation.

Weiland, P. 2000. Anaerobic waste digestion in Germany: Status and recent development. *Proceedings of the 4th International Symposium on Biotechnology*, S. Hartmans and P. Lens, eds. Noordwijkerhout, The Netherlands. 10–12 April.

Wheatley, A. 1991. *Anaerobic digestion: A waste treatment technology*, Essex, United Kingdom: SCI and Elsevier Applied Science.

Glossary of Terms and Abbreviations

A

Absorption: Entrapment of a component within the volume of another component.

ABF: Activated biofilter.

ABNR: Advanced biological nutrient removal.

Acclimation period: Period of adjustment of a microbial cell or culture to a particular carbon source, substrate, or environmental condition after inoculation.

Acid: Hydrogen-containing substance that donates hydrogen ions upon disassociation.

Acid mine drainage: Seepage of acid solutions from mining wastes that result from the interaction of sulfide minerals and water.

Acidogenesis: Conversion of fatty acids, amino acids, and monosaccharides to lower intermediate compounds.

Activated sludge: Active biological solids consisting of bacteria, protozoa, and other microorganisms in a suspended floc in an aerated wastewater treatment tank.

Adsorption: Sorption of a substance from water or air phases onto the surface of a solid.

Aerated lagoon: A wastewater treatment pond with mechanical or diffused air aeration but without sludge return.

Aerobic: In the presence of free oxygen.

Aerobic bacteria: Bacteria that require oxygen for respiration and metabolism.

Aerobic digestion: Use of aerobic microbes to digest suspended organic matter.

Air sparging: Pumping of air into the groundwater.

ALD: Anoxic limestone drain.

Algae: Single or multicellular microorganisms capable of photosynthesis.

Alkalinity: Presence of bicarbonate, carbonate, and hydroxide ions in water which are capable of neutralizing acids. Expressed as mg /L or meq/L of $CaCO_3$.

Ammonia: Nitrogen in the form of NH_3 or NH_4^+.

Anaerobic: In the absence of free oxygen.

Anaerobic bacteria: Bacteria that require oxygen in the form of SO_4^{2-}, NO_3^-, PO_4^{3-}, or other similar forms for respiration.

Anaerobic digestion: Digestion of suspended organic matter by anaerobic bacteria.

Anion: Negatively charged ion.

Anoxic: Lack of free oxygen.

Anthropogenic: Contaminant or material resulting from human activity such as spills or discharges.

Aquaclude: Geologic forms that transmit water with difficulty.

Aquifer: Geologic formation of porous rock or sand saturated with water in the subsurface that transmits water easily for extraction by wells.

ATAD: Autogenous thermophilic aerobic digestion.

ATAD™:   Autothermal Thermophilic Aerobic Digestion system.

ATP: Adenosine triphosphate.

Attenuation: Reduction of contaminants over time and distance via physical, chemical, and biological means.

B

Bacilli: Rod-shaped or cylindrical bacterial cells.

Bacteria: Unicellular, microscopic organisms belonging to prokaryotes that use organic or inorganic matter and that are incapable of photosynthesis. They can live alone, in pairs, or in chains.

Bacteriophage: Virus that infects bacteria.

Base: Substance that produces one or more hydroxyl ions upon disassociation.

BATS: BioTrol aqueous treatment system.

Batch reactor: Reactor into which all reactants are added, the reaction takes place, and the products are then removed.

Bioaccumulation: The accumulation of contaminants in the fatty tissue of an organism.

Bioaugmentation: Enhancement of specific bacterial cultures to biodegrade specific contaminants.

Bioavailable: Available for biodegradation by microorganisms.

Biochemical oxygen demand (BOD): The amount of oxygen required by microorganisms to degrade organic waste. Used as a measurement of wastewater strength.

Biodegradation: Biological breakdown of organic contaminants by microbial action to simple, less toxic molecules.

Biofilm: Living microorganisms, polysaccharides, and other materials in a thin film.

Biofilter: Device containing filter media used for immobilizing bacteria that digest contaminants captured from the air or water that passes through the media.

Biomass: Quantity of biological microbial cells in a population in water, sediment, or soil.

Biopile: Piling of contaminated soil in heaps with an air distribution system.

Bioreactor: Reactor for the promotion of biological reactions.

Bioremediation: Cleanup of contaminated sites by microorganisms.

Bioscrubber: Unit consisting of a spray or trickling tower that contacts and removes contaminants from air by dissolving and biodegradation.

Bioslurry reactor: Tanks where microorganisms, wastes, and nutrients are added to water to obtain a suspension that is mixed.

Biosolids: Organic solids from a wastewater treatment process.

Biosorption: Uptake of an element or compound by biological cell membranes.

Biosurfactants: Surfactants produced by bacteria, yeast, or fungi.

Biotrickling filter: Unit consisting of a packed material that supports microorganisms to enable the biodegradation of pollutants from air passing through the material.

Biotransformation: Conversion of a compound by biological means.

Bioventing: Stimulation of biological activity by controlled aeration of an *in situ* contaminated site.

BMP: Biochemical methane production potential.

BMSR: Bacteria Metal Slurry Reactor.

BNR: Biological nitrogen removal.

BOD: Biological oxygen demand. Measure of the amount of oxygen required to biodegrade organic material. The standard test is performed for five days at 20°C.

BTA: BioTechnische Abfalverweertung GmbH.

BTEX: Benzene, toluene, ethylbenzene, and xylene, major components in gasoline.

Bulking agent: Material such as straw or woodchips added to another material to increase porosity, reduce compaction, and increase air flow.

C

Carbohydrates: Organic compounds containing carbon, hydrogen, and oxygen such as glucose and other sugars, starch, and cellulose.

Carcinogen: Agent capable of causing cancer.

Catalytic oxidation: Oxidation at lower temperatures than thermal oxidation due to a catalyst such as platinum or palladium.

CEC: Cation Exchange Capacity.

CERCLA: Comprehensive Environmental Response Compensation and Liability Act.

CFU: Colony Forming Units.

Chelant: An organic chemical that forms organo-metallic complexes with metal cations.

Chemical Oxygen Demand (COD): Amount of oxygen needed to chemically oxidize organic matter. Expressed as mg/L and used to measure wastewater strength.

Clarification: Settling of solids by gravity at the bottom of a clarifier to obtain a clear liquid.

Cocci: Spherical-shaped bacterial cells.

COD: Chemical oxygen demand. Amount of oxygen needed to chemically oxidize organic matter. Expressed as mg/L and used to measure wastewater strength.

Coliform bacteria: Bacterial group that live in human and other mammal intestines, water, wastewater, and in the soil. Includes aerobic and facultative anaerobic bacteria, gram-negative, non-sporeforming bacilli that ferment lactose to gas. Used to indicate contamination.

Co-metabolism: Degradation of a compound without gaining energy, particularly for chlorinated compounds, while another compound is used for carbon or energy.

Completely mixed system: Reactor with immediate dispersion of all entering elements within the reactor volume.

Compost: Stabilized organic residues after biological decomposition. Used as soil additive.

Composting: Stabilization of organic wastes by biological means under controlled conditions. Wastes such as straw and hay (bulking agents) are added to increase porosity of the pile.

Consortium: Group of microorganisms that work together to degrade a contaminant.

Constructed wetland: Use of plants for biological wastewater treatment.

Contact stabilization: An activated sludge process including a contact tank for sorption of organic material and a sludge stabilization for biooxidation of sorbed organic material.

Contaminated site: Place where hazardous waste has been dumped.

Continuous flow reactor: Reactor with continuous streams of influent reactants and effluent products.

Conventional digester: A low-rate anaerobic digester.

CSTRs: Continuous flow stirred tank reactors.

Culture: Microbial growth developed through the addition of nutrients and the appropriate environment.

D

DBR: Ethylene dibromide.

DCE: Dichloroethylene.

Denitrification: Conversion of nitrate to nitric oxide, nitrous oxide, and nitrogen gas by anoxic cell growth.

Detention time: Average time that water or sewage remains in a tank or basin.

Digested sludge: Sludge digested aerobically or anaerobically that is stable due to the lower volatile content.

Digester: Tank used for sludge digestion.

Digestion: Biological oxidation of organic matter to stabilize sludge.

Dispersed plug-flow reactor: Rectangular reactor with longitudinal mixing over its length.

Dissolved oxygen: Oxygen dissolved in water, expressed as mg/L or percent saturation.

Dissolved solids: Solid matter dissolved in water that passes through a 1.2 micron filter.

DMAc: N,N-dimethylacetamide.

DO: Dissolved oxygen. Oxygen dissolved in water, expressed as mg/L or percent saturation.

Doubling time: Time required for microbial cells to double their numbers.

Domestic wastewater: Wastewater generated by homes, businesses, institutions, etc.

Downflow: Water flowing downward through a column by gravity.

DNAPL: Dense non-aqueous phase liquid.

DTPA: Diethylenetriamine pentaacetic acid.

E

EDTA: Ethylenediaminetetraacetic acid

Effluent: Wastewater or flow from a reactor, tank, process, industrial plant, etc.

EGSB: Expanded granular sludge bed.

Electron acceptor: Compound, also known as an oxidizing agent, that accepts electrons in an oxidation-reduction reaction.

Electron donor: Compound, also known as a reducing agent, that donates electrons in an oxidation-reduction reaction.

Endogenous decay: Continuous process of microbial cell decay.

Environment: Air, water, or land surroundings.

Enzyme: Substance that catalyses reactions within living cells.

Eutrophication: Increase in nutrient (nitrogen and phosphorus) concentrations in an aquatic ecosystem that causes high levels of plant growth and sediment accumulation.

Ex situ: Requiring the removal or excavation of soil or groundwater from its original place.

Extended aeration activated sludge process: Activated sludge process with sufficient detention to allow endogenous decay.

Extracellular: On the outside of the cell.

F

Facultative bacteria: Bacteria that can use molecular oxygen if available or combined oxygen if not.

Fermentation: Biochemical utilization by bacteria or yeast of an organic compound as an electron donor and acceptor.

Fill and draw reactor: Batch operated activated sludge process in a single reactor.

Five-day biochemical oxygen demand (BOD$_5$): Oxygen required for stabilization of organic matter over a five day period at 20°C. Expressed in mg/L.

Fixed bed: A bed that is stationary in a column.

Flowsheet: Representation in a diagram of unit operations and processes.

Flushing: Pumping of water through an *in situ* contaminated site to remove the contaminants from the soil.

Fluidized bed: Particles in a bed suspended by upward flow of water.

F/M: Food to microorganism ratio; ratio of BOD organic load to amount of activated sludge (microorganisms) in system.

Fresh sludge: Undigested organic sludge.

Fresh wastewater: Raw, untreated wastewater.

FRP: Fiber reinforced plastic.

Fungi: Multicellular, heterotrophic, non-photosynthetic microorganisms in the eukaryote group and which have a cell wall; feed on organic matter.

G

GAC: Granular activated carbon, frequently used for adsorption of compounds for air or water.

GASS: Granular aerobic sludge system.

Garbage: Food wastes from homes, restaurants, and food processing facilities.

GC-MS: Gas chromatography-mass spectrometry.

Groundwater: Subsurface water that is used for drinking water, is under a greater pressure than atmospheric pressure, and moves due to gravity.

H

HAF: Hyundai anaerobic filter reactor.

Half-life: Time required for a compound to degrade to one-half its initial amount. Often used for radioactivity or biodegradation.

Hazardous waste: Waste that can lead to sickness, injury, death and/or environmental damage.

HDPE: High density polyethylene.

Heap leaching: *Ex situ* technology for the removal of metals from soil or residues by passing an extraction fluid through the pile and then collecting the fluid to extract the metals.

Heavy metals: High molecular weight metals with atomic numbers higher than 38 such as lead, silver, mercury, and cadmium but often includes those with atomic numbers greater than 20 including zinc, copper, nickel, cobalt, chromium, etc.

HEDTA: -N(2-Hydroxyethyl) ethylenediamine tetraacetic acid.

Henry coefficient: Indicator of the volatility of a compound. The ratio of the vapor pressure of the contaminant to its solubility in water.

Herbicide: Agent for the control of plant growth.

Heterotrophs: Bacteria that feed on organic matter.

High-rate digester: Digester with continuous feed, mixing, and heating.

HRT: Hydraulic retention time or hydraulic detention time. Defined as reactor volume divided by wastewater flow rate or average time that the water remains in the reactor.

Humus: Product of garbage or compost.

Hydraulic conductivity: Water flow rate per cross-sectional area. Common units are m^3/m^2-day or m/day.

Hydraulic gradient: Change in head (water pressure) per unit length in the direction of flow.

Hydraulic head: Fluid potential which drives fluid flow from places of high hydraulic head to where hydraulic head is lower. Consists of the total of elevation head, pressure head, and velocity head.

Hydraulic loading rate: Amount of wastewater treated per reactor volume or area per time.

Hydrolysis: Breaking of organic bonds by reaction with water.

Hydrophilic: Capable of dissolving in water since there is an affinity for water. Also called water loving.

Hydrophobic: Low solubility in water due to an aversion to water. Also called water hating.

I

Incineration: Conversion at high temperatures of organic sludges or wastes to carbon dioxide, water, ash, and other stable forms for disposal and volume reduction.

Indigenous microorganisms: Microorganisms living at the site.

Industrial wastewater: Effluent wastewater from industrial plants.

Influent: Water or other streams flowing into a plant, reactor, or process.

Injection well: Wells used to pump water, nutrients, or other solutions into the ground.

Inoculate: To add living, active microbial cells into a medium.

Inorganic: Containing no carbon such as metals, ions such as phosphate, sulfate, etc.

Insecticide: Agent for the killing and control of insects.

In situ: Without movement or excavation of soil; in the place where it is found.

Ion: Charged atom, molecule, or radical.

Ion exchange: Process of reversibly exchanging ions between liquid and solid media.

L

Lag phase: Initial phase of a slow bacterial growth.

Lagoon: Pond for the biological stabilization of organic waste.

Landfarming: Spreading of excavated contaminated soil for treatment on a liner or pad to collect leachate. Also called land treatment.

Landfill: Site where unwanted waste is placed.

Leachate: Liquid percolating through solid material that extracts or dissolves materials.

Lignin: Major complex, aromatic polymeric component in wood.

LTU: Land Treatment Unit.

Low-rate conventional system: Anaerobic digestion with intermittent mixing and feeding.

LNAPL: Light non-aqueous phase liquid.

M

Mass transfer: Transfer of a substance from one phase to another.

MAT: Müll und Abralltechnik GmbH.

Mean cell residence time (θ_c): The average time that a microbial cell remains in an activated sludge or sludge digestion reactor. Determined by dividing the mass of cells by the rate of cell wastage.

Media: Packing material such as stone in a trickling filter, fluidized bed, or biofilter.

Mesophilic: Anaerobic digestion at a temperature of 20 to 45°C .

Metabolism: Biochemical reactions to produce energy necessary for cell life.

Methane: Gaseous product of anaerobic digestion.

Methanogen: Strict anaerobes that convert carbon material to methane.

Methanotroph: Bacteria that can use methane as a carbon source.

Microorganism: Organisms visible by optical microscope only. Also known as microbe.

Microtox®:   A method used to evaluate microbial toxicity.

Milliequivalent (meq): The weight in milligrams of a substance that combines with or displaces 1 mg of hydrogen. Calculated by dividing the formula weight by its valence.

Mineralization: Complete biodegradation of an organic molecule into carbon dioxide, water, ions, and cellular material.

Mixed liquor: The mixture of wastewater and activated sludge in a reactor or aeration basin.

MLSS: Mixed liquor suspended solids. Suspended microbiological and inert solids in the mixed liquor (usually 1,500 to 4,000 mg/L).

MLVSS: Mixed liquor volatile suspended solids. The volatile fraction of the suspended solids in the mixed liquor. Often used to represent active biological solids.

Moisture content: The amount of water in a sludge or soil. Often expressed as weight percentage.

Monod equation: Equation describing the rate of growth that depends on substrate concentration.

MTBE: Methyl tert butyl ether.

Municipal solid waste: Waste generated by residential and commercial sources.

Municipal wastewater: Wastewater generated in dwellings, businesses, institutions, etc.

N

NAPL: Nonaqueous phase liquid.

Nitrification: Conversion of ammonium ions into nitrite and then nitrate.

Nitrobacter: Bacteria that oxidizes ammonia to nitrates.

Nitrosomonas: Bacteria that oxidizes ammonia to nitrites.

Nutrient: Mineral substance required for life.

Nutrient removal: Tertiary treatment to remove ammonia, nitrates, phosphates, and sulfates not removed in secondary treatment processes.

O

OCDD: Octachlorodibenzo-p-dioxin.

Octanol water partition coefficient (K_{ow}): ratio of organic chemical in water compared to water.

Organic matter: Substances of animal or vegetable nature consisting of carbon, hydrogen, oxygen, and other elements.

Organotroph: Organism that utilizes organic materials for energy during respiration.

Oxidation: Conversion of an organic compound to carbon dioxide and water in the presence of oxygen.

Oxidation ditch: Aeration basin in the form of a race-track for extended aeration processes.

Oxidation pond: Pond for the biological oxidation of organic matter in wastewater.

Ozone: Oxygen in the form of three atoms of oxygen (O_3).

P

PAH: Polycyclic aromatic hydrocarbon or polynuclear aromatic compound. Persistent long chain compounds.

Particulate matter: Small particles in a gas or liquid stream of organic or inorganic origin.

Partition coefficient: Ratio of concentration of a component in two adjacent phases at equilibrium.

Pathogen: Disease causing micro organism.

PCB: Polychlorinated biphenyl. Chlorinated isomeric compounds used for dielectric fluid, heat transfer fluid, and turbine lube; banned in 1979.

PCP: Pentachlorophenol.

PCE: Perchloroethylene.

Permeability: Rate of fluid flow through porous media such as soil. Also known as hydraulic conductivity.

Pesticide: Agent to control or kill insects, weeds, algae, rodents, and other pests.

pH: The reciprocal of the logarithm of the hydrogen ion concentration in moles per liter.

PHB: Poly-ß-hydroxybutyrate

Photoautotroph: Organisms that use light for energy.

Phytoremediation: Use of photosynthetic plants to remediate sites.

Pilot test/pilot scale: Field process or operation at a larger scale than laboratory but not full scale.

PNA: Polynuclear aromatics hydrocarbons, also known as PAHs.

Polishing ponds: Ponds used as a finishing treatment after other treatment processes to meet or exceed regulatory limits.

Pollution: Condition caused by the presence of harmful or objectionable matter in air, water or soil.

Pore volume: Volume of water occupying the pore space of soil or other geologic material.

Porosity: Percentage of pore space within soil.

POTW: Publicly owned treatment works for wastewater treatment.

ppb: Parts per billion. In water, equal to mg/m^3.

ppb$_v$:   Parts per billion on a volume basis; usually for the gaseous components.

ppm: Parts per million. Also equal to mg/L in water.

ppm$_v$:   Parts per million on a volume basis; usually for the gaseous components.

Preliminary or pretreatment: Treatment to prepare the wastewater for further treatment, including screening, grit removal, and communition.

Primary treatment: Removal of setteable suspended solids by sedimentation.

Prokaryote: Cell with genetic material not enclosed in a nuclear membrane.

Protozoa: Single-celled microorganisms including ciliates, amoebas, and flagellants that consume bacteria and algae.

Psychrophilic: Adapted to low temperatures (0 to 20°C).

Pump and treat: Removal of groundwater to treat on the surface followed by reinjection or disposal.

Pure culture: Microbial culture comprised of a single species.

Pure oxygen activated sludge process: Activated sludge process using pure oxygen instead of atmospheric oxygen.

R

RAS: Returned activated sludge.

Raw sludge: Untreated sludge.

Raw wastewater: Untreated wastewater.

RCRA: Resource Conservation and Recovery Act which controls hazardous waste.

Recalcitrant or refractory: Organic material that is not easily biodegraded.

Recycle rate: Ratio of recycled flow to influent flow in an activated sludge or trickling filter process.

Refuse: Solids waste from a community for disposal that includes garbage, rubbish, and trash.

Residence time: Time required for a fluid (liquid or gas) to pass through a vessel.

Respiration: Energy for growth and metabolism.

Returned sludge: Sludge that has been settled and mixed with raw or primary influent wastewater.

RNA: Ribonucleic acid.

Rotary biological contactor (RBC): A biological reactor consisting of partially submerged circular discs that contact and treat wastewater during rotation. Oxygen is obtained from the air.

Rotifer: Microscopic, multicellular invertebrate animal that feeds on organic matter.

Runoff: Precipitation that does not leach through the soil or waste and runs over land to streams, rivers, and lakes.

S

SBR: Sequencing batch reactor.

Secondary sludge: Sludge from the final clarifier in wastewater treatment process.

Secondary treatment: Treatment of wastewater to remove organics (usually biologic) after primary treatments.

Sedimentation: Removal of suspended solids by gravity. Also known as settling or clarification.

Septic tank: Underground concrete or steel tank for primary settling, anaerobic digestion, and storage of sanitary sewage.

Sequencing batch reactor (SBR): Timed batch reactor for the biological treatment of wastewater.

Sewage: Wastewater containing sanitary or industrial wastes from domestic, commercial, or industrial sources.

Sloughing: Breaking off of the slime layer from a trickling filter due to lack of food or anaerobic conditions.

Sludge: Solid accumulation from wastewater treatment or industrial processes.

Sludge age: Mean cell residence time.

Sludge digestion: Biological stabilization of sludge for volume reduction and pathogen destruction.

Slurry: Suspension of solids in a liquid.

Slurry-phase bioremediation: Combination of water, contaminated solids, and additives mixed in a bioreactor to enhance microbial activity.

Solid-phase bioremediation: Above the ground treatment of soil with collection systems to prevent escape of contaminants.

Solid waste: Unwanted waste of solid form from human, animal, industrial, agricultural, or commercial activity. Also called refuse.

SRB: Sulfur-reducing bacteria.

Stabilization pond: Oxidation pond without additional aeration.

Static pile composting: Compost in piles with aeration by blowers or vacuum pumps.

Substrate: Food for microorganisms that serves as carbon, energy, or nutrient source.

Superfund: Money fund from the U.S. Federal Government to pay for the remediation of abandoned hazardous waste dump sites.

Supernatant liquor: Liquid from aerobic or anaerobic digestion.

Surfactant: Surface-active compound that increases solubility of organics in water due to their hydrophobic and hydrophilic portions. Also known as detergents or soaps.

Suspended solids: Solids suspension in water that can be removed by filtration.

SVE: Soil vapor extraction. Blowing of air through contaminated, unsaturated soil to remove the contaminants.

SVI: Sludge volume index.

T

TCA: Tetrachloroethane.

TCE: Trichloroethylene.

Tertiary treatment: Physical, chemical, or biological treatment after secondary treatment. Also called advanced wastewater treatment.

Thermal oxidation: Oxidation at high temperature with no catalyst.

Thermophilic: Anaerobic digestion at a temperature between 45 and 75°C .

Thickened sludge: Sludge with partial water removal to obtain four percent solids.

TKN: Total Kjeldahl nitrogen; ammonia nitrogen plus inorganic and organic nitrogen.

TMB: Trimethylbenzene.

TOC: Total organic carbon.

TPH: Total petroleum hydrocarbons. Hydrocarbons measured by a flame ionization detector.

TS: Total solids. Suspended plus dissolved solids in water; usually expressed as mg/L.

TSS: Total suspended solids.

Treatment: Technique, process, or method to change the properties of a waste so that it is less hazardous or toxic.

Trickling filter: Biological reactor consisting of a bed of material (rock or plastic) with water sprayed over it that is allowed to trickle through the bed covered with microorganisms growing as a slime.

Turbidity: Suspended material in water that scatters or absorbs light; used to indicate the presence of colloidal material.

U

Upflow: Water flowing upward through a column by pressure.

UASB: Upflow anaerobic sludge blanket.

UST: Underground storage tank.

V

Vadose zone: Zone of unsaturated soil above the water table. Soil pores are not completely saturated with water.

VC: Vinyl chloride.

Virus: Smallest living form capable of causing disease in humans, animals, and plants.

VIP: Virginia Initiative Process.

VISITT: Vendor information system for innovative technologies. Now replaced by EPA Reachit.

Vitrification: Conversion of waste at high temperatures to glassy, impermeable state.

Volatile: Can be converted to vapor easily.

VFA: Volatile fatty acids. Water soluble fatty acids with no more than six carbon atoms.

VOC: Volatile organic compounds. Organic compounds with high vapor pressure that evaporate readily. Henry's law coefficient is greater than 0.01.

Volatile solids: Solids lost at temperatures above 550°C .

Volumetric loading rate: Volumetric air flow rate through a bed such as a biofilter; often expressed as cubic metres of air per second per cubic metre of bed volume.

VSS: Volatile suspended solids; suspended solids lost at temperature above 550°C; indication of biosolids content.

W

Waste activated sludge: Sludge removed from the activated sludge system equal to the rate of sludge production.

Wastewater: Used water from domestic, institutional, and industrial sources.

Water quality: Physical, chemical, and biological characteristics of water.

Water table: Water level in the subsurface where the pore water pressure equals the atmospheric pressure, and soil pores are completely saturated with water.

Water treatment: Treatment of water for a specific use.

Wetland: Flooded area of land that forms a transition from land to aquatic systems and supports vegetation suited to saturated conditions.

Windrow: Form of composting where the waste is laid out in long heaps in a triangular cross section and turned regularly by tractors or other similar types of equipment.

WWTP: Wastewater treatment plant.

X

Xenobiotic: Man-made chemical that does not occur naturally. Biodegradation is difficult since natural degradation pathways are not developed.

Introduction

Based on information obtained from the Internet, industry articles, current buyer's guides, advertisements, and company literature, vendors were selected and classified according to the technology supplied. If technologies for the treatment of more than one type of media is supplied by the company, then the company is listed in all relevant categories. For selection of the companies, special emphasis is placed on those that have developed technologies discussed in this book. Others were selected based on their presence in more than one listing. This was to ensure that the company supplied at least one technology based on biological treatment, and to determine the category that the company should be placed.

Some of the listings that have been consulted to compile the following vendor lists include buyer's guides in magazines such as Hazardous Materials Management and Biocycle. Internet resources for company listings include EPA Reachit (www.epareachit.org), Water Online (www.wateronline.com), Waterlink (www.waterlink.com), Solid Waste Online (www.solidwasteonline.com), Enviro Access (www.enviroaccess.ca/eng/index.html), Green Pages (www.eco-web.com), and Industry Canada's Canadian Environmental Solutions (strategis.ic.gc.ca/SSG/es00001e.html).

Air Treatment

Air Science Technologies
1751 Richardson St. Suite 3525
Montreal, Quebec
Canada H3K 1G6
Tel: (514) 937-4614
Fax:(514) 937-4820
Email:102374.3051@compuserve.com

Ambient Engineering Inc.
P.O. Box 279
Rocky Hill, New Jersey 08553-0279
Tel: (609) 279-6888
Fax: (609) 279-9444
Email: sales@chemtainer.com
Web site: www.chemtainer.com

Ambio Biofiltration
2983 Baseline Road
St. Pascal Baylon, Ontario
Canada K0A 3N0
Tel: (613) 488-2743
Fax:(613) 488-3333

Ametek Roton Biofiltration
North Street
Saugerties, New York 12477
Tel: (914) 246-3711
Fax: (914) 246-3802
Email: info@biocube.com
Web site: www.biocube.com

B.B. Environmental Inc.
704 Mara St., Suite 201
Port Edward, Ontario
Canada N7V 1X4
Tel: (519) 337-0228
Fax:(519) 337-9178
Email: bbmccrie@ebtech.net

BBK Bioaircleas A/S
Linnerupvej 5
Hjortsvang
DK-7160 Torring
Denmark
Tel: +45 7567 6066
Fax: +45 7567 6580
Internet: www.bbk.dk

Biocube, Inc.
100 Rawson Rd. Suite 230
Victor, New York 14564
Tel: (716) 924-2220
Fax: (716) 924-8280
Email: info@biocube.com
Web site: www.biocube.com

Biofiltec GmbH
Rudolf-Diesel-Strasse 1
D-37308 Heilbad
Heiligenstadt
Germany
Tel: + 49 (0) 360655-25-0
Fax: + 49 (0) 360655-25-20

Biogenie Inc.
350 Franquet Street
Sainte-Foy, Quebec
Canada G1P 4P3
Tel: (418) 653-4422
Fax: (418) 653-3583
Email: info@biogenie.com
Internet: www.biogenie-env.com

Bio Reaction Industries Inc.
9673 S.W. Tualatin-Sherwood Rd.
Tualatin, Oregon 97062
Tel: (503) 691-2100
Fax: (503) 691-8051

Biorem Technologies Inc.
7496 Wellington Road 34
R.R. #3
Guelph, Ontario
Canada N1H 6H9
Tel: (519) 767-9100
Fax: (519) 767-1824
Email: biorem@intrnear.com
Web site: www.bioremtechnologies.com

Biothane Corp.
2500 Broadway, Drawer 5
Camden, New Jersey 08104
Tel: (609) 541-3500
Fax: (609) 541-3366
Email: sales@biothane.com
Web site: www.biothane.com

Biothane Systems International BV
Postbus 5068 GB
Tanthofdreef 21
2600 GG Deft
The Netherlands
Tel: +31 15 27001 11
Fax: +31 15 25609 27
Email: 101672.2231@compuserve.com
Internet: www.biothane.com

H. Brechbühl AG
Tel: 41 33 437 65 65
Fax: 41 33 437 65 26
Internet: www.brechbuehl.ch

Bohn Biofilter Corp.
P.O. Box 44235
Tucson, Arizona 85733
Tel: (520) 624-4664
Fax: (520) 621-1647
Email: hbohn@ag.ariz.edu

Braintech Planung und Bau von Industrieanlagen GmbH
Kaiserstrasse 100
D-52134 Herzogenrath
Germany
Tel: + 49 (0) 2407-96389
Fax: + 49 (0) 2407-80120

Camp Dresser & McKee
Raritan Plaza I, Raritan Center
Edison, New Jersey 08818
Tel: (732) 225-7000
Fax: (732) 225-7851
Email: webmaster@cdm.com
Web site: www.cdm.com

Ceilcote Luftreinhaltung Air Cure GmbH
Brunnenweg 1, D-64584 Biebesheim
Germany
Tel: (44) 6258 991-0
Fax: (44) 6258 6079

Centre de recherche industrielle du Quebec (CRIQ)
Parc technologique du Quebec
333, rue Franquet
Sainte-Foy G1P 4C7
Canada
Email: info@criq.qc.ca
Internet: www.criq.qc.ca

CH2M Hill
C:N Composting Systems
P.O. Box 91500
Bellevue, WA 98009-2050
Tel: (425) 453-5000
Fax (425) 462-5957
Email: CNcompostsystems@ch2m.com
Web site: www.ch2m.com

ClairTech BV (CT)
Postbus 65
Stationsweig Oost 279
3931 ER Woudenberg
The Netherlands
Tel: +31 33 22267376
Fax: +31 33 2865736
Email: cvanlith@pi.net

CMS Group Inc.
185 Snow Boulevard, Suite 200
Concord, Ontario
Canada L4K 4N9
Tel: (905) 660-7580
Fax:(905) 660-0242
Email: avaidila@rotordisk.com
Internet: www.rotordisk.com

Corain Impianti Engineering & Contracting
S.r.l.
Via Balso degli Ubaldi
250-00167 Rome
Italy

Dessau-Soprin, Inc.
1441 Rene-Levesque Blvd. W, Suite 500
Montreal, Quebec
Canada H3G 1T7
Tel: (514) 281-1010
Fax: (514) 875-2666
Email: jean-pierre.pelletier@dessausoprin.com
Internet: www.dessausoprin.com

EG&G Biofiltration
North St.
Saugerties, New York 12477
Tel: (914) 246-3711
Fax: (914) 246-3802
Email: apa-v@egginc.com
Web ste: www.egginc.com

Envirogen Inc.
Princeton Research Center
4100 Quakerbridge Road
Lawrenceville, New Jersey 08648
Tel: (609) 936-9300
Fax: (609) 936-9221
Email: fucich@envirogen.com
Internet: www.envirogen.com

ETA Process Plant Ltd.
The Levels, Brereton
Rugeley, Staffs, England WS15 1RD
Tel: +44 1889-5764501
Fax: +44 1889-579856

Europe CVT Bioway
Posbus 166
6740 AD Lunteren
The Netherlands

Integrated Explorations Inc.
67 Watson Rd., Unit 1
Guelph, Ontario
Canada N1H 6H8
Tel: (519) 822-2608
Fax: (519) 822-3076
Email: ieinc@istar.ca

Isolation Air Systems, Inc.
P.O. Box 99
Minoa, New York 13116
Tel: (315) 656-3884
Fax: (315) 656-8078

Keramchemie GmbH
Postfach 1163
Berggarten 1
D-56425 Siershahn
Germany
Tel: + 49 (0) 2623 600-0
Fax: + 49 (0) 2623 600-513
Email: info@kch.de
Web site: www.sglcarbon.com/cp/index.html

Kessler + Luch GmbH
Postfach 10 05 54
Rathenaustrasse 8
D-35394 Giessen
Germany
Tel: +49 641 707 00
Fax: +49 641 707 316

Kruger A/S
Gladsaxevej, 363
DK-2860 Soborg
Denmark
Tel: +45 39 690222
Fax: +45 39 690806

Geoenergy International Corp.
7617 South 180th Street
Kent, Washington 98032
Tel: (425) 251-0407
Fax: (425) 251-0414

H. Seus GmbH & Ci,
Systemtechnik KG
Banter Weg 13
D-2940 Wilhelshaven
Germany
Tel: +49 21-2008-0

MBI International
3900 Collins Road
Lansing, Michigan 48910
Tel: (517) 337-3181
Fax: (517) 337-2122
Email: serverin@mbi.org
Internet: www.mbi.org

Media and Process Technology Inc.
1155 William Pitt Way
Pittsburgh, Pennsylvania 15238
Tel: (412) 826-3716

Monsanto Enviro-Chem Systems Inc.
P.O. Box 14547
14522 South Outer 40 Road
St. Louis, Missouri 63178-4547
Tel: (314) 275-5700
Fax: (314) 275-5967
Email: enviroch@monsanto.com
Internet: www.enviro-chem.com

Paques ADI Inc.
389 Main St.
Salem, New Hampshire 03079
Tel: (603) 893-2134
Fax: (603) 898-3991
Email: sdigroup@adi.ca
Web site: www.adi.ca/sys

PPC Biofilter
3000 E. Marshall. Ave.
Longview, Texas 75601
Tel: (903) 758-3395
Fax: (903) 758-6487
Email: info@ppcbio.com
Internet: www.ppcbio.com

Remediation Technologies, Inc.
1011 S.W. Klickitat Way, Suite 207
Seattle, Washington 98134
Tel: (206) 624-9349
Fax: (206) 624-2839

SNC Research Corp.
455 Rene-Levesque W.
Montreal, Quebec
Canada H2Z 1Z3
Tel: (514) 866-6635
Fax: (514) 866-0600
Email: Catherine.Mulligan@snclavalin.com
Internet: www.snc-lavalin.com

SRE Inc.
510 Franklin Ave.,
Nutley, New Jersey 07110
Tel: (973) 661-5192
Fax: (973) 661-3713
Email: sreinc.erols.com
Web site: www.srebiotech.com

Sulzer Chemtech Ltd.
P.O. Box 65
8404 Winterthur
Switzerland
Tel: +41 (0)52 262-11-22
Fax: +41 (0)52 262-0051
Email: peter.huber@sulzer.ch
Web site: www.sulzer.ch

Thiopaq Sulfur Systems BV.
T. De Boerstraat 13
P.O. Box 53
8560 AB Balk
The Netherlands
Tel: +31 514-60 8500
Fax: +31 514-60 8666
Internet: www.paques.nl

U.S. Filter Corp.
40-004 Cook Street
Palm Desert, CA 92211
Tel: (760) 340-0098
Fax: (760) 341-9368
Internet: www.usfilter.com

VTT Chemical Technologies
P.O. Box 1403
FIN 02044 VTT
Finland

Weststates Biocarb
1501 East Woodfield Road Suit
Schamburg, Illinois 60173-5417
Tel: (708) 706-6900
Zander Umwelt GmbH
Nordring 69
90409 Nuremberg
Germany
Tel: + 49 911-3503-0
Fax: + 49 911-3503-100

Soil and Groundwater Treatment

2 The 4 Technology Solutions
2621 Cutler Ave., N.E.
Albuquerque, New Mexico 87106
Tel: (505) 254-7738
Fax: (505) 208-3030
Web site: www.2the4.net

ABB Environmental Services Inc.
107 Audubon Rd.
Corporate Place 128
Wakefield, Massachusetts 01880
Tel: (617) 245-6606
Fax: (617) 246-5050
Web site: www.us.abb.com

Action Environmental, Inc./WK
17 Green Street
Waltham, Massuchetts 02451
Tel: (781) 893-9922
Fax: (781)893-6622
Email: actionenv@aol.com
Web site: www.actionenviromental.net

Adrian Brown Consultants, Inc.
333W Bayaud Ave.
Denver, Colorado 80223
Tel: (303) 698-9080
Fax: (303) 698-9241

ADS Environnement inc.
1441 Rene Levesque Blvd., Suite 500
Montreal, Quebec
Canada H3G 1T7
Tel: (514) 875-1441
Fax: (514) 875-2666

Advanced Geoservices Corp.
Rts. 202& 1
Brandwine One, Suite 202
Chadds Ford, Pennsylvania 19317-9676
Tel: (610) 558-3300
Fax: (610) 558-2620
Email: agc@agcinfo.com
Web site: www.agcinfo.com

Alabaster Corp.
6921 Olson
Pasadena, Texas 77505
Tel: (281) 487-5470
Fax: (281) 487-9014
Email: bratinc@neosoft.com
Web site: www.alabastercorp.com

AMEC
3232 West Virginia Avenue
Phoenix, AZ 85009
Tel: (602) 272-6848
Fax: (602) 272-7239
Web site: ee.amec.com

American Bioremediation Services, LLC
347 Elizabeth Avenue, Suite 100
Somerset, New Jersey 08873
Tel: (732) 469-9190
Fax: (732) 469-1120
Email: mail@ambio.com

American Compliance Technologies, Inc.
1875 W. Main St.
American Way & State Rd 60W
Bartow, Florida 33830
Tel: (863) 533-2000
Fax: (863) 534-1133
Email: info@act-environmental.com
Web site: www.act-environmental.com

Anderson Columbia Environmental
Constructors, Inc.
P.O. Box 357490
Gainesville, Florida 32635-7490
Tel: (352) 384-0272
Fax: (352) 384-0282
Email: ac@atlantic.net

Applied Natural Sciences Inc.
4129 Tonya Trail
Hamilton, Ohio 45011
Tel: (513) 895-6061
Fax: (513) 895-6062
Email: ans@fuse.net
Web site: www.treemediation.com

Applied Remedial Technologies
220 Montgomery Street Suite 432
San Francisco, California 94104
Tel: (415) 986-1284
Fax: (415) 986-1359

ARCADIS Geraghy & Miller
1099 18th St. , Suite 2100
Denver, Colorado 80202-1921
Tel: (303) 294-1200
Fax: (303) 294-1221
Email: arcadisgm@gmgw,com
Web site: www.gmgw.com

ARS Technologies Inc.
271 Cleveland Ave.
Highland Park, New Jersey 08904
Tel: (908) 739-6444
Fax: (908) 739-0451
Web site: www.arstechnologies.com

ARUSI AR Utility Specialists, Inc.
2840 South 36th Street
Building E, Suite 1
Phoenix, AZ 85034-7238
Tel: (602) 431-2175
Fax: (602) 431-2163
Web site: www.arusi.net

August Mack Environmental, Inc.
8007 Castleton Road
Indianopolis, Indiana 46250
Tel: (317) 579-7400
Fax: (317) 579-7410
Email: BPetroko@aol.com
Web site: www.augustmack.com

Baker Environmental, Inc.
420 Rouser Rd., Bldg. 3
Airport Office Park
Corapolis, Pennsylvania 15108
Tel: (412) 269-6000
Fax: (412) 269-6097

Batelle Memorial Institute
505 King Avenue
Department 3171
Columbus, Ohio 43201
Tel: (614) 424-6424
Fax: (614) 424-3667
Web site: www.batelle.org
Email: environmental-client-
services@batelle.org

B.B. Environmental Inc.
704 Mara Street, Suite 201
Point Edward, Ontario
Canada N7V 1X4
Tel: (519) 337-0228
Fax: (519) 337-9178
Email: bbmccrie@ebtech.net

Beak International
42 Arrow Rd.
Guelph, Ontario
Canada N1K 1S6
Tel: (519) 763-2325
Fax: (519) 763-2378
Web site: www.beak.com

Bennett Engineering Inc.
301 Center St.
North Muskegon, Michigan 49455-3105
Tel: (231) 744-8989
Fax: (231) 744-0916
Email: dbennett@nspemail.com

Billings and Associates, Inc.
6808 Academy Parkway E. N.E.
Suite A-4
Albuquerque, New Mexico 87109
Tel: (505) 345-1116
Fax: (505) 345-1756

BioGenesis Enterprises, Inc.
7420 Alban Station Boulevard, Suite B 208
Springfield, Virginia 22150
Tel: (703) 913-9700
Fax: (703) 913-9704

Biogenie Inc.
350 Franquet Street
Sainte-Foy, Quebec
Canada G1P 4P3
Tel: (418) 653-4422
Fax: (418) 653-3583
Email: info@biogenie.com
Internet: www.biogenie-env.com

Bioinotech Inc.
100 6th Street
Suite 501
Sacramento, California 95814
Tel: (916) 491-0450
Fax: (916) 491-0463

Biorem Technologies Inc.
7496 Wellington Rd. 34 RR#3
Guelph, Ontario
Canada N1H 6H9
Tel: (519) 767-9100
Fax: (519) 767-1824

Bioremediation Systems
1106 Commonwealth Ave.
Boston, Massachusetts 02215
Tel: (617) 232-2207

Bioscience Inc.
1550 Valley Centre Parkway,
Suite 140
Bethlehem, Pennsylvania 18017
Tel: (610) 974-9693
Fax: (610) 691-2170
Email: bioscience@aol.com
Website: www.bioscienceinc.com

Bio-Rem Inc.
P.O. Box 116
Butler, Indiana 46721
Tel: (219) 868-5823
Fax: (219) 868-5851

Biotechnik Incorporated
1315 Finch Avenue, West Suite 410
North York, Ontario
Canada M3J 2G6
Tel: (416) 633-6308
Fax: (416) 633-0432

Biotrol Inc.
10300 Valley View Road, Suite 107
Eden Prairie, Minnesota 55344-3456
Tel: (612) 942-8032
Fax: (612) 942-8426

BMG Engineering AG
Ifangstrasse 11
8057 Schlieren
Switzerland
Tel: 41/1-732-9258
Fax: 41/1-730-6622
Email: mathias.schluep@bmgeng.ch

Boojum Technologies Ltd.
468 Queen St. E.
Suite 400, Box 19
Toronto, Ontario
Canada M5A 1T7
Tel: (416) 861-1086
Fax: (416) 861-0634
Email: kalin@ecf.toronto.edu

Canadian Crude Separators Inc.
Suite 1400, 815-8th Avenue SW
Calgary, Alberta
Canada T2P 3P2
Tel: (403) 233-7565
Fax: (403) 261-5612
Email: ramirault@cdncrude.com

CH2M HILL 825 NE Multnomah, Suite 1300
Portland, Oregon 97232
Tel: (503) 235-5000
Email: mmadison@ch2m.com
Web site: www.ch2m.com

Chempete Inc.
405 East Pierce Street
Elburn, Illinois 60119
Tel: (630) 365-2007
Fax: (630) 365-2064

Clayton Environmental Consultants
3611 South Harbor Blvd.
Suite 260
Santa Ana, California 92704
Tel: (714) 431-4106
Fax: (714) 825-0685

Dames & Moore
2325 Maryland Road
Willow Grove, Pennsylvania 19090
Tel: (215) 657-5000
Fax: (215) 657-5454

D. Glass Associates, Inc.
124 Bird Street
Needham, Massachusetts 02192
Tel: (781) 449-7940
Fax: (781) 449-8045
Email: DGlassAssc@aol.com
Web site: www.channel1.com/dglassassoc/

Earthcare, Inc.
P.O. Box 1922
Denison, Texas 75021-1922
Tel: (903) 463-3087
Fax: (903)463-3087
Email: gbrar@netscape.net

Ecology Technologies International, Inc.
3941 Park Drive
Suite 20-322
El Dorado Hills, California 95762
Tel: (916) 939-2397
Fax: (916) 939-2449
Email: sales@fyrezym-se.com

Ecology Technologies International, Inc.
Suite 21225 South 48th Street
Tempe, Arizona 85281
Tel: (602) 985-5524
Fax: (602) 985-2988

ECO-TEC Inc./Ecology Technology
P.O. Box 1113
Issaquah, Washington 98027-1113
Tel: (425) 392-0304
Fax: (425) 392-0575
Web site: www.eco-tec.com

Ecolotree
3017 Valley View Lane
North Liberty, Iowa 52317
Tel: (319) 665-3547
Fax: (319) 665-8035
Email: Ecolotree@aol.com
Web site: www.ecolotree.com

Edenspace Systems Corp.
15100 Enterprise Court
Dulles, Virginia 20151
Tel: (703) 961-8700
Fax: (703) 961-8939
Email: Info@edenspace.com
Web site: www.edenspace.com

Eimco Process Equipment Co,
P.O. Box 300
Salt Lake City, Utah 84110
Tel: (801) 526-2000
Fax: (801) 526-2425

Eko Tec AB
Nasuddsvagen 1o
93221 Skelleftehamn
Sweden
Tel: 46/910-33366
Fax: 46/910-33375
Email:erik.backlund@ebox.tninet.se

Electrokinetics Inc.
11552 Cedar Park Ave.
Baton Rouge, Louisianna 70809
Tel: (504) 753-8004
Fax: (504) 753-0028

ENSR Consulting and Engineering
3000 Richmond and Engineering
Houston, Texas 77098
Tel: (713) 520-9900
Fax: (713) 520-6802
Email: ramsden@ensr.com

Envirogen Inc.
4100 Quakerbridge Road
Lawrenceville, New Jersey 08648
Tel: (609) 936-9300
Fax: (609) 936-9221

Environmental Remediation Consultants, Inc.
677 Washington Blvd.
Sarasota, Florida 34236
Tel: (941) 952-0076
Fax: (941) 957-3530

EnviroCare Environmental Services Ltd.
19 McNaughton Avenue
Regina, Saskatchewan
Canada S4R 4L9
Tel: (306) 545-1021
Fax: (306) 545-3411
Email: jdw@cableregina.com

Envirogen Inc.
Princeton Research Center
4100 Quakerbridge Road
Lawrenceville, New Jersey 08648
Tel: (609) 936-9300
Email: webmaster@envirogen.com
Web site: www.envirogen.com

Environmental BioTechnologies Inc
255C South Guild Avenue
Lodi, California 95220-0844
Tel: (209) 333-4570
Fex: (209) 333-4572
Email: webmaster@e-b-t.com
Web site: www.e-b-t.com

Environmental Remediation Consultants Inc.
677 N. Washington Boulevard
Sarasota, Florida 34236
Tel: (941) 952-5825
Fax: (941) 362-298
Email: manhardr@lfci.com

Environmental Resources Management (ERM)
855 Springdaale Drive
Exton, Pennsylvania 19341
Tel: (800) 544-3117
Fax: (610) 524-7335
Web site: www.erm.com

Envirosite, inc.
855 Rue Pepin
Sherbrooke, Quebec
Canada J1L 2P8
Tel: (819) 829-0101
Fax: (819) 829-2717

Enzyme Technologies Inc.
2233 NE-244th #C1
Troutdale, Oregon 97060
Tel: (503) 254-4331
Fax: (503) 254-1722
Email: info@enzymetech.com
Web site:www.enzymetech.com

EWS Services Inc.
12804 N. Lake Dr.
Thornton, Colorado 80241
Tel: (303) 254-6641
Fax: (303) 254-4916
Email: theronjohn@hotmail.com

First Environment Inc.
90 Riverdale Rd.
Riverdale, New Jersey 07457
Tel: (973) 616-9700
Fax: (973) 616-1930
Email: contactus@firstenvironment.com
Web site: www.firstenvironment.com

Fluor Daniel GTI
310 Horizon Centre Drive
Trenton, New Jersey 08691
Tel: (609) 587-0300
Fax: (609) 587-7908

FOREMOST Solutions Inc.
350 Indiana Street
Suite 415
Golden, Colorado 80401
Tel: (303) 271-9114
Fax: (303) 216-0362
Email: foremost@earthlink.com

Foster Wheeler Environmental Corp.
P.O. Box 479
Livingston, New Jersey 07039-0479
Tel: (800) 580-3765
Fax: (973) 597-7590
Web site: www.fwenc.com

Fyrezyme Southeast Inc.
14229 Cypress Circle
Tampa, Florida 33624-2710
Tel: (800) 975-5336
Email: sales@fyrezyme-se.com
Web site: www.fyrezyme-se.com

GAI Consultants Inc.
570 Beatty Rd,
Monroeville, Pennsylvania 15146
Tel: (412) 856-6400
Fax: (412) 856-4970
Email: marketing@aiconsultants.com
Web site: www.gaiconsultants.com

Geo-Microbial Technologies Inc.
East Main Street
P.O. Box 132
Ochelata, Oklahoma 74051
Tel: (918) 535-2281
Fax: (918) 535-2564

Golder Associates
940-6 Ave. S.W., 10th floor
Calgary, Alberta
Canada T2P 3T1
Tel: (403) 299-5600
Fax: (403) 299-5606
Emai solutions@golder.com
Web site: www.golder.com.

Grace Bioremediation Technologies
3451 Erindale Station Rd
Mississauga, Ontario
Canada L5C 2S9
Tel: (905) 273-5374
Fax: (905) 273-7422
Web site: www.gracebioremediation.com

GSI Environmental, Inc
965 Newton, Suite 270
Quebec, Quebec
Canada G1P 4M4
Tel:(418) 872-3600
Fax: (418) 872-0149
Email:quebec@serrener.ca

Harding ESE
743 Horizon Court, Suite 334
Grand Junction, Colorado 81506
Tel: (970) 242-4749
Fax: (970) 242-5784
Web site: www.mactec.com

Hayward Baker Inc.
1130 Annapolis Road
Suite 202
Odenton, Maryland 21113
Tel: (410) 551-8200
Fax: (410) 551-1900
Web site: www.haywardbaker.com

Heritage Environmental Services Inc.
7901 W. Morris St.
Indianapolis, Indiana 46231
Tel: (317) 243-0811
Fax: (317) 486-5085
Email: heritage@heritage-enviro.com
Web site: www.heritage-enviro.com

Inland Environmental Inc.
3921 Howard St.
Skokie, Illinois 60076
Tel: (847) 677-7500
Fax: (847) 677-7533
Email: Internet@InlandEnv.com
Web site: www.inlandenv.com

In-Situ Fixation Inc.
P.O. Box 516
Chandler, Arizona 85244-0516
Tel: (602) 821-0409
Fax: (602) 786-3184

Institute of Gas Technology (IGT)
1700 S. Mt. Prospect Rd.
Des Plaines, Illinois 60018
Tel: (847) 768-0500
Fax: (847) 769-0262
Web site: www.igt.org

INTECH 180 Corporation
1770 N. Research Parkway, Suite 100
North Logan, Utah 84341
Tel: (801) 753-2111
Fax: (801) 753-8321

Integrated Explorations Inc.
67 Watson Rd., Unit 1
Guelph, Ontario
Canada N1H 6H8
Tel: (519) 822-2608
Fax: (519) 822-3076
Email: ieinc@istar.ca

International Daleco Corporation
33 Journey
Aliso Viejo, California 92656
Tel: (949) 360-4288 or (800) 432-5326 (U.S.)
Fax: (949) 360-8774
Email: comments@daleco.com
Web site: www.daleco.com

IT Corporation
312 Directors Drive
Knoxvlle, Tennessee 37923
Tel: (423) 690-3211
Fax: (423) 690-9573
Web site: www.itcorporation.com

ITT Night Vision
7635 Plantation Road
Roanoke, Virginia 24019-3257
Tel: (540) 362-7356
Fax: (540) 362-7370

Jacques Whitford
3 Spectacle Lake Dr.
Dartmouth, Nova Scotia
Canada B3B 1W8
Tel: (902) 468-7777
Fax: (902) 468-9009

J. R. Simplot Company
P.O. Box 198
Lanthrop, California 95330
Tel: (209) 858-2511
Fax: (209) 85802519

Komex H2O Science Inc.
5500 Bolsa Avenue
Suite 105
Huntingdon Beach, California 92649
Tel: (714) 379-1157
Fax: (714) 379-1160

Larsen Engineers
700 West Metro Park
Rochester, New York 14623-2678
Tel: (716) 272-7310
Fax: (716) 272-0159

Lawler Matusky & Skelly Engineers LLP
One Blue Hill Plaza
Pearl River, New York 10965
Tel: (914) 735-8300
Fax: (914) 735-7466
Email: lms@lmseng.com
Web site: www.lmseng.com

MBI International
3900 Collins Road
P.O. Box 27609
Lansing, Michigan 48909
Tel: (517) 336-4626
Fax: (517) 337-2122
Email: jain@mbi.org

Media and Process Technology Inc.
1155 William Pitt Way
Pittsburgh, Pennsylvania 15238
Tel: (412) 826-3716

Micro-Bac International Inc.
3200 N. H-35
Round Rock, Texas 78681-2410
Tel: (512) 310-9000
Fax: (512) 310-8800
Email: mail@micro-bac.com
Web site: www.micro-bac.com

Microbe Inotech Laboratories
12133 Bridgeton Square Drive
St. Louis, Missouri 63044
Tel: (800) 688-9144
Fax: (314) 344-3031

Microbial International
463 North Shatluck Place
Orange, California 92866
Tel: (714) 666-0924
Fax: (714) 538-5134

Midwest Bio-Systems
380 Oak Street
Elmhurst, Illinois 60126
Tel: (630) 279-8346
Fax: (630) 279-9593

Monsanto Company
800 N. Lindbergh Boulevard
St. Louis, Missouri 63167
Tel: (314) 694-5179
Fax: (314) 694-1531

O'Brien & Gere Companies
5000 Brittonfield Pkwy
Syracuse, New York 13221
Tel: (315) 437-6100
Fax: (315) 463-7554
Email: capponmpj@obg.com
Web site: www.obg.com

OHM Remediation Services
10 Ward Road
North Tonawanda, New York 14120
Tel: (716) 693-8800
Fax: (716) 693-8001

Oppenheimer Biotechnology Inc.
P.O. Box 5919
Austin, Texas 78763
Tel: (512) 474-1016
Fax: (512) 472-2909
Email: Jan.Neve@obio.com
Web site: www.obio.com

Parsons Engineering Services Inc.
1700 Broadway, Suite 900
Denver, Colorado 80290
Tel: (303) 831-8100
Fax: (303) 831-8208

Phytokinetics Inc.
1770 North Research Park Way, Suite 110
North Logan, Utah 84341-1941
Tel: (435) 750-0985
Fax: (435) 755-6296
Web site: www.phytokinetics.com

Phytotech Inc.
1 Deer Park Drive Suite I
Monmouth Junction, New Jersey 08852
Tel: (732) 438-0900
Fax: (732) 438-1209
Email: soilrx@aol.com

Praxair Inc.
39 Old Ridgebury Road (K-1)
Danbury, Connecticut 06810-5113
Tel: (203) 837-2174
Fax: (203) 837-2540
Email: webmaster@Praxair.com
Web site: www.praxair.com

Regenesis Bioremediation Products
1011 Calle Sombra
San Clemente, California 92672
Tel: (949) 366-8000
Fax: (949) 366-8090
Email: orc@regenesis.com
Website: www.regenesis.com

Remediation Technologies, Inc.
1011 S.W. Klickitat Way, Suite 207
Seattle, Washington 98134
Tel: (206) 624-9349
Fax: (206) 624-2839

R.E. Wright Environmental Inc,
3240 Schoolhouse Road
Middletown, Pennsylvania 17057-3595
Tel: (717) 944-5501
Fax: (717) 948-9398

Roy F. Weston Inc.
One Weston Way
West Chester, Pennsylvania 19380
Tel: (610)701-3000
Fax: (610)701-3186
Email: webmaster@rfweston.com
Web site: www.rfweston.com

Sanexen Environmental Services
57 Rue Le Breton
Longueuil, Quebec
Canada J4G 1R9
Tel: (450) 646-7878
Fax: (450) 646-5127
Email: info@sanexen.com
Web site: www.sanexen.com

SCG Industries Ltd.
250 King William Rd.
Saint John New Brunswick
Canada E2M 5Y5
Tel: (506) 674-1081
Fax: (506) 674-1082
Email: scgind@nbnet.nb.ca
Web site: www.remedi8.com

Services Environnement AES, inc.
3500 Chemin du Plateau S.
Latterriere, Quebec
Canada G0V 1K0
Tel: (418) 677-3247
Fax: (418) 677-3279

Sevenson Environmental Services Inc.
2749 Lockport Rd.
Niagara Falls, New York 14302
Tel: (716) 284-0431
Fax: (716) 284-1796
Email: jmdamon@sevenson.com
Web site: www.sevenson.com

Silt
Haven 1025
Scheldedijk 30
Zwijndrecht, Belgium 2070
Tel: +33 (0) 3250-54-11
Fax: +33 (0) 3250-52-53
Email: silt@dredging.com
Web site: www.silt.be

Sodexen Group
2519 Chomedy Boul.
Laval, Quebec
Canada H7T 2R2
Tel: (450) 973-7757
Fax: (450) 973-7758
Web site: www.sodexen.com

Soil Enrichment Systems Inc.
10800 Weston Road
Vaughan, Ontario
Canada L4L 1A6
Tel: (905) 832-2166
Fax: (905) 832-0751
Email: ssoil@aol.com

SRE Inc.
510 Franklin Ave.,
Nutley, New Jersey 07110
Tel: (973) 661-5192
Fax: (973) 661-3713
Email: sreinc.erols.com
Web site: www.srebiotech.com

Sybron Chemicals Inc.
Birmingham Road
P.O. Box 66
Birmingham, New Jersey 08011
Tel: (800) 678-0020
Fax: (609) 894-8641

Tauw B.V.
Handelskade 11
P.O. Box 133
744 AC Deventer
The Netherlands
Tel: +31-57-06-99-911
Fax: +31-57-06-99-666
Email: jfh@tauw.nl
Web site: www.tauw.nl

Terra Vac Corp.
92 North Main Street, Building 15
P.O. Box 468
Windsor, New Jersey 08561-0468
Tel: (609) 371-0070
Fax: (609) 371-9446
Email: tturner@terravac.com
Web site: www.terravac.com

ThermoRetec
7011 North Chaparral
Tucson, Arizona 85718
Tel: (520) 377-8323
Fax: (520) 377-7455
Web site: www.thermoretec.com

Thomas Consulting
P.O. Box 54924
Cincinnatti, Ohio 45254
Tel: (513) 271-9923
Fax: (513) 271-0036
Email: tc@iac.net
Web site: www.thomasconsultants.com

Troy Corp.
8 Vreeland Rd.
Florham Park, New Jersey 07932
Tel: (973) 443-4200
Fax: (973) 443-0258
Email: marketing@troy.com

U.S. Air Force
Mail Stop ASC-EMR
1801 10th Street
Building 8, Suite 200, Area B
Wright Patterson Air Force Base, Ohio 45433
Tel: (513) 255-7716, ext. 302
Fax: (513) 255-4155

Vega Power Resources
3401 Custer Road, Suite 115
Plano, Texas 75023
Tel: (972) 612-1103
Fax: (972) 612-1132

Verdant Technologies Inc.
12600 8th Avenue NE
Seattle, Washington 98125
Tel: (206) 365-3440
Fax: (206) 365-4957
Email: info@verdanttech.com
Web site: verdanttech.com

Walsh Environmental Scientists & Engineers
4888 Pearl E. Circle, Suite 108
Boulder, Colorado 80301-2475
Tel: (303) 443-3282
Fax: (303) 443-0367

Wasatch Environmental Inc.
2410 West California Avenue
Salt Lake City, Utah 84109-4109
Tel: (801) 972-8400
Fax: (801) 972-8459
Web site: www.wasatch-environmental.com

Waste Stream Technology Inc.
302 Grote Street
Buffalo, New York 14207
Tel: (716) 876-5290
Fax: (716) 876-2412

Westford Chemical Corp.
P.O.Box 798
Westford, Massachusetts 01886
Tel: (508) 885-1113
Fax: (508) 885-1114
Email: info@biosolve.com
Website: www.biosolve.com

Westinghouse Savannah River Co.
P.O. Box 616
Building 773-42A
Aiken, South Carolina 29802
Tel: (803) 725-5178

YES Technologies
320 South Wilson Ave.
Bozeman, Montana 59715
Tel: (406) 586-2002
Fax: (406) 586-8818
Email: yes@yestech.com
Web site: www.yestech.com.

Waste Treatment

AAA New Buoyancy/Gravity Mixer Co.
724 W. Pine Ave.
El Segundo, California 90245-2929
Tel: (310) 322-3257
Fax: (310) 322-3457
Web site: www.gwvandrie.com

AABio
Hauptstrasse 37
CH-4450 Sissach
Switzerland
Tel: +41 (0)61 976 96 00
Fax: +41 (0) 61 976 96 09
Email: euromaier@compuserve.com
Web site: www.euromaier.ch

ABB Environmental Services Inc.
107 Audubon Rd.
Corporate Place 128
Wakefield, Massachusetts 01880
Tel: (617) 245-6606
Fax: (617) 246-5050
Web site: www.us.abb.com

AC Compressor
401 E. South Island St.
Appleton, Wisconsin 54915
Tel: (414) 738-5968
Fax: (414) 738-3141

Aeration Products Inc.
P.O. Box 42216
Cincinnati, Ohio 45242
Tel: (513) 884-9907
Fax: (513) 891-0721

Ag-Bag Environmental
2320 S.E. Ag Bag Ln
Warrenton, Oregon 97146
Tel: (800) 334-7632
Email: compost@ag-bag.com
Web site: www.ag-bag.com

Agstar Progam
Tel: (800) 952-4782
Web site: agstar.com

Agronomic Management Goup
P.O. Box 012006
Arlington, Texas 76012
Tel: (817) 571-9391
Fax: (817) 571-6783

Alpha Alpha Umwelttechnik AG
Schloss-Strassee 15
CH-2560 Nidau
Schweitz, Switzerland
Tel: +41 (0) 32 331 54 54
Fax: +41 (0) 32 331 23 37
Email: info@alphat.ch
Web site: www.alphaut.ch

Alpha-Biotek Environmental, Inc.
1530 LaPalco Blvd., Suite 28
Harvey, Lousianna 70058
Tel: (504) 366-2800
Fax: (504) 366-3959
Web site: www.abenvironmental.com

American Bulk Conveying Inc.
564 Central Ave.
Murray Hill, New Jersey 07974
Tel: (908) 464-0700
Fax: (908) 464-0703

AN Maschinenbau und Umweltschutzanlagen
Waterbergstraße 11
D-28237 Bremen
Germany
Tel: +49 (0) 4 21 / 6 94 58-51
Fax: +49 (0) 4 21 64 22 83

Antico Olindo & Cesare Srl
Via Savona 26
Milan, Italy 20144
Tel: +39-2-83361
Fax: +39-2-89402788
Email: antico@energy.it
Web site: www.anticomix.it

Aqua Alliance
30 Harvard Mill Square
Wakefield, Massachusetts 01880-5371
Tel: (781) 246-5200
Fax: (781) 245-0823

Aquacare Systems Inc.
11820 N.W. 37th Street
Coral Springs, Florida 33065
Tel: (954) 796-3390

Arcadis
US Office
88 Duryea Road
Melville, New York 11747
Tel: (516) 391-5262
Fax: (516) 249-7610
Web site: www.arcadis.nl/arcadis

Arcadis NV
Utrechtseweg 69
Arnhem, The Netherlands
P.O. Box 33 6800 LE Arnhem
Tel: +31 26 3778911
Fax: +31 26 3515235

Arkenol
26001 Pala St.
Mission Viejo, California 92691
Tel: (949) 588-3737
Fax: (949) 588-3972
Email: arkenol@aol.com
Web site: www.arkenol.com

Augspurger Engineering
15455 N Greenway-Hayden Loop Suite C14
Scottsdale, Arizona 85260-1609
Tel: (480) 483-5966
Fax: (480) 483-0070
Email: engineering@aeincaz.com
Web site: www.aeincaz.com

Baker Process Chemical Group
14990 Yorktown Plaza Drive
Houston, Texas 77040
Tel: (936) 321-2244
Fax: (936) 321-2336
Email: houston.bakerhughes
    @bakerhughes.com
Web site: www.bakerhughes.com

BC International Corp.
990 Washington Street, Suite 104
Dedham, Massachusetts 02026
Tel: (781) 461-5700
Fax: (781) 461-2626
Web site:www.bcintlcorp.com

Bedminister Bioconversion Corp.
3220 Tillman Dr. Suite 107
Bensalem, Pennsylvania 19020
Tel: (215) 639-6644
Fax: (215) 639-7673
Email: roderrusso@bedminister.com

Bercan Environmental Resources Inc.
1702, 924-14 Avenue S.W.
Calgary, Alberta
Canada T2R 0N7
Tel: (403) 244-5081
Fax: (403) 244-5081
Email: candlshb@cadvision.ca

Best Sand Corp.
11830 Ravenna Road
Cleveland, Ohio 44024
Tel: (800) 237-4986
Fax: (216) 285-4109

Beulah Tec Ltd.
110-10524-170 Street
Mayfield Business Centre
Edmonton, Alberta
Canada T5P 4W2
Tel: (403) 578-3299
Fax: (403) 578-3313
EmailL btecprrc@telusplanet.net

Bio Gro.Div.
Wheelabrator Water Technologies
1110 Benfield Blvd. Suite B
Millersville, Maryland 21108
Tel: (410) 729-1440
Fax: (410) 729-0857

Bio-Mate Technologies, Inc.
3939 Grant St.
Burnaby, British Columbia
Canada V5C 3N4
Tel: (604)298-2143
Fax: (604)298-3262
Email: primespot@zoolink.com

Biomax
764 St. Joseph Est Suite 124
Quebec, Quebec
Canada G1K 3C4
Tel: (418) 529-585
Fax: (418) 529-9413

Bion Technologies
555 17 th St.
Denver, Colorado 80202
Tel: (303) 738-0845

Biorem Technologies Inc.
7496 Wellington Rd. 34 RR #3
Guelph, Ontario
Canada N1H 6H9
Tel: (519) 767-9100
Fax: (519) 767-1824

Bio-System
Gesellschaft fur Anwendungen
biologischer Verfahren mbH
Lohnerhofstr. 7
78467 Konstanz
Germany
Tel: +49 (0) 7531 69 06 50
Fax: +49 (0) 7531 69 06 60
Email: info@bio-system.de
Web site: bio-system/de

Bio-Terre Systems Inc.
12 Aviation Boulevard
St. Andrew's, Manitoba
Canada R1A 3N5
Tel: (204) 334-8846
Fax: (204) 334-6965
Email: dgheng@mb.sympatico.ca

Biothane Corporation
2500 Broadway/Dwr #5
Camden, New Jersey 08104
Tel: (609) 541-3500
Fax: (609) 541-3366
Email: sales@biothane.com
Web site: www.biothane.com

Boojum Technologies Ltd.
468 Queen St. E.
Suite 400, Box 19
Toronto, Ontario
Canada M5A 1T7
Tel: (416) 861-1086
Fax: (416) 861-0634
Email: kalin@ecf.toronto.edu

Brown Bear Corp.
P.O. Box 29
Corning, Iowa 50841-0029
Tel: (515) 322-4220
Fax: (515) 322-3527
Email: brnbear@mddc.com
Web site: www.brownbearcorp.com

BTA Biotechnische Abfallverwertung GmbH
& CO KG
Rottmannstr. 18
D-80333 Munchen
Germany
Tel: +49 89 52 04 60-6
Fax: +49 89 523 23 29
Email: post@bta-technologie.de
Web site: www.bta-technologie.de

BW Organics, Inc.
150 E Houston,
Sulphur Springs, Texas 75482
Tel: (903) 438-2525
Fax: (903) 438-2626
Email: Dairyland@neto.com
Web site: www.neto.com/bworgani/top.htm

Canada Composting Inc./BTA Process
390 Davis Dr., Suite 301
Newmarket, Ontario
Canada L3Y 7T8
Tel: (905)830-1160

CBI Walker
601 W. 143rd St.
Plainfield, Illinois 60544
Tel: (815) 439-3100
Fax: (815) 439-3130

CEMCORP. Ltd.
2170 Stanfield Road
Mississauga, Ontario
Canada L4Y 1R5
Tel: (905) 566-7227
Fax: (905) 566-7228
Email: cemcorp@cemcorp.com
Web site: www.cemcorp.com

Centre de recherche industrielle du Quebec
(CRIQ)
Parc technologique du Quebec
333 Rue Franquet
Sainte-Foy, Quebec
Canada G1P 4C7
Email: info@criq.qc.ca
Internet: www.criq.qc.ca

Chief Industries Inc.
1808 Raymond Dr.
P.O. Box 2078
Grand Island, Nebraska 68802-2078
Tel: (308) 389-7403
Fax: (308) 389-7448
Email: russelld@chiefind.com

Celto Canadian Enviro Systems, Ltd
1515 56th St.
Delta, British Columbia
Canada V4L 2A9
Tel: (604) 946-2414

CH2M Hill
C:N Composting Systems
P.O. Box 91500
Bellevue, Washington 98009-2050
Tel: (425) 453-5000
Fax: (425) 462-5957
Email: CNcompostsytems@ch2m.com
Web site: www.ch2m.com/composting

Coastal BioAgresearch Ltd.
RR #5
Truro, Nova Scotia
Canada B2N 5B3
Tel: (902) 893-9139
Fax: (902) 893-4523
Email: cbaltd@fox.nstn.ca

Columbian Steel Tank Co.
5400 Kansas Ave.
P.O. Box 2907
Kansas City, Kansas 66110
Tel: (913) 621-3700
Fax: (913) 621-2145
Web site: www.columbiantank.com

Les Composts du Quebec
415 Chemin Plaisance
C.P. 448
St. Henri de Levis, Quebec
Canada G0R 3E0
Tel: (418) 882-2736
Fax: (418) 882-2255
Web site: www.composts.com

Conporec/S&W Process
One Remington Park Dr.
Cazenovia, New York 13035
Tel: (315) 655-4953
Web site: www.stearnswheler.com

Consito SRL
Piazza 4 Novembre-1
Milan, Italy 20124
Tel: +39-2-66-98-1651
Fax: +39-2-66-98-4739
Email: consito@tin.it

Consolidated Environwaste Industries, Inc.
27715 Huntingdon Rd.
Abbotsford, British Columbia
Canada V4X 1B6
Tel: (800) 667-1942

Crom Corporation
250 S.W. 36th Terrace
Gainesville, Florida 32607
Tel: (352) 372-3436
Fax: (352) 372-6209
Web site: www.cromcorp.com

D & D Chemical Inc.
P.O. Box 57
Grifton, North Carolina 28530
Tel: (252) 524-3323
Fax: (252) 524-4576
Web site: www.ddchem.com

Deep Shaft Technology Inc.
700, 1207 11th Ave S.W..
Calgary, Alberta
Canada T3C 0M5
Tel: (403) 244-5340
Fax: (403) 245-5726

Dorr Oliver Inc.
612 Wheelers Farm Road
Milford, Connecticut 06460-8719
Tel: (203) 876-5400
Fax: (203) 876-5432
Email: stgeorge@dorr-oliver.com
Web site: www.glv.com/english/process

Double T Equipment
P.O. Box 3637
#2 E. Lake Way
Airdrie, Alberta
Canada T4B 2B8
Tel: (403) 948-5618
Fax: (403) 948-4780
Email solutions@double-t.com

EarthCare Technologies
820 Industrial Dr.
P.O. Box 998
Lincoln, Arizona 72744
Tel: (501) 824-5511

Eimco Process Equipment Co,
669 West 200 South
Salt Lake City, Utah 84101-1020
Tel: (801) 526-2000
Fax: (801) 526-2943
Website: www.eimcoprocess.com

EKO Systems
P.O. 14026
Lakewood, Colorado 55110
Tel: (303) 233-8440

Electrokinetics Inc.
11552 Cedar Park Ave.
Baton Rouge, Louisianna 70809
Tel: (504) 753-8004
Fax: (504) 753-0028

Engineering Resource Inc.
P.O. Box 808
Winchester, Massachusetts 01890
Tel: (781) 729-6777
Fax: (781) 721-7445
Email: eri@gis.net
Web site: www.gis.net/~eri

ENSR Consulting and Engineering
3000 Richmond and Engineering
Houston, Texas 77098
Tel: (713) 520-9900
Fax: (713) 520-6802
Email: ramsden@ensr.com

Entek BioSystems, LC
P.O. Box 372
Smithfield, Virginia 23431
Tel: (757) 357-6500

Entec Environmental Technology
Umwelttechnik GmbH
Schielfweg 1
RSB-Haus A-6972 Fussach
Austria
Tel: +43 55787946
Fax: +43 557873638
Email: entec@biogas.at
Web site: www.biogas.at

Envirodyne Systems Inc.
50 Utley Drive
Camp Hill, Pennsylvania 17011
Tel: (717) 763-0500
Fax: (717) 763-9308

Enviro-Ganics
4505 Baker Rd., RR. #3
Niagara Falls, Ontario
Canada L2E 6S6
Tel: (905) 383-0330

Environmental Dynamics Inc.
5601 Paris Road
Columbia, Missouri 65202-9399
Tel: (573) 474-9456
Fax: (573) 474-6988
Web site: www.wastewater.com

Environmental Energy Company
6007 Hill Road NE
Olympia, Washington 98516
Tel: (360) 923-2000
Website: www.makingenergy.com
Email: engineer@makingenergy.com

Environmental Equipment & Systems, Inc.
35 Corporate Drive
Trumbull, Connecticut 06611
Tel: (203) 452-0299
Fax: (203) 452-0313

Environmental Products & Technologies Corp
(EPTC)
5380 N. Sterling Center Dr.
Westlake Village, California 91361
Tel: (818) 865-2205
Fax: (818) 865-2210
Email: informed@eptciorp.com
Web site: www.eptcorp.com

Environmental Recycling Alternatives
P.O. Box 6417
High Point, North Carolina 27262
Tel: (803) 788-0477

Enviroquip
2404 Rutland Drive Suite 200
P.O. Box 9069
Austin, TX 78758-8519
Tel: (512) 834-6000
Fax: (512) 834-6039
Web site: www.enviroquip.com

EnviroSystems Supply, Div. AquaCare Inc.
11820 N.W. 37th St.
Coral Springs, Florida 33065
Tel: (954) 796-3390
Fax: (954) 796-3405

EPM Inc.
P.O. Box 1295
Cottage Grove, Oregon 97424
Tel: (541) 895-5990

ESG Manufacturing
P.O. Box 431
30070 Dan Pierson Road
Holden, Louisiana 70744
Tel: (225) 567-9200
Fax: (225) 567-9858

Fairfield Engineering Co.
P.O. Box 526
240 Boone Ave.
Marion, Ohio 43302
Tel: (740) 387-3327
Fax: (740) 387-4869
Email: sales@fairfieldengineering.com
Web site: www.fairfieldengineering.com

Farmer Automatic
P.O. Box 39
Register, Georgia 30452
Tel: (912) 681-2763
Fax: (912) 681-1096
Web site: www.farmerautomaticusa.com

Fiberglass Structures and Tank Co, Inc.
P.O. Box 583
Wyoming, Minnesota 55092
Tel: (651) 462-2605
Fax: (651) 462-3789
Web site: www.fiberglassstructures.com

Flowerfield Enterprises
10332 Shaver Rd.
Kalamazoo, Michigan 490024
Tel: 616-327-0108

Fusion Tanks and Silos
Eye 1P237HS
Suffolk, United Kingdom
Tel: 44-1379 870723
Fax: 44 1379 870530
Web site: www.fusiontanks.com

Geo-Microbial Technologies Inc.
East Main Street
P.O. Box 132
Ochelat, Oklahoma 74051
Tel: (918) 535-2281
Fax: (918) 535-2564

Global Earth Products
R.R. 2.
Utopia, Ontario
Canada L0M 1T0
Tel: 705-726-1339

GL&V Process Equipment Group
174 West St. S.
Orillia, Ontario
Canada L3V 6L4
Web site: www.glv.com

Grace Bioremediation Technologies
Div. W.R. Grace & Co. of Canada
341 Erindale Station Rd.
P.O. Box 3060
Station A
Mississauga, Ontario
Canada   L5A 3T5
Tel: (905) 273-5374
Fax: (905) 272-7472
Web site: www.daramend.com

Green Mountain Technologies
P.O. Box 17
Whitingham, Vermont 05361
Tel: (802) 368-7291
Web site: www.gmt-organic.com

Groupe Conporec
3125 Joseph-Simard Street
Tracy, Quebec
Canada J3P 5N3
Tel: (450) 746-9996
Fax: (450) 746-7587
Email: info@conporec.com
Web site: www.conporec.com

Harza Environmental Services, Inc.
233 Wackler Drive
Chicago, Oklahoma 60606
Tel: (312) 831-3806
Fax: (312) 831-3999

Heil Engineered Systems
205 Bishops Way, #201
Brookfield, Wisconsin 53005
Tel: (707) 894-7724

Infilco Degremont Inc.
P.O. Box 71390
Richmond, Virginia 23255-1390
Tel: (804) 756-7600
Fax: (804) 756-7643
Email: idi@idi-online
Web site: www.infilcodegremont.com

Institute of Gas Technology (IGT)
1700 S. Mt. Prospect Rd.
Des Plaines, Illinois 60018
Tel: (847) 768-0783
Fax: (847) 769-0262

International Bio-Recovery Corp.
52 Riverside Drive
North Vancouver, British Columbia
Canada V7H 1T4
Tel: (604) 924-1023
Fax: (604) 924-1024
Email: ibrcorp@direct.ca
Website: www.ibrcorp.com

Iogen Corp.
300 Hunt Club Rd. East
Ottawa, Ontario
Canada K1V 1C1
Tel: (613) 733-9830

Ionics Italba SpALivraghi
1/BMilan, Italy 20126
Tel: +39-02-270791
Fax: +39-02-27079291
Email: office@ionic-isitalba.it

JDV Equipment Corporation
26 Commerce Rd. Unit G
Fairfield, New Jersey 07004
Tel: (973) 244-1633
Fax: (973) 244-1055

Jones & Attwood/Waste Tech inc.
1931 Industrial Drive
Libertyville, Illinois 60048
Tel: (847) 367-5480
Fax: (847) 367-3983

Kankyo System Engineering Co. Ltd.
3-9-25 Danbara, Minami-ku
Hiroshima City
Hiroshima, 732-0811
Japan
Tel: 81 82-261-9676
Fax: 81 82-06-1553
Email: kse@md.neweb.ne.jp

KC Environmental Group Ltd.
15619-112 Ave.
Edmonton, Alberta
Canada T5M 2V8
Tel: (780) 488-7926
Fax: (780) 452-8284
Email: kcgroup@connect.ac.ca

Kelly Green Environmental Services
59 Columbus Ave.,
Exeter, New Hampshire 03833
Tel: (603) 772-6490

Knight Industrial Division
1501 West 7th Ave.
Brodhead, Wisconsin 5320-0167
Tel: (608) 897-2131
Fax: (608) 897-2561
Web site: www.knightmfg.com

KOMPGAS AG
Rohrstrasse 36
CH-8152 Glattbrugg
Switzerland
Tel: +41 1 809 71 00
Fax: +41 1 809 71 10
Email: info@kompogas.ch
Web site: kompogas.ch

Kruger A/G
Klamshgervej 2-4, 8230
Aabyhoj, Denmark
Tel: +458-746-3300
Web site: www.kruger.com

Kruger Inc.
401 Harrison Oaks Blvd., Suite 100
Cary, North Carolina 27513
Tel: (919) 677-8310
Fax: (919) 677-0082
Email: webmaster@krugerworld.com
Web site: www.krugerworld.com

Labconco Corp.
8811 Prospect Ave.
Kansas City, Missouri 64132-2696
Tel: (816) 333-8811
Fax: (816) 363-0130
Email: labconco@labconco.com
Web site: www.labconco.com

Larsen Engineers
700 West Metro Park
Rochester, New York 14623-2678
Tel: (716) 272-7310
Fax: (716) 272-0159

Liquid Waste Technology
P.O Box 250
422 Main St.
Somerset, Wisconsin 54025
Tel: (715) 247-5464
Fax: (715) 247-3934
Email: info@lwtpithog.com
Web site: www.wtpithog.com

Longwood Manufacturing
816 E. Baltimore Pike,
Kennett Square, Pennsylvania 19348-1890
Tel: (610) 444-4200
Fax: (610) 444-9552
Email: Mail@LMConline.com
Web site: www.LMConline.com

LOTEPRO Corp.
115 Stevens Ave.
Valhalla, New York 10695
Tel: (914) 749-5228
Fax: (914) 747-3422

Lumenite Control Technology, Inc.
2331 N. 17th ve.
Franklin, Park, Illinois 60131
Tel: (847) 455-1450
Fax: (847) 455-0127
Web site: www.lumenite.com

MacLeod & Miller (Engineers) Ltd.
P.O. Box 4
Logans Rd.
Motherwell, Scotland
ML1 3NP
Tel: +44 (0) 1698-230300
Fax: +44 (0) 1698-263178

Madison Chemical Industries
490 McGeachie Drive
Milton, Ontario
Canada L9T 3Y5
Tel: (905) 878-8863
Fax: (905) 878-1449
Web site: www.madisonchemical.com

Masada Resource Group, LLC
950 22nd Avenue North, Suite 850
Birmingham, Alabama 35203
Tel: (205) 320-1888
Fax: (205) 320-0055

Micro-Bac International, Inc.
3200 N. 1 H35
Round Rock, Texas 78610-2410
Tel: (512) 310-9000
Fax: (512) 310-8800
Web site: www.micro-bac.com

Micronair
11259 Phillipps Plwy Dr. Easte
Jacksonville, Florida 32256
Tel: (904) 268-5457
Fax: (904) 268-5457
Web site: www.micronairusa.com

Midwest Bio-Systems
Route 1, Box 121
Tampico, Illinois 61283
Tel: (630) 279-8396
Web site: www.aeromasterequipment.com

Miller Waste Systems/Ebara
8050 Woodbine Ave.
Markham, Ontario
Canada L3R 2N8
Tel: (905) 475-6356
Web site: www.millergroup.ca

Mixing Systems Inc.
7058 Corporate Way
Dayton, Ohio 45459
Tel: (937) 435-7227
Fax: (937) 435-9200
Email: mixing@mixing.com
Web site: www.mixing.com

Natgun Corp.
11 Teal Road
Wakefield, Massachusetts 01880
Tel: (617) 246-1133
Web site: www.natgun.com

National Technical Services Inc.
33881 Wastgate Circle
Corvallis, Oregon 97333
Tel: (541) 754-0670
Fax: (541) 758-0207

NaturTech Composting Services, Inc.
44 28th Ave. North, Suite J
St. Cloud, Minnesota 56303
Tel: (320) 253-6255
Fax: (320) 253-4976
Web site: www.composter.com

NewBio Inc.
P.O. Box 711
Hopkins, Minnesota 55343
Tel: (612) 949-1331
Fax: (612) 934-4067
Web site: www.newbio.com

Norton Environmental
6200 Rockside Woods Blvd.
Independence, Ohio 44131
Tel: (216) 447-0070

Nu Environ Tech Inc.
124 East "F" Street Suite 14
Ontario, California 91764
Tel: (909) 391-6654
Fax: (909) 391-6453
Email: nuenviron@nuenviron.com
Web site: www.nuenviron.com

N-Viro International Corp.
3450 West Central Ave. Suite 328
Toledo, Ohio 43606
Tel: (800) 66-NVIRO
Fax: (419) 535-7008
Web site: www.nviro.com

Orca Environmental Corp.
Suite 400, 3 Fan Tan Alley
Victoria, British Columbia
Canada V8W 3G9
Tel: (250) 360-0476
Fax: (250) 380-6791
Email: info@orcaenviro.com
Web site: www.orcaenviro.com

Original Vermitech Systems, Ltd.
2328 Queeb St. Easte
Toronto, Ontario
Canada M4E 1G9
Tel: (416) 693-1027

Organic Waste Systems/Dranco
Dok Noord 4
B-9000 Gent
Belgium
Tel: +32(0) 9 233 02 04
Fax: +32(0) 9 233 28 55
Web site: www.ows.be

Paques ADI Inc.
389 Main St.
Salem, New Hampshire 03079
Tel: (603) 893-2134
Fax: (603) 898-3991
Email: adigroup@adi.ca
Web site: www.adi.ca

Petrozyme Technologies Inc.
R.R. #3
Guelph, Ontario
Canada N1H 6H9
Tel: (519) 767-2299
Fax: (519) 767-9435
Email: hdimock@petrozyme.com
Web site: www.petrozyme.com

Philadelphia Mixers
1221 E. Main
Palmyra, Pennsylvania 17078-9518
Tel: (717) 838-1341
Fax: (717) 832-1740
Email: philamixers@philamixers.com
Web site: www.philamixers.com

Phytotech Inc.
1 Deer Park Drive Suite I
Monmouth Junction, New Jersey 08852
Tel: (732) 438-0900
Fax: (732) 438-1209
Email: soilrx@aol,com or
ericmuhr@mars.superlink.net

Pinnacles Biotechnologies International, Inc.
1667 Cole Blvd., Suite 400
Golden, Colorado 80401
Tel: (303) 235-0613
Fax: (303) 235-0603
Email: chris_rivard@nrel.gov
Web site: www.pinnaclebiotech.com

Pintail Systems Inc.
11801 East 33rd Avenue, Suite C
Aurora, Colorado 80010
Tel: (303) 367-8443
Fax: (303) 364-2120

Pitt Des Moines Inc.
979 Lincoln Village Drive, Suite 301
Sacramento, California 95827
Tel: (916) 366-6663
Fax: (916) 366-8547

P. J. Hannah Equipment Sales Corp.
#10, 8528 123rd St.
Surrey, British Columbia
Canada V3W 3V6
Tel: (604) 591-5999
Fax: (604) 591-9925

Plenty Ltd.
Hambridge Rd.
Newbury, Berkshire
England RG14 5TR
Tel: +44-1635-42363
Fax: +44-1635-49758
Email: cbartiam@plenty.co.uk
Web site: www.plenty.co.uk

Pollution Control Systems Inc.
5827 Happy Hollow Rd. Suite 1B
Milford, Ohio 45150-1839
Tel: (513) 831-1165
Fax: (513) 965-4812
Email: polconsys@aol.com
Web site: www.pollutioncontrolsystem.com

Preload Inc.
60 Commerce Drive
Hauppauge, New York 11788
Tel: (888) PRELOAD
Fax: (516) 222-0528
Email: sales@preload.com
Web site: www.preload.com

Prism Resource Management Ltd.
4833 Tufford Road
Beamsville, Ontario
Canada L0R 1B1
Tel: (905) 563-3533
Fax: (905) 563-3463

Purestream Inc.
P.O. Box 68
Florence, Kentucky 41042-0068
Tel: (606) 371-9898
Fax: (606) 371-3577

R. Cave and Associates Engineering Ltd.
Suite 404, 345 Lakeshore Rd. E.
Oakville, Ontario
Canada L6J 1J5
Tel: (905) 825-8440
Fax: (905) 825-8446
Email: rcave@rcave.com

R.E Wright Environmental Inc.
3440 Schoolhouse Road
Middletown, Pennsylvania 17057-3595
Tel: (717) 944-5501
Fax: (717) 944-5642

Resource Optimization Technologies
R.R. #2, Box 495
Cornish, New Hampshire 03745
Tel: (603) 542-5291

Resource Recovery Systems of Nebraska
KW Composter
Route 4, 511 Pawnee Dr.
Sterling, Colorado 80751-8698
Tel: (970) 522-0663
Fax: (970) 522-3387
Email: rrskw@kci.net
Web site: www.rrskw.com

Rieckermann Thai Enviro-Chem Co. Ltd.
19th Floor, Banga Tower C
2/3 Moo 14, Bangna-Trad Road
Km 6.5 Bangkaew, Bangplee
Samutprakarn 10541
Thailand
Tel: +662 7519111
Fax: +662 7519112-6

R.O.M. AG
Mattstrasse
8502 Frauenfeld
Switzerland
Tel: +41 (0) 52 22 46 60
Fax: +41 (0) 52 722 40 42
Email: info.rom@zucker.ch
Web site: www.rom.ch

Salmet Poultry Systems
P.O. Box 24
West Mansfield, Ohio 43358
Tel: (937) 355-3021

Sanexen Environmental Services
57 Rue Le Breton
Longueuil, Quebec
Canada J4G 1R9
Tel: (450) 646-7878
Fax: (450) 646-5127
Email: info@sanexen.com
Web site: www.sanexen.com

Seepex Inc.
511 Speedway Drive
Enon, Ohio 45323
Tel: (937) 864-7150
Fax: (937) 864-7157
Web site: www.seepex.com

SHW-GmbH
Wilhemstrasse 67
73433 Aalen, Germany
Tel: 49-761-502-430
Fax: 49-7361-502-288
Web site: www.shw.de

Smith & Loveless Inc.
14040 Santa Fe Trail Drive
Lenexa, Kansas 66215-1284
Tel: (913) 888-5201
Fax: (913) 888-2173
Email: answers@smithandloveless.com
Web site: www.smithandloveless.com

Spencer Turbine Co.
600 Day Hill Road
Windsor, Connecticut 06095
Tel: (860) 688-8361
Fax: (860) 688-0098
Email: marketing@spencer-air.com
Web site: www.spencerturbine.com

Steinmüller Valorga Sarl
Rue Albert Einstein
F-34935 Montpellier
Cedex, France
Tel: +33 (0) 33/4 67 99 41 00
Fax: +33 (0) 33/4 67 99 41 01

Steinmüller Valorga/Waste Recovery Systems
33655 Marlin Spike Dr.
Monarch Beach, California 92629-3328
Tel: (949) 219-4555

Synagro
1800 Bering Drive, Suite 1000
Houston, Texas 77057
Tel: (800) 370-0035
Fax: (713) 369-1750
Web site: www.synagro.com

Taylor Environmental Products
628 Old Robinson Rd.
Louisville, Missouri 39339
Tel: (601) 773-3421
Fax: (601) 773-7139
Web site: www.taylorenviro.com

Teg Environmental Plc
c/o RKB Enterprises Inc.
1322 DeBree Ave.
Norfolk,Virginia 23517
Tel: (757) 622-0692
Web site: teg-environmental.com

TEMCOR
24724 S. Wilmington Ave.
Carson, California 90745
Tel: (310) 549-4311
Fax: (310) 549-4588
Web site: www.temcor.com

Texel/Compostex
245 Ten Stones Circle
Charlotte, Vermont 05455
Tel: (802) 425-5556
Fax: (802) 425-5557

Thermodyne Corp.
2 Smedley Drive
Newtown Square, Pennsylvania 19073-1012
Tel: (610) 644-5406
Fax: (610) 644-3122

Thermo Tech Technologies Inc.
Suite 101
20436 Fraser Hwy.
Langley, British Columbia
Canada V3A 4G2
Tel: (604) 514-8390
Web site: www.ttrif.com

Transform Compost Systems Lt.
34642 Mierau St.
Abbotsford, British Columbia
Canada V2S 4W8
Tel: (604) 504-5660
Web site: wwww.transformcompost.com

TurboSonic Inc.
550 Parkside Dr., Suite A-14
Waterloo, Ontario
Canada N2L 5V4
Tel: (519) 885-5513
Fax: (519) 885-6992
Email: infor@turbosonic.com
Web site: www.turbosonic.com

U.S. Filter/CPC
441 Main St.
P.O. Box 36
Sturbridge, Massachusetts 01566
Tel:(508) 347-4560
Web site: www.water.usfilter.com

U.S. Filter Davis Process Products
2650 Tallevast Road
Sarasota, Florida 34243
Tel: (800) 345-3982
Fax: (941) 351-4756

U.S. Filter Envirex Products
P.O. Box 1604
Waukesha, Wisconsin 53187-1604
Tel: (414) 547-0141
Fax: (414) 547-4120
Web site: www.usfilterenvirex.com

U.S. Filter Industrial Wastewater Systems
181 Thorn Hill
Warrendale, Pennsylvania 15086
Tel: (724) 772-0044
Fax: (724) 772-1360
Web site: www.usfilter.com

U.S. Filter Microfloc Products
441 Main Street
P.O. Box 36
Sturbridge, Massachusetts 01566-0036
Tel: (508) 347-7344
Fax: (508) 347-7049
Website: www.water.usfilter.com

VermiCo
P.O. Box 1134
Merlin, Oregon 97532
Tel: (541) 476-9626
Web site: www.vermico.com

Vermitechnology Unlimited, Inc.
P.O. Box 130
Orange Lake, Florida 32681
Tel: (352) 591-1111
Web site: www.vermitechnology.com

Vulcan Chemicals
P.O. Box 530390
Birmingham, Alabama 35253-0390
Tel: (205) 877-3021
Fax: (205) 877-3805

Walker Process, Div. Of McNish Corp.
840 N. Russell Ave.
Aurora, Illinois 60506
Tel: (630) 892-7921
Fax: (630) 892-7951
Web site: www.walker-process.com

Waste Management, Inc.
1001 Fannin, Suite 4000
Houston, Texas 77002
Tel: (713) 512-6200
Fax: (713) 512-6299
Email: webmaster@wm.com
Web site: www.wastemanagement.com

Westech Engineering Inc.
3605 Southwest Temple
Salt Lake City, Utah 84115
Tel: (801) 265-1000
Fax: (801) 265-1080
Email: info@westech-inc.com
Web site: www.westech-inc.com

Wright Environmental Management
9050 Yonge St., Suite 300
Richmond Hill, Ontario
Canada L4C 9S6
Tel: (905) 881-3950
Fax: (905) 881-2334

ZMI/Portec Chemical Processing Group
1102 Egret Drive
Sibley, Iowa 51249
Tel: (712) 754-2595
Fax: (712) 754-3607

Water and Wastewater Treatment

AAA New Bouyancy/Gravity Mixer Co.
724 W. Pine. Ave.
El Segundo, California 90245
Tel: (310) 322-3258
Fax: (310) 322-3457
Web site: www.gwvandrie.com

AABio
Hauptstrasse 37
CH-4450 Sissach
Switzerland
Tel: +41 (0)61 976 96 00
Fax: +41 (0) 61 976 96 09
Email: euromaier@compuserve.com
Web site: www.euromaier.ch

ABJ Water Pollution Control Corp.
9333 N. 49th Street
Brown Deer, Wisconsin 53223
Tel: (414) 365 2200
Fax: (414) 365 2210

AC Compressor
401 E. South Island St.
Appleton, Wisconsin 54915
Tel: (414) 738-5968
Fax: (414) 738-3141

ADI Systems Inc.
1133 Regent Street, Suite 300
Fredicton, New Brunswick
Canada E3B 3Z2
Tel: (506) 452-7307
Fax: (506) 452-7308
E-mail: systems@adi.ca
Web site: www.adi.ca

Advanced Bio-Catalysts Corporation
3451 Airway Ave., Suite U-2
Costa Mesa, California 92626
Tel: (714) 427-6333
Fax: (714) 427-0380

Advanced Environmental Systems, Inc.
P.O. Box 50356
Sparks, Nevada 89435
Tel: (775) 425-0911
Fax: (775) 425-0212

Aeration Industries Inc.
4100 Peavey Road
Chaska, Minnesota 55318
Tel: (612) 448-6789
Fax: (612) 448-7293

Aeromix Systems Inc.
2611 N. Second St.
Minneapolis, Minnesota 55411
Tel: (612) 521-8519
Fax: (612) 521-1455
Web site: www.aeromix.com

Aero-Mod Inc.
7927 U.S. Highway 24
Manhattan, Kansas 66502
Tel: (800) 352-2376
Fax: (785) 537-0813

Aerostrip Corp.
P.O. Box 1168
Old Saybrook, Connecticut 06475
Tel: (860) 388-6686
Fax: (860) 388-6948

Aga Gas Inc.,
6055 Rockside Woods, Blvd.,
Independence, Ohio 44131
Tel: (216) 573-7851
Fax: (216) 573-7873

Alfa Laval Thermal Inc.
4093 Gettysburg St.
Ventura, California 93003
Tel: (805) 642-9721
Fax: (805) 642-8222

Alken-Murray Corp.
P.O. Box 400
New Hyde Park, New York 11040
Tel: (540) 636-1236
Fax: (516) 775-6597
Web site: alken-murray.hypermart.net

Allied Colloids
2301 Wilroy Road
Suffolk, Virginia 23434
Tel: (504) 355-2800
Fax: (504)366-3959

E. Roberts Alley & Associates Ltd.
230 Wilson Pike Circle
Brentwood, Tennessee 37027
Tel: (615) 373-1567
Fax: (615) 373-3697
Web site: www.eralley.com

Alpha-Biotek Environmental, Inc.
1530 LaPalco Blvd., Suite 28
Harvey, Louisianna 70058
Tel: (504) 366-2800
Fax: (504) 366-3959
Web site: abenvironmental.com

Alpha Umwelttechnik AG
Schloss-strasse 15
CH-2560 Nidau
Schweitz, Switzerland
Tel: +41 32 331 54 54
Fax: +41 32 331 23 37
Email: info@alphaut.ch
Web site: www.alphaut.ch/index.html

Alpheus Enviromental Ltd.
Cambridge Road
Bedford, England
MK42 0LL
Tel: +44-1234-270-344
Fax: +44-1234-357-088

American Compliance Technologies, Inc.
1875 W. Main St.
American Way & State Rd, 60 W
Bartow Florida 33830
Tel: (941) 533-2000
Fax: (941) 534-1133
Email: info@act-environmental.com
Web site: www.act-environmental.com

American Engineering Services Inc.
5912 Breckenridge Pkwy., Suite F
Tampa, Florida 33610
Tel: (813) 621-3932
Fax: (813) 621-4085
Email: watermaker@aesh2o.com
Web site: www.aesh2o.com

AnAerobic
P.O. Box 307
Aurora, New York 13026
Tel: (315) 364-5062
Fax: (315) 364-8664
Web site:www.anaerobics.com

Applied Biosciences
Tel: (800) 280-7852
Fax: (801) 809-9089
Web site: www.bioprocess.com

Applied Process Technology Inc.
3333 Vincent Road, Suite 222
Pleasant Hill, CA 94523
Tel: (925) 977-1811
Fax: (925) 977-1818
Web site: www.aptwater.com

Aqua-Aerobic Systems Inc.
6306 N. Alpine Rd.
P.O. Box 2026
Rockford, Illinois 61130 USA
Tel: (815) 654-2501
Fax: (815) 654-2508
Email: solutions@aqua-aerobic.com
Web page: www.aqua-aerobic.com

Aqua-Biotechnology
U6-8 Robin St.
Menora, WA Australia 6050
Tel: +61-8-9371-5657
Fax: +61-8-9-2011103
Web site: www.aqua-bio.com

AquaCare International, Inc.
14 Green River Rd.
Morris, Minnesota 56267-0593
Tel: (320) 589-0900
Fax: (320) 589-2814
Web site: www.aquacareinternational.com

Aquaflo Inc.
6244 Frankford Ave.
Baltimore Maryland 21206
Tel: (410) 485-7600
Fax: (410) 488-2030
Email: aquaflow@erols.com
Web site: www.aquafloinc.com

Aqua-Plant Construction Co.. Ltd.
329-1508 W. Broadway
Vancouver, British Columbia
Canada V5V 4J7
Tel: (604) 874-3514
Fax: (604) 874-3520
Email: aqua@istar.ca

Aqua-System AG
Postfach 29
CH-8410 Winterthur
Switzerland
Tel: +41(0)52-214 27 00
Fax: +41(0)52-214 27 59
Email: as@aquasystem.ch
Web site: www.aquasystem.ch

Aquaturbo Systems Inc.
1754 Ford Ave.
Springdale, Arizona 72765
Tel: (501) 927-1300
Fax: (501) 927-0700

Atara Corp.
9700 Henri-Bourassa W.
Saint-Laurent, Quebec
Canada H4S 1R5
Tel: (514) 331-8332
Fax: (514) 335-9346

Babcock Water Engineering
Badminton Court
Church Street
Amersham, Buckinghamshire
United Kingdom HP7 0DD
Tel: (44) 1494-727-296
Fax: (44) 1494-721-909
Web site: www.babcock.co.uk

Baker Process Chemical Group (EIMCO Process
Equipment)
669 West 200 South
Salt Lake City, Utah 84101-1020
Tel: (801) 526-2000
Fax: (801) 526-2943
Email: eimco.info@eimcoprocess.com
Web site: www.eimcoprocess.com

B.B. Environmental Inc.
704 Mara Street, Suite 201
Point Edward, Ontario
Canada N7V 1X4
Tel: (519) 337-0228
Fax: (519) 337-9178
Email: bbmccrie@ebtech.net

BIFS Technologies Corp.
2341 Porter Lake Dr. #101
Sarasota, Florida 34240
Tel: (941) 343-9300
Fax: (941) 343-0404
Email: bfs@gte.net
Web site: www.biofiltrationsystems.com

Big Dipper-Thermaco Inc.
646 Gransborost
Asheboro, North Carolina 27203
Tel: (800) 633-4204
Fax: (336) 626-5739
Web site: www.big-dipper.com

BioChem Technology, Inc.
100 Ross Road, Suite 201
King of Prussia, Pennsylvania 19406-2100
Tel: (610) 768-9360
Fax: (610) 768-9360
Web site: www.biochemtech.com

Bio Matrix Technologies, Inc.
26 Albion Road
Lincoln, Rhode Island 02865
Tel: (401) 334-1280
Fax: (401) 334-1285

BIOMAX Systems
9902 NE Glisan St.
Portland Oregon 97220
Tel: (503) 256-7324
Fax: (503) 256-7325
Web site: www.biomaxwa.com

Bio-Microbiotics, Inc.
8271 Melrose Drive
Lenexa, Kansas 66214
Tel: (913) 753-3278
Fax: (913) 492-0808
E-mail: onsite@biomicrobics.com
Web site: www.biomicrobics.com

Bioscience Inc.
1550 Valley Center Pkwy, Suite 140
Bethleham, Pennsylvania 18017-2263
Tel: (610) 974-9693
Fax: (610) 691-2170
Email: bioscience@aol.com
Web site: www.bioscienceinc.com

Bio-System
Gesellschaft fur Anwendungen
biologischer Verfahren mbH
Lohnerhofstr. 7
78467 Konstanz
Germany
Tel: +49 (0) 7531 69 06 50
Fax: +49 (0) 7531 69 06 60
Email: info@bio-system.de
Web site: biosystem/de

Bioteg Umwelttechnik
von Linde Str. 16
95326 Kulmbach
Germany
Tel: +49(0) 9221 9053 80
Fax: +49(0) 9221 9053 99
Email: service@bioteg.de
Web site: www.bioteg.de

Biothane Corporation
2500 Broadway/Dwr #5
Camden, New Jersey 08104
Tel: (609) 541-3500
Fax: (609) 541-3356
Email: sales@biothane.com
Web site: www.Biothane.com

Biotrol
10300 Valley View Road, Suite 107
Eden Prairie, Minnesota 55344-3546
Tel: (612) 942-8032
Fax: (612) 942-8526

BOC Gases
575 Mountain Ave.
Murray Hill, New Jersey 07974
Tel: (800) 742-4726
Fax: (908) 508-3814
Web site: www.boc.com/gases.index.htm

Brentwood Industries Inc.
P.O. Box 605
Reading, Pennsylvania 19603
Tel: (610) 236-1100
Fax: (610) 236-1199
Web site: brentw.com

Brown-Minneapolis Tank, Div. ITEQ Storage
Systems
2875 Highway 55 S
St. Paul, Minnesota 55121
Tel: (612) 454-6750
Fax: (612) 454-1987

BTA Biotechnische Abfallverwertung GmbH
& CO KG
Rottmannstr. 18
D-80333 Munchen
Germany
Tel: +49 89 52 04 60-6
Fax: +49 89 523 23 29
Email: post@bta-technologie.de
Web site: www.bta-technologie.de

Burns & McDonnell
9400 Ward Parkway
Kansas City, Missouri 64114
Tel: (816) 822-3226
Fax: (816) 822-3413
Web site: www.burnsmcd.com

BYO-GON PX 109
P.O. Box 912
Kermit, Texas 79745
Fax: (915) 586-6496

C-A Engineering
P.O. Box 190332
San Juan, Puerto Rico 00919-0332
Tel/Fax: (787) 748-6106
Email: ancallec@caribenet
Web site: netdial.caribe.net/~ancallec/
    home.html

CBI Walker
151 N. Division St.
Plainfield, Illinois 60544-8984
Tel: (815) 439-4000
Fax: (815) 439-4010

CH2M Hill
6060 S. Willow Drive
Greenwood Village, Colorado 80111
Tel: (303) 771-0900
Fax: (303) 754-0199
E-mail: feedback@ch2m.com
Web site: www.ch2m.com

Chemineer Inc.
P.O. Box 1123
Dayton, Ohio 45401
Tel: (513) 454-3200
Fax: (513) 454-3379
Web site: www.chemineer.com

CMS Group
185 Snow Blvd., Suite 200
Concord, Ontario
Canada L4K 4N9
Tel: (905) 660-7580
Fax: (905) 660-0243
Email: cms.group@aims.on.ca
Web site: www.rotordisk.com

Cromaglass Corp.
2902 N. Reach Road
P.O. Box 3215
Williamsport, Pennsylvania 17701
Tel: (717) 326-3396
Fax: (717) 326-6426

Crom Corporation
250 S.W. 36th Terrace
Gainesville, Florida 32607
Tel: (352) 372-3436
Fax: (352) 372-6209
Web site: www.cromcorp.com

D & D Chemical Inc.
P.O. Box 57
Grifton, North Carolina 28530
Tel: (252) 524-3323
Fax: (252) 524-4576
Web site: www.ddchem.com

Deep Shaft Technology Inc.
700, 1207 11th Ave S.W.
Calgary, Alberta
Canada T3C 0M5
Tel: (403) 244-5340
Fax: (403) 245-5726

Diffused Gas Technologies Inc.
1776 Mentor Dr.
Cincinnati, Ohio 45212
Tel: (513) 531-4426
Fax: (513) 531-4436

Dorr-Oliver
612 Wheelers Farm Road
Milford, Connecticut 06460
Tel: (203) 876-5400
Fax: (203) 876-5432
Web site: www.dorroliver.thomasregister.com
    /olc/dorroliver/home.htm

Dry Biofilter (DBF) Inc.
271 Glidden Road, #15
Brampton, Ontario
Canada L6W 1H9
Tel (905) 796-9653
Fax: (905) 803-8836
Email: dbf@aibn.com
Web site: www.drybio.com

Ducks Unlimited Inc.
One Waterfowl Way
Memphis, Tennessee 38120
Tel: (901) 758-3825
Website: www.ducks.org

Eco-Cycle Inc.
P.O. Box 228
Manchester, Maine 04351
Tel: (207) 622-7800
Fax: (207) 622-5197
Website: www.eco-cycle.com

Eco Equipment & Processes
3330 boul des Entreprises
Terrebonne, Quebec
Canada J6X 4J8
Tel: (514) 477-7879
Fax: (514) 477-7880
Email: ecofep@cam.org
Web site: www.sequencertech.com

Ecological Laboratories, Inc.
70 N Main St.
Freeport, New York 11520
Tel: (516) 379-3441
Fax: (516) 379-3632
Web site: www.microbelift.com

Environmental Enterprises USA, Inc.
58485 Pearl Acres Road, Suite D
Slidell, Louisianna 70461
Tel: (985) 646-2787
Fax: (985) 646-2810
Email: info@eecusa.com
Web site: www.eeusa.com

EIMCO Canadian Region Headquarters
5155 Creekbank Road
Mississauga, Ontario
Canada L4X 1X2
Tel: (905) 625-6821
Fax: (905) 625-3519
Eamil: canada.eimco@eimcoprocess.com
Web site: www.eimcoprocess.com

EIMCO Europe, Africa and Mid-East
Headquarters
Swift House
Cosford Lane Ruby,
Warwickshire CV21 1QN
England
Tel: +44-1-788-555777
Fax: +44-1-788-555778

EIMCO-KCP
Ramakrishna Buildings
183, Anna Salai
Chennai 600 006
India
Tel: +91-044 8555171
Fax: +91-044 8555863
Web site: www.ekcp.com

EIMCO Process Equipment
669 West 200 South
P.O. Box 300
Salt Lake City, Utah 84110-0300
Tel: (801) 526-2000
Fax: (801) 526-2911
Web site: www.eimcoprocess.com

Enprotec
4465 Limaburg Rd.
Hebron, Kentucky 41048
Tel: (606) 689-4300
Fax: (606) 689-4322

Envirodyne Systems Inc.
50 Utley Drive
Camp Hill, Pennsylvania 17011
Tel: (717) 763-0500
Fax: (717) 763-9308

Envirogen, Inc.
Prince Research Center
4100 Quakerbridge Rd.
Lawrenceville, New Jersey 08648
Tel: (609) 936-0075
Fax: (609) 936-0085
Web site: www.envirogen.com

Environmental Dynamics Inc.
5601 Paris Road
Columbia, MO 65202-9399
Tel: (573) 474-9456
Fax: (573) 474-6988
Web site: www.wastewater.com

Environmental Energy Company
6007 Hill Road NE
Olympia, Washington 98516
Tel: (360) 923-2000
Fax: (360) 923-1642
Web site: www.makingenergy.com
Email: engineer@makingenergy.com

Enviroquip
P.O. Box 9069
Austin, Texas 78728-8519
Tel: (512) 834-6010
Fax: (512) 834-6039
Web site: www.enviroquip.com

Environmental Resources Management (ERM)
855 Springdale Drive
Exton, Pennsylvania 19341
Tel: (800) 544-3117
Fax: (610) 524-7335
Web site: www.erm.com

EnviroSystems Supply, Div. Of AquaCare Inc.
11820 N.W. 37th St.
Coral Springs, Florida 33065
Tel: (954) 796-3390
Fax: (954) 796-3405

Environmental Treatment Systems, Inc.
1500 Wilson Way, Suite 100
Smyrna, Georgia 30082
Tel: (770) 384-0602
Fax: (770) 384-0603

Enviro-Zyme International
P.O. Box 169
Stormville, New York 12582
Tel: (800) 882-9904
Fax: (914) 878-7917
Web site: www.envirozyme.com

ETA Process Plant Ltd.
The Levels, Brereton,
Rugeley Staffs, England WS15 1RD
Tel: +44-1889-576501
Fax: +44-1889-579856
Email: sales@etapp.com

Export Technologies
3955 Leapheart Rd. #1A
West Columbia, South Carolina 29169-2418
Tel: (803) 794-2543
Fax: (803) 796-0999

FE3 Inc.
P.O. Box 808
Celina, Texas 75009
Tel: (800) 441-2659
Fax: (972) 382-3211
Web site: www.fe3.com

Ferro Corporation, Filtros Plant
603 W. Commercial St.
P.O. Box 389
East Rochester, New York 14445
Tel: (716) 586-8770
Fax: (716) 586-7154

Fiberglass Structures and Tank Co. Inc.
P.O. Box 582
Wyoming, Minnesota 55092
Tel: (651) 462-2605
Fax: (651) 462-3789
Web site: www.fiberglassstructures.com

Fluidyne Corp.
35 Wellington
Conroe, Texas 77304
Tel: (319) 266-9967
Fax: (319) 277-6034
Web site: www.fluidynecorp.com

Foster Wheeler Environmental Corp.
P.O. Box 479
Livingston, New Jersey 07039-0479
Tel: (800) 580-3765
Fax: (973) 597-7590
Web site: www.fwenc.com

Frontier Technology Inc.
11601 Drive
Rolla, MO 65401
Tel: (573) 364-2200
Web site: umr.edu/~volner/

Fusion Tanks and Silos
Eye 1P237HS
Suffolk, United Kingdom
Tel: +44-1379 870723
Fax: +44 1379 870530
Web site: www.fusiontanks.com

Geoform Inc.
227 Hathaway Street
Girard, Pennsylvania 16417
Tel: (814) 774-5020
Fax: (814) 774-8459
Email: geoform@ncinter.net
Web page: www.geoforminc.com

GL&V / Dorr-Oliver Inc.
612 Wheeler's Farm Road
Milford, Connecticut 06460-8719
Tel: (203) 876-5400
Fax: (203) 876-5599
Email: Mike.Smith@glv.com
Web site: www.glv.com

G.M.H. Associates of America Inc.
5 Cheften Way, Bldg. 15
Trenton, New Jersey 08638
Tel: (609) 396-4751
Fax: (609) 396-1067
Web site: www.gmhassociates.com

Groth Equipment Corporation
1202 Hahlo
Houston, Texas 77020
Tel: (713) 675-6151
Fax: (713) 675-7528
Web site: www.grothcorp.com
Email: GPG@grothcorp.com

Harza Environmental Services, Inc.
233 Wackler Drive
Chicago, Oklahoma 60606
Tel: (312) 831-3806
Fax: (312) 831-3999

Hibernia Export Trading House Ltd.
Knoknagin House
Balbriggan, Co.
Dublin, Ireland
Tel: +353-1-841 4560
Fax: +353-1-841 4564
Web site: indigo.ie/~hibernia/index.html

Hoffland Environmental
303 Silver Spring S. Road
Conroe, Texas 77303
Tel: (409) 856-4515
Web site: www.hofflandenv.com

Hudson Industries
Department F
P.O. Box 2212
Hudson, Ohio 44236
Tel: (216) 487-0668
Fax: (216) 487-0811

Humboldt Water Resources
P.O. Box 165
Arcata, California 95518
Tel: (707) 826-2869
Fax: (707) 826-2165
Email: water@humboldt1.com
Web site: humboldt1.com/~water/

Hyandai Engineering Co., Ltd
45 Bangi-dong, Songpa-ku
Seoul, Korea
Tel: 2 410-8114
Fax: 2 410-8118
Email: jonhlee@hec.co.kr
Web site: www.hec.co.kr

Hyder North America Inc.
270 Granite Run Drive
Lancaster, Pennsylvania 17601
Tel: (717) 569-7021
Fax: (717) 560-0577
Web site: www.thearrogroup.com/
Environmental%20Services.htm

Hydro-Aerobics, Inc.
P.O. Box 16327
Houston, Texas 77222-6327
Tel: (281) 449-0322
Fax: (281) 987-2134

Hydrocal Inc.
227332 Granite Way, Suite A
Laguna Hills, California 92653
Tel: (949) 455-0765
Fax: (949) 455-0764
Web site: www.hydrocal.com

Hydroxyl Systems Inc.
9800 McDonald Park Road
P.O. Box 2278
Sidney, British Columbia
Canada V8L 3S8
Tel: (250) 655-3348
Fax: (250) 655-3349
Web site: www.hydroxyl.com

Hydro-Thermal Corporation
400 Pilot Court
Waukesha, Wisconsin 53188
Tel: (414) 548-8900
Fax: (414) 548-8908
Web site: www.hydro-thermal.com

IMR Corporation
901 Fox Hollow
Catoosa, Oklahoma 74015-2304
Tel: (918) 266-1050
Fax: (918) 266-1628

Industrial Municipal Equipment
1430 Progress Way, #105
Eldersburg, Maryland 21784
Tel: (800) 858-4857
Fax: (410) 795-1373

Infilco Degremont Inc.
P.O. Box 71390
Richmond, Virginia 23255-1390
Tel: (804) 756-7600
Fax: (804) 756-7643
Email: idi@idi-online
Web site: www.infilcodegremont.com

In-Pipe Technology
100 Bridge St.
Wheaton, Illinois 60187
Tel: (630) 871-9010
Fax: (630) 871-0303
Web site: www.milieu-nomics.com

Integrated Explorations Inc.
67 Watson Rd., Unit 1
Guelph, Ontario
Canada N1H 6H8
Tel: (519) 822-2608
Fax: (519) 822-3076
Email: ieinc@istar.ca

JDV Equipment Corporation
26 Commerce Rd.
Fairfield, New Jersey 07004
Tel: (973) 244-1633
Fax: (973) 244-1055

John Meunier Inc.
6290 Perineault St.
Montreal, Quebec
Canada H4K 1K5
Tel: (514) 334-7230
Fax: (514) 334-5010
Web site: www.johnmeunier.com

J.W. Salm Engineering Inc.
Bishopville, Maryland 21813
Tel: (410) 213-0805
Email: comments@jwse.com
Web site: www.jwse.com

Kruger Inc.
401 Harrison Oaks Boulevard
Suite 100
Cary, North Carolina 27513
Tel: (919) 677-8310
Fax: (919) 677-0082
Email: webmaster@krugerworld.com
Web site: krugerworld.com

Kvaerner Werner a.s.
Pindslevn. 1c
N-3204 Sandefjord
Norway
Tel: +47 33 48 88 00
Fax:+47 33 48 88 99
Email: water-sandefj@kvaerner.com

L & J Technologies
5911 Butterfield Rd.
Hillside, Illinois 60162
Tel: (708) 236-6000
Fax: (708) 235-6006
Web site: www.ljtechnologies.com

Lakeside Equipment Co.
P.O. Box 8448
Bartlett, Illinois 60103
Tel: (603) 837-5640
Fax: (603) 837-5647

Lantec Products Inc.
5308 Derry Ave., Unit E
Agora Hills, California 91301
Tel: (818) 707-2285
Fax: (818) 707-9367

LAS International Ltd.
3811 Lockport Street
Bismark, North Dakota 58501
Tel: (701) 222-8331
Fax: (701) 222-2773
Email: info@lasinternational.com
Web site: www.lasinternational.com

Lemna Technologies Inc.
1408 Northland Drive, Suite 310
St. Paul, Minnesota 55120
Tel: (651) 688-0836
Fax: (651) 688-8813
Email: sales@lemna.com
Web site: www.lemnatechnologies.com

Linked, Inc.
2069 East Corsen
Tempe, Arizona 85282

Living Machines, Inc.
125 La Posta Road
8018 NDCBU
Taos, New Mexico 87571
Tel: (505) 751-4448
Fax: (505) 751-4449
Email: erik@goodwater.com
Web site: www.livingmachines.com

LOTEPRO Corp.
115 Stevens Ave.
Valhalla, New York 10695
Tel: (914) 749-5228
Fax: (914) 747-3422

Lurgi AG
Lurgiallee 5
Frankfurt am Main
Germany 60295
Tel: +49 69 5808 0
Fax: +49 695808 3888
Email: sabine_biewer@lurgi.de
Web site: www.lurgi.com

Martin Marietta Magnesia Specialities
2323 Eastern Blvd.
P.O. Box 15470
Baltimore, Maryland 21220-0470
Tel: (410) 780-5500
Fax: (410) 780-5555
Web site: www.magspecialties.com

Mazzei Injector Corp.
500 Rooster Drive
Bakersfield, California 93307
Tel: (805) 363-6500
Fax: (805) 363-7500
Web site: www.mazzei-injector.com

Membrane Technologies
5601 Paris Road
Columbia, Missouri 65202-9399
Tel: (573) 474-9456
Fax: (573) 474-6988
Web site: www.wastewater.com

Metcalf & Eddy
P.O. Bpx 4071
Wakefield, Massachusetts 01880-5371
Tel: (781) 245-6293
Fax: (781) 245-6293
Email: janelle_coleman@air-water.com
Web site: www.m-e.com

Met-Pro Corp., Systems Division
P.O. Box 144
Harleysville, Pennsylvania 19438
Tel: (215) 723-6751
Fax:(215) 723-6758
Web site: www.met-pro.com

Micro-Bac International Inc.
3260 N. 1H35
Round Rock, Texas 78610-2410
Tel: (512) 310-9000
Fax: (512) 310-8800
Web site: www.micro-bac.com

Mixing Systems Inc.
7058 Corporate Way
Dayton, Ohio 45459
Tel: (937) 435-7227
Fax: (937) 435-9200
Web site: www.mixing.com

Napier-Reid Ltd.
10-2 Alden Rd.
Markham, Ontario
Canada L3R 2S1
Tel: (905) 475-1545
Fax: (905) 475-2021
Email: info@napier-reid.com
Web site: www.napier-reid.com

Neozyme International Inc.
33 Journey #300
Alisa Viejo, California 92656
Tel: (800) 982-8676
Fax: (714) 360-8774

NewBio
Corporate Office
P.O. Box 771
Hopkins, Minnesota 55343
Tel: (612) 949-1331
Fax: (612) 934-4067
Web site: www.newbio.com

NGK Insulators Ltd.
Head Office
2-56, Suda-cho,
Mizuho-ku, Nagoya 467
Japan
Tel: 81 52 872-7171
Email: pr-office@ngk-co.jp
Web site: www.ngk.co.jp

Nitrate Removal Technologies
1667 Cole Blvd. #400
Golden, Colorado 80401
Tel: (303) 274-1426

NORAM Biosystems Inc.
200 Granville Street, Suite 400
Vancouver, British Columbia
Canada V6C 1S4
Tel: (604) 681-2020

NordBeton North America
P.O. Box 470858
Lake Monroe, Florida 32747
Tel: (407) 322-8122
Fax: (407) 322-8159
Web site: www.nordbeton.com

Orenco Systems Inc.
814 Airway Ave.
Sutherlin, Oregon 97479
Tel: (541) 459-4449
Fax: (541) 459-2884
Web site: www.orenco.com

Osmonics Inc.
5951 Clearwater Drive
Minnetonka, Minnesota 55343
Tel: (612) 933-2277
Fax: (612) 933-0141
Web site: www.osmonics.com

Osprey Biotechnics
1833 57th Street
Sarasota, Florida 34243
Tel: (941) 351-2700
Fax: (941) 351-0026
Email: Info@OspreyBiotechnics.com
Web site: www.ospreybiotechnics.com

Pall Corp.
2200 Northern Blvd.
East Hills, New York 11548
Tel: (516) 484-5400
Fax: (516) 484-3216
Web site: www.pall.com

Paques ADI Inc.
389 Main Street
Salem, New Hampshire 03079
Tel: (603) 893-2134
Fax: (603) 898-3991
Email: sdigroup@adi.ca
Web site: www.adi.ca/sys

Paques BV.
P.O. Box 52
8560 AB Balk
The Netherlands
Tel: 31 514 608500
Email: info@paques.nl
Internet: www.paques.nl

Parkson Corp.
P.O. Box 408399
2727 N.W. 62nd St.
Ft. Lauderdale, Florida 33340-8399
Tel: (954) 974-6610

PDR Engineers
800 Corporate Drive, Suite 100
Lexington, Kentucky 40503
Tel: (606) 223-8000
Fax: (606) 224-1025

Philadelphia Mixers
1221 E. Main
Palmyra, Pennsylvania 17078
Tel: (717) 838-1341
Fax: (717) 832-1740
Email: philamixers@philamixers.com
Web site: www.philamixers.com

P. J. Hannah Equipment Sales Corp.
#10, 8528 123rd St.
Surrey, British Columbia
Canada V3W 3V6
Tel: (604) 591-5999
Fax: (604) 591-9925
Email: pjhannah@direct.ca

Pollution Control Systems, Inc.
5827 Happy Hollow Rd., Suite 1-B
Milford, Ohio 45150
Tel:(513) 831-1165
Fax: (513) 965-4812
Email: polyconsys@aol.com
Web site: www.pollutioncontrolsystem.com

Praxair Inc.
39 Old Ridgebury Road (K2-476)
Danbury, Connecticut 06810-5113
Tel: (203) 837-2039
Fax: (203) 837-2454
Email: webmaster@Praxair.com
Web site: www.praxair.com

Premier Tech Ltd.
1 Avenue Premier
P.O. Box 3500
Rivière-du-Loup, Quebec
Canada G5R 4C9
Tel: (418) 867-8883
Fax: (418) 862-6642
Email:agri@premiertech.com
Web site: www.premiertech.com

Pro-Equipment, Inc.
1010 Ridge Rd.
Waukesha, Wisconsin 53186
Tel: (414) 542-6331
Fax: (414) 542-2297
Web site: www.proequipment.com

PURAC AB
Box 1146
Clemenstorget 15
S-221 05
Lund, Sweden
Tel: +46 46 1919 00
Fax: +46 46 19 19 19
Email: lh@purac.se
Web site: www.purac.se

Purestream Inc.
P.O. Box 68
Florence, Kentucky 41042-0068
Tel: (606) 371-9898
Fax: (606) 371-3577

RBC Services
4031 W. Kiehnau Ave.
Milwaukee, Wisconsin 53209-3022
Tel: (800) 232-7011
Fax: (414) 228-9150
Web site: www.allaneng.com

R. Cave and Associates Engineering Ltd.
Suite 404, 345 Lakeshore Rd. E.
Oakville, Ontario
Canada L6J 1J5
Tel: (905) 825-8440
Fax: (905) 825-8446
Email: rcave@rcave.com

Remco Engineering
410 Bryant Circle
Ojai, California 93023
Tel: (805) 646-3706
Fax: (805) 646-3923
Web site: www.remco.com

Resource Management & Recovery
4980 Baylor Canyon Rd.
Las Cruces, New Mexico 88011
Tel: (505) 382-9228
Fax: (505) 382-9228

RGF Environmental Systems Inc.
3875 Fiscal Court, Suite 100
West Palm Beach, Florida 33404
Tel: (561) 848-1826
Fax: (561) 848-9454
Web site: www.rgf.com

Ringlace Products Inc.
9902 N.E. Gilson
Portland, Oregon 97220
Tel: (503) 251-1295
Fax: (503) 256-7325

Roediger Pittsburgh Inc.
3812 Route 8
Allison Park, Pennsylvania 15101
Tel: (412) 487-6010
Fax: (412) 487-6005
Web site: www.roediger.com

Samsung USA R &D Center
2 Executive Drive, Suite 555
Fort Lee, New Jersey 07024
Tel: (201) 461-5064
Fax: (201) 461-5721

Sanexen Environmental Services
295 The West Mall, Suite 205
Etiboke, Ontario
Canada M9C 4Z4
Tel: (416) 622-5011
Fax: (416) 622-5823
Email: info@sanexen.com
Web site: www.sanexen.com

Sanitaire
9333 N. 49th St.
Brown Deer, Wisconsin 53223
Tel: (414) 365-2200
Fax: (414) 365-2210
Web site: www.sanitaire.com

Sanitherm Engineering Ltd.
431 Mountain Highway, Suite 4
North Vancouver, British Columbia
Canada V7J 2L1
Tel: (604) 986-9168
Fax: (604) 986-5377
Email: saneng@sanitherm.com
Web site: www.sanitherm.com

Santec Corp.
220 Malibu St.
Castle Rock, Colorado 80104
Tel: (303) 660-9211
Fax: (303) 660-2180
Web site: www.santeccorporation.com

Schreiber Corp. Inc.
100 Schreiber Drive
Trussville, Alabama 35122
Tel: (205) 655-7466
Fax: (205) 655-7669
Web site: www.schreiber-water.com

Seghers Better Technology
344 Emes Circle
Austell, Georgia 30001
Tel: (770) 739-4205
Fax: (770) 739-0117
Web site: www.seghers-water.com

Semblex
1635 W. Walnut St.
Springfield, Missouri 65806
Tel: (417) 866-0235

Smith & Loveless Inc.
14040 Santa Fe Trail Drive
Lenexa, Kansas 66215-1284
Tel: (913) 888-5201
Fax: (913) 888-2173
Email: answers@smithandloveless.com
Web site: www.smithandloveless.com

SNC-Lavalin Group Inc.
455 Rene Levesque Blvd. W.
Montreal, Quebec
Canada H2Z 1Z3
Tel: (514) 866-6635
Fax: (514) 866-0600
Email: Catherine.Mulligan@snclavalin.com
Web site: www.snc-lavalin.com

Soil Enrichment Systems Inc.
10800 Weston Road
Vaughan, Ontario
Canada L4L 1A6
Tel: (905) 832-2166
Fax: (905) 832-0751
Email: ssoil@aol.com

Solmar Corp.
P.O. Box 2329
Carlsbad, California 92018
Tel: (760) 734-1685
Fax: (760) 734-1778
B.V. Sorbex
3610 University Street
Montreal, Quebec
Canada H3A 2B2
Tel: (514) 398- 4494
Fax: (514) 398-6678
Email: volesky@chemeng.lan.mcgill.ca

Southwest Wetlands Group, Inc
901 W. San Mateo, Suite M
Santa Fe, New Mexico 87505
Tel: (505) 988-7453
Fax: (505) 988-3720
Email: swgroup@uswest.net
Web site: www.swg-inc.com

SRE Inc.
510 Franklin Ave.
Nutley, New Jersey 07110
Tel: (973) 661-5192
Fax: (973) 661-3713
Web site: srebiotech.com
Email: sre@srebiotech.com

Stormtreat Systems Inc.
3179 Main Street
Barnstable, Massachusetts 02630
Tel: (508) 778-4449
Fax: (508) 362-5335

Sybron Chemicals Inc.
P.O. Box 66
Birmingham, New Jersey 08011
Tel: (609) 893-1100
Fax: (609) 894-8641
Web site: http://63.241.177.140

Syneco Systems Corp.
4930 West 35th Street
St. Louis Park, Minnesota 55416
Tel: (952) 927-9215
Fax: (952) 927-9224
Web site: www.synecosystems.com

Systems Ecotechnologies Inc.
222-111 Research Drive
Saskatoon, Saskatchewan
Canada S7N 3R2
Tel: (306) 955-0872
Fax: (306) 975-7011
Email: lakshmang@innovationplace.com

Tanks-A-Lot Ltd.
1810 Yellowhead Trail N.E.
Edmonton, Alberta
Canada T6S 1B4
Tel: (780) 472-8265
Fax: (780) 478-6699
Web site: www.tanks-a-lot.com

TASKEM OMEGA Water Treatment Group
4639 Van Epps Road
Brooklyn Hts, Ohio 44131
Tel: (216) 351-1500
Fax: (216) 351-5677
Web site: www.taskem.com

Tauw B.V.
Handelskade 11
P.O. Box 133
744 AC Deventer
The Netherlands
Tel: +31-57-06-99-911
Fax: +31-57-06-99-666
Email: jfh@tauw.nl
Web site: www.tauw.nl

Taylor Environmental Products
628 Old Robinson Rd
Louisville, Mississippi 39339
Tel: (601) 773-3421
Fax: (601) 773-7139
Web site: www.taylorenviro.com

Technologia Intercontinental S.A. de C.V.
Rio Lerma 171-4
Col. Cuauhtemoc,
Mexico D.F. 06500
Tel: (52) 5514-0321
Fax: (52) 5207-2478
Web site: www.ticsa.com.mx

Tetra Technologies Inc.
6302 Benjamin Road
Tampa, Florida 33634
Tel: (813) 886-9331
Fax: (813) 886-0651

Thermodyne Corp.
2 Smedley Dr.
Newton Square, Pennsylvania 19073-1012
Tel: (610) 644-6406
Fax: (610) 644-3122

TurboSonic Inc.
550 Parkside Dr.,
Suite A-14
Waterloo, Ontario
Canada N2L 5V4
Tel: (519) 885-5513
Fax: (519) 885-6992
Email: info@turbosonic.com
Web site: www.turbosonic.com

Unisol
1810 W. Drake Dr.
Tempe, Arizona 85283
Tel: (800) 440-2121
Fax: (602) 491-7185

United Industries Inc.
P.O. Box 3838
Baton Rouge, Louisianna 70821
Tel: (504) 292-5527
Fax: (504) 293-1655
Web site: www.ui-inc.com
U.S. Filter, Davco Products
1828 Metcalf Ave.
Thomasville, California 31792
Tel: (912) 226-5733
Fax: (912) 228-0312
Web site: www.davcoproducts.com

U.S. Filter/Envirex
1901 S. Prairie Avenue
Waukesha, Wisconsin 53186
Tel: (262) 547-0141
Fax: (262) 547-4120
Web site: www.usfilterenvirex.com

U.S. Filter, Industrial Wastewater Systems
181 Thorn Hill
Warrendale, Pennsylvania 15086
Tel: (724) 772-0044
Fax: (724) 772-1360
Web site: www.usfilter.com

U.S. Filter, Jet Tech Products
1051 Blake
P.O. Box 13306
Edwardsville, Kansas 66111
Tel: (913) 422-7600
Fax: (913) 422-7667
Email: reynoldss@usfilter.com
Web site: www.usfilter.com

U.S. Filter/Zimpro
301 W. Military Rd.
Rothschild, Wisconsin 54474
Tel: (715) 359-7211
Fax: (715) 355-3335
Email: mattmiller@usfilter.com
Web site: www.zimpro.usfilter.com

VA TECH WABAG Ltd.
P.O. Box 414
CH-8401 Winterthur
Switzerland
Tel: +41-52-262 4904
Fax: +41-52-262 00 74
Email: martin.baggenstos@vatech.ch
Web site: www.vatech.ch

Walden Inc.
P.O. Box 1378
Waterville, Maine 04903-1378
Tel: (207) 873-4234
Fax: (207) 872-6273
Web site: www.waldeninc.com

Walker Process, Div. Of McNish Corp.
840 N. Russell Ave.
Aurora, Illinois 60506
Tel: (630) 892-7921
Fax: (630) 892-7951
Email: walker.process@walker-process.com
Web site: www.walker-process.com

Washington Group International, Inc.
1993 S. Centennial Avenue SE
Aiken, South Carolina 29803
Tel: (803) 502-9950
Fax: (803) 502-9795
Web site: www.wgint.com

Wastewater Solutions
13422 Clayton Rd.
Mason Woods Village, Ste 201
St. Louis, Missouri 63131-1008
Tel: (314) 434-3554
Fax: (314) 434-2883

Waterlink/Biological Wastewater Systems
630 Current Road
Fall River, Massachusetts 02720-4732
Tel: (508) 679-6770
Fax: (508) 672-5779
Email: biological@waterlink.com
Web site: www.mtsjets.com

Waterlink/Mass Transfer Systems
630 Current Road
Fall River, Massachusetts 02720-4732
Tel: (508) 679-6770
Fax: (508) 672-5770
Email: mtsjets@waterlink.com
Web site: www.waterlink.com

Waterloo Biofilter Systems Inc.
143 Denis Street
Rockwood, Ontario
Canada N0B 2K0
Tel: (519) 856-0757
Fax: (519) 856-0759
E-mail: wbs@waterloo-biofilter.com
Website: www.waterloo-biofilter.com

Westech Engineering Inc.
3625 Southwest Temple
Salt Lake City, Utah 84115
Tel: (801) 265-1000
Fax: (801) 265-1080
Email: info@westech-inc.com
Web site: www.westech-inc.com

Wetland Ecosystems
1/126 Nepean Hwy
Aspendale VIC 3195
Australia
Web site: www.wetlandecosystems.com.au

Zenon Environmental Inc.
3239 Dundas Street West
Oakville, Ontario
Canada L6M 4B2
Tel: (905) 465-3030
Fax: (905) 465-3050
Web site: www.zenonenv.com

Index

Environmental factors
  moisture, 20, 25
  osmotic pressure, 25
  oxygen, 5, 20, 25
  temperature, 1, 5, 20, 25
Enzyme, 16, 20, 26, 29-30, 46, 178, 232
Equilibrium, 41, 224
Escherichia coli, 24, 232, 244
Europe, 5, 7-8, 11, 105, 111, 147, 191, 241, 253, 262-263, 271, 274, 290, 311
Ethanol, 9, 50, 59, 65-67, 158-159, 304-306, 308-309, 312
Ether, 27, 46
Ethylbenzene, 30, 48, 53, 212
Ethylenediaminetetraacetic acid (EDTA), 180, 199
Excavating, 2-3, 6-7, 25, 44, 143, 174, 186, 188, 193-194, 202, 204, 206-207, 209-210, 215, 217-219
Expanded Granular Sludge Blanket Reactor (EGSB), 121, 133-134
Explosives
  HMX, 217
  RDX, 217
  TNT, 34, 37, 217
Exponential growth, 21, 213
Ex situ bioremediation
  biopiles, 204-205, 210, 212, 217-219, 233
  biopit, 205
  biovaults, 205
  composting, 210-217
  landfarming, 296, 298
  slurry-phase, 218-221, 234, 238-239, 241, 244, 298-299, 302

F

Facultative pond, 96, 98, 152
Feasibility studies, 1, 73, 80, 180, 189, 250, 297
Fenton's reagent, 223
Fermentation
  ethanol production, 9, 304, 306
  high solids (dry), 145, 263-264
  low solids (wet), 265
Fertilizer, 9, 152, 204, 260, 272, 280, 305-306, 310
Filter, 40, 42-43, 45, 50, 53-54, 58-61, 64, 68, 71, 74-77, 84, 98-103, 105, 114-115, 123, 126-127, 129, 134-135, 146, 150, 164, 166, 170, 187, 192-193, 249, 253, 262, 269, 272, 279, 288, 294-295, 311
Filtration, 39, 65, 81, 95, 105, 119, 123, 192, 288

Field-scale, 178, 202
First-order reaction, 95
Fixed film reactor, 80, 105, 121, 123, 125, 134, 153, 159, 167
Fixed cover, 252
Flagella, 20
Floating covers, 252, 254, 256, 259
Fluidized bed reactors, 101, 124, 133, 135, 156, 165
Fly ash, 250
Foam, 58, 61, 99, 101, 114, 187, 223, 227, 252, 301
Food processing, 55, 63, 79, 90, 120, 145, 213, 277, 280, 305, 307
Food waste, 246, 267, 272, 277, 281-283, 286, 291, 311
Formaldehyde, 32, 55, 71, 90, 101, 103
F/M ratio, 85-87, 89, 91
France, 90, 105, 117, 120, 135, 263, 290
Frost, 152, 154, 184, 206
Full scale
  composting, 264, 267
  slurry-phase, 218, 298
Fungi
  classification, 17, 274
  fungi, 2, 4, 11, 15-17, 23, 28, 30, 32, 42, 70, 73, 124, 150, 237, 274, 296
  mold, 16
  yeast, 16, 119, 132, 154
Future trend, 9

G

GAC, 48, 215
Gambusia fish, 145
Gas aphron, 187
Gas composition, 139, 295
Gas flux, 44
Gas production, 8, 119, 124, 130, 135, 252-254, 257-258, 262, 264, 292, 295
Gasoline, 6, 30, 37, 39, 49, 105, 177, 186, 188-189, 194-195, 204, 238, 244
Genetically engineered, 2, 232, 305
Geochemistry, 182
Geology, 1-2, 182
Germany, 5, 11, 39, 61-62, 65-66, 104, 217, 260, 263, 265, 296, 311-312
Glucose, 29, 257, 304
Granular activated carbon (GAC), 38, 48, 124, 215
Granulation, 48, 124, 155
Grasses, 17, 141-142, 145-146, 199, 201, 209, 213, 281, 284

Government Institutes Mini-Catalog

PC #	ENVIRONMENTAL TITLES	Pub Date	Price*
629	ABCs of Environmental Regulation	1998	$65
672	Book of Lists for Regulated Hazardous Substances, 9th Edition	1999	$95
4100 ◉	CFR Chemical Lists on CD ROM, 1999-2000 Edition	1999	$125
512	Clean Water Handbook, Second Edition	1996	$115
581	EH&S Auditing Made Easy	1997	$95
673	E H & S CFR Training Requirements, Fourth Edition	2000	$99
825	Environmental, Health and Safety Audits, 8th Edition	2001	$115
548	Environmental Engineering and Science	1997	$95
643	Environmental Guide to the Internet, Fourth Edition	1998	$75
820	Environmental Law Handbook, Sixteenth Edition	2001	$99
688	EH&S Dictionary: Official Regulatory Terms, Seventh Edition	2000	$95
821	Environmental Statutes, 2001 Edition	2001	$115
4099 ◉	Environmental Statutes on CD ROM for Windows-Single User, 1999 Ed.	1999	$169
707	Federal Facility Environmental Compliance and Enforcement Guide	2000	$115
708	Federal Facility Environmental Management Systems	2000	$99
689	Fundamentals of Site Remediation	2000	$85
515	Industrial Environmental Management: A Practical Approach	1996	$95
510	ISO 14000: Understanding Environmental Standards	1996	$85
551	ISO 14001: An Executive Report	1996	$75
588	International Environmental Auditing	1998	$179
518	Lead Regulation Handbook	1996	$95
608	NEPA Effectiveness: Mastering the Process	1998	$95
582	Recycling & Waste Mgmt Guide to the Internet	1997	$65
615	Risk Management Planning Handbook	1998	$105
603	Superfund Manual, 6th Edition	1997	$129
685	State Environmental Agencies on the Internet	1999	$75
566	TSCA Handbook, Third Edition	1997	$115
534	Wetland Mitigation: Mitigation Banking and Other Strategies	1997	$95

PC #	SAFETY and HEALTH TITLES	Pub Date	Price*
697	Applied Statistics in Occupational Safety and Health	2000	$105
547	Construction Safety Handbook	1996	$95
553	Cumulative Trauma Disorders	1997	$75
663	Forklift Safety, Second Edition	1999	$85
709	Fundamentals of Occupational Safety & Health, Second Edition	2001	$69
612	HAZWOPER Incident Command	1998	$75
662	Machine Guarding Handbook	1999	$75
535	Making Sense of OSHA Compliance	1997	$75
718	OSHA's New Ergonomic Standard	2001	$95
558	PPE Made Easy	1998	$95
683	Product Safety Handbook	2001	$95
598	Project Mgmt for E H & S Professionals	1997	$85
658	Root Cause Analysis	1999	$105
552	Safety & Health in Agriculture, Forestry and Fisheries	1997	$155
669	Safety & Health on the Internet, Third Edition	1999	$75
668	Safety Made Easy, Second Edition	1999	$75
590	Your Company Safety and Health Manual	1997	$95

Government Institutes

4 Research Place, Suite 200 • Rockville, MD 20850-3226
Tel. (301) 921-2323 • FAX (301) 921-0264
Email: giinfo@govinst.com • Internet: http://www.govinst.com

Please call our customer service department at (301) 921-2323 for a free publications catalog.

CFRs now available online. Call (301) 921-2355 for info.

*All prices are subject to change. Please call for current prices and availablity.

Government Institutes Order Form

4 Research Place, Suite 200 • Rockville, MD 20850-3226
Tel (301) 921-2323 • Fax (301) 921-0264
Internet: http://www.govinst.com • E-mail: giinfo@govinst.com

4 EASY WAYS TO ORDER

1. Tel: **(301) 921-2323**
Have your credit card ready when you call.

2. Fax: **(301) 921-0264**
Fax this completed order form with your company purchase order or credit card information.

3. Mail: **Government Institutes Division**
ABS Group Inc.
P.O. Box 846304
Dallas, TX 75284-6304   USA

Mail this completed order form with a check, company purchase order, or credit card information.

4. Online: Visit http://www.govinst.com

PAYMENT OPTIONS

❏ **Check** *(payable in US dollars to ABS Group Inc. Government Institutes Division)*

❏ **Purchase Order** *(This order form must be attached to your company P.O. Note: All International orders must be prepaid.)*

❏ **Credit Card** ☐ VISA ☐ MasterCard ☐ AMERICAN EXPRESS

Exp. ____ /____

Credit Card No. _____

Signature _____

(Government Institutes' Federal I.D.# is 13-2695912)

CUSTOMER INFORMATION

Ship To: (Please attach your purchase order)

Name _____

GI Account # (*7 digits on mailing label*) _____

Company/Institution _____

Address _____
(Please supply street address for UPS shipping)
_____

City _____ State/Province _____

Zip/Postal Code _____ Country _____

Tel ( ) _____

Fax ( ) _____

E-mail Address _____

Bill To: (if different from ship-to address)

Name _____

Title/Position _____

Company/Institution _____

Address _____
(Please supply street address for UPS shipping)
_____

City _____ State/Province _____

Zip/Postal Code _____ Country _____

Tel ( ) _____

Fax ( ) _____

E-mail Address _____

Qty.	Product Code	Title	Price

Subtotal _____
MD Residents add 5% Sales Tax _____
Shipping and Handling (see box below) _____
Total Payment Enclosed _____

30 DAY MONEY-BACK GUARANTEE

If you're not completely satisfied with any product, return it undamaged within 30 days for a full and immediate refund on the price of the product.

SOURCE CODE: BP03

Shipping and Handling	Sales Tax
Within U.S:	Maryland 5%
1-4 products: $6/product	Texas 8.25%
5 or more: $4/product	Virginia 4.5%
Outside U.S:	
Add $15 for each item (Global)	